DEVELOPMENT

The S203 Course Team

Chairman and General Editor
Michael Stewart

Academic Editors
Caroline M. Pond (Book 1)
Norman Cohen (Book 2)
Michael Stewart (Book 3)
Irene Ridge (Book 4)
Brian Goodwin (Book 5)

Authors
Mary Bell (Book 4)
Sarah Bullock (Book 2)
Norman Cohen (Book 2)
Anna Furth (Book 2)
Brian Goodwin (Book 5)
Tim Halliday (Book 3)
Robin Harding (Book 5)
Stephen Hurry (Book 2)
Judith Metcalfe (Book 5)
Pat Murphy (Book 4)
Phil Parker (Book 4)
Brian Pearce (Book 3)
Caroline M. Pond (Book 1)
Irene Ridge (Book 4)
David Robinson (Book 3)
Michael Stewart (Book 3)
Margaret Swithenby (Book 3)
Frederick Toates (Book 3)
Peggy Varley (Book 3)
Colin Walker (Book 2)

Course Managers
Phil Parker
Colin Walker

Editors
Perry Morley
Carol Russell
Dick Sharp
Margaret Swithenby

Design Group
Diane Mole (Designer)
Pam Owen (Graphic Artist)
Ros Porter (Design Group Coordinator)

External Course Assessor
Professor John Currey (University of York)

General Course Consultant
Peggy Varley

Course Secretaries
Valerie Shadbolt
Christine Randall

DEVELOPMENT

Edited by Brian Goodwin

BIOLOGY: FORM AND FUNCTION

Hodder & Stoughton
The Open University

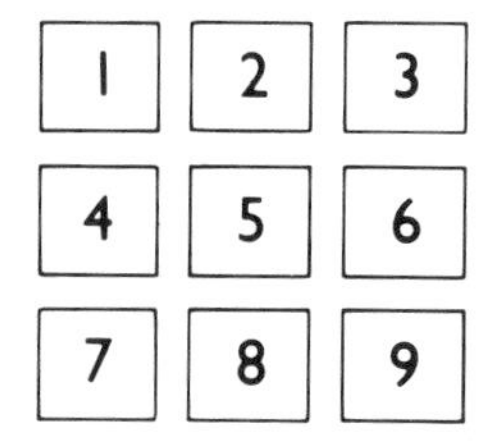

Cover illustrations

1 Scanning electron micrograph of the floral meristem of *Aquilegia*, a member of the buttercup family.

2 The tentacles of *Hydra*, stained with an immunofluorescent antibody.

3 Spiral shell of a marine gastropod.

4 Mirror-image duplication of a frog limb caused by vitamin A (retinoic acid).

5 A flower of *Echeveria fulgens* (*Crassulaceae*) showing the five-fold symmetry.

6 Embryos of the frog *Xenopus laevis*, immunostained to show myotomes.

7 Whorls and developing cap of *Acetabularia acetabulum*.

8 Colour graphic of *Acetabularia* whorl formation, showing predicted calcium distribution.

9 Spiral arrangement of elements viewed from the base of a pine cone.

Back cover: Colour graphic of a model of pair-rule transcript pattern in *Drosophila*.

British Library Cataloguing in Publication Data
Development.
1. Organisms. Development
I. Goodwin, B. C. (Brian Carey) II. Series
574.3

ISBN 0–340–53190–8

First published 1991.

Designed by the Graphic Design Group of the Open University.

The text forms part of an Open University course. Further information on Open University courses may be obtained from the Admissions Office, The Open University, P.O. Box 48, Walton Hall, Milton Keynes, MK7 6AB.

Typeset by Wearside Tradespools, Fulwell, Sunderland, printed in Great Britain by Thomson Litho Ltd, East Kilbride for the educational division of Hodder and Stoughton Ltd, Mill Road, Dunton Green, Sevenoaks, Kent TN13 2YA, in association with the Open University, Walton Hall, Milton Keynes, MK7 6AB.

1.1

CONTENTS

PREFACE

Development is the fifth in a series of five volumes that provide a general introduction to biology. It is designed so that it can be read on its own (like any other textbook) or studied as part of *S203, Biology: Form and Function*, a second level course for Open University students. As well as the five books, the course consists of five associated study texts, 30 television programmes, several audiocassettes and a series of home experiments. As is the case with other Open University courses, students of S203 are required both to complete written assignments during the year and to sit an examination at the end of the course.

In this book, each subject is introduced in a way that makes it readily accessible to readers without any specific knowledge of that area. The major learning objectives are listed at the end of each chapter, and there are questions (with answers at the end of the book) which allow readers to assess how well they have achieved these objectives. Key words are identified in **bold** type both in the text where they are explained and also in the index, for ease of reference. A further reading list is included for those who wish to pursue certain topics beyond the limits of this book.

INTRODUCTION

Organisms come in an immense diversity of sizes and forms, but all species have a common characteristic — their existence depends upon a cyclic process of development and reproduction, each species having a distinctive life cycle. Although, strictly speaking, a cycle has neither a beginning nor an end, life cycles are often thought of as starting with a fertilized egg and finishing with the mature organism that produces the gametes for the next generation. There are many variations on this basic theme, and it is important to remember that many species (for example most algae and fungi, all Bryophytes, and ferns) have two major stages to this cycle, one in which the organism is haploid and the other in which it is diploid. The life cycle in such species therefore involves an alternation of generations. Furthermore, many species reproduce by an asexual process in which a new individual arises directly from the parent. This can occur either by growth and division of single cells, as happens in many unicellular species, or by the emergence of a protuberance or a bud on the parental body. The bud develops into an individual that detaches from the parent and wends its independent way to maturity when it, too, can make another of its kind.

The deep and abiding questions that arise from the contemplation of this remarkable range of biological forms focus on questions of origins. Since Darwin, it is axiomatic that the diversity of species emerged historically by diversification from a few primary types that started the living adventure. However, history does not explain origins; it simply describes them. We want to know why and how organisms take the shapes and forms that we see about us, and that are reconstructed from the fossil record. A common answer to these questions is given in functional terms. The morphologies that arise in the organic realm are there because they give organisms the wherewithal to make a living in a particular habitat. So plants have roots that gather water and nutrients from the soil, leaves for photosynthesis, and flowers that attract insects for pollination. Animals have guts for digestion of food, limbs for locomotion, and eyes to see their way about in the world. This must be true: nothing can persist unless it is dynamically stable. The form of each organism contributes to the stability of its own life cycle — that is, the ability to reproduce its kind and perpetuate the species in particular environments.

But this still fails to get to the root of the question of why and how organisms take the particular forms they do. We still want to know what makes some morphologies possible and others impossible, and how the possible ones are generated. These are the questions to which developmental biologists seek answers. How does something apparently very simple, such as a microscopic spherical egg, develop into a complex form such as a water lily or a frog, with a repetition of the same performance in the next generation? Such processes define the generative basis of all life. Development, one might say, is where all aspects of the study of biological form and function come together. Are the processes that occur in the development of an alga or a plant similar to or totally different from those that occur in a frog or a human? What is the role of the genes in determining similarities and differences between species? How are different cells and tissues generated during embryonic development and what processes underlie their coherent organization into an integrated whole, the adult organism? How do organisms regenerate their parts, what processes are responsible for healing? What causes morphological abnormalities, and cancer? These are some of the questions we shall be considering in this book.

In Chapter 1 the basic principles of developmental biology are introduced by examining three very different species, a unicellular green alga, a simple invertebrate (*Hydra*) and a vertebrate (the newt). What emerges from this is evidence for spatial gradients of substances that are directly involved in morphogenesis through their action as components of morphogenetic fields, the generators of form. Furthermore, despite the differences between these species, certain basic aspects of cell organization and behaviour emerge as fundamental to the generation of characteristic form in each of these species.

These cellular properties and their morphological consequences are examined in more detail in Chapters 2 and 3, first in relation to the production of specialized male and female gametes and then, through the processes of fertilization and activation, to the detailed cellular events that underlie morphogenesis in amphibians. What this reveals is the coordinated large-scale organization of the developing embryo resulting from cell interactions, movements and changes of shape. This leads to questions about the specific processes and molecular distributions that underlie the emergence of organized spatial patterns such as segments and limbs.

These questions are pursued in Chapter 4, taking the developmental story to the level of genes and the spatial distributions of their products in a study of *Drosophila* development. Spatial gradients and periodicities of gene products in the developing embryo emerge as basic contributors to pattern formation. The role of gene activity in the specification of form and behaviour is examined further in Chapters 5 and 6 in connection with sex determination. Here a remarkable diversity of mechanisms that act in the specification of sex is revealed across a range of species. Environmental temperature turns out to be a major sex-determining factor in a variety of species, showing that developmental processes are as much influenced by environmental stimuli as by gene activity, a point that emerges repeatedly throughout the book. Furthermore, the complexity of the sex-determination process has the interesting consequence that there are not just two distinct sexual states, male and female, but a whole range of intersexes as well.

Finally, the last chapter illustrates the relevance of developmental studies in a human context by considering two case studies. The first examines the cause of a developmental abnormality, spina bifida (partial failure of spinal cord formation) and considers ways of preventing it. The second looks at the dynamic organization of the human skin and its capacity for healing. The cellular and molecular processes involved recapitulate in a new context the principles of spatial and temporal organization already studied in other chapters, and provide a basis for understanding cancer and its treatment. Thus this final chapter reviews fundamental developmental processes and shows their relevance to the understanding and prevention of abnormalities and misadventures of the human body.

1.1 *ACETABULARIA*: THE MERMAID'S CAP

The organism shown in Figure 1.1 is a green alga, a member of the *Chlorophyceae*. It grows in shallow waters around the shores of the Mediterranean, anchored to rocks by the base which is known as a hold-fast or **rhizoid**. The stalk is 3–5 cm long and the cap is about 0.5 cm in diameter. The cap is the distinctive part which differs between species. The one shown has the name *Acetabularia acetabulum*, commonly known as the mermaid's cap. On first acquaintance, the alga would certainly be taken to be multicellular, the cap appearing to be made of separate cells. However, the whole organism is in fact one giant cell, whose specific form arises in a distinctive manner during development. The capacity of a single cell to grow to such a size and into such a shape is what makes these organisms so fascinating to a developmental biologist. As we shall see, they provide an unusual opportunity to examine particular aspects of development, though other aspects are not easily investigated in these algae. To understand the full range of developmental phenomena it is necessary to study many different species, each one providing an opportunity to explore in detail certain features of the overall process of development.

The life cycle of *Acetabularia acetabulum* is shown in Figure 1.2. The starting point can be taken to be the fusion or conjugation of two haploid flagellated gametes to form the **zygote**, the diploid cell about 50 μm in diameter that will develop into the mature plant. This process begins with the formation of an axis, a stalk extending upward by growth at the tip, while root-like processes grow out from the base, forming the rhizoid. The nucleus grows to a diameter

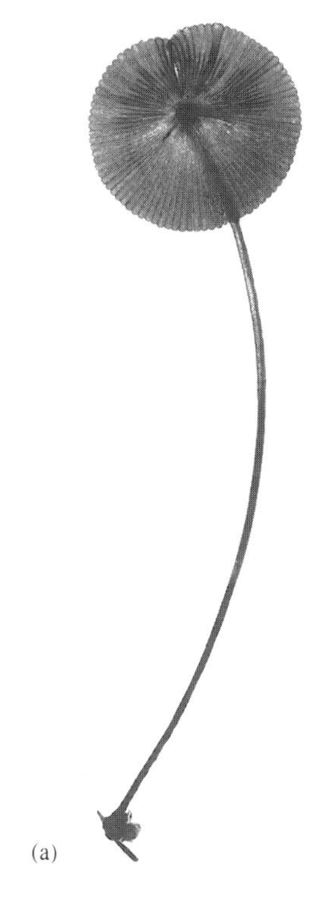

Figure 1.1 Acetabularia acetabulum, the mermaid's cap alga.

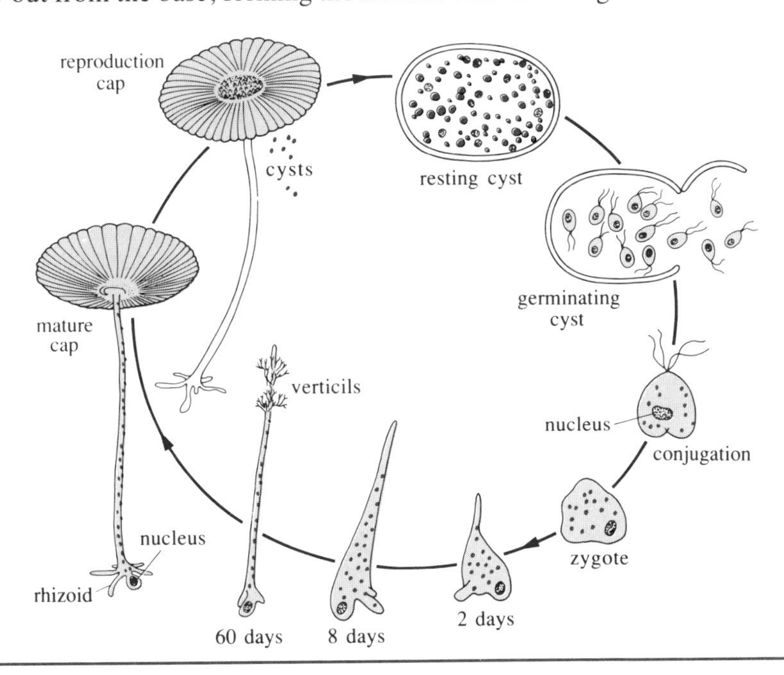

Figure 1.2 Acetabularia life cycle.

of about 150 μm as the plant develops, but remains in one branch of the rhizoid. Since the nucleus is the source of the species-specific macromolecules that influence the formation of the distinctive structure produced at the tip, these macromolecules must somehow be transported from the rhizoid at the base to the apex.

Recall that the DNA in the nucleus produces messenger RNA (mRNA) molecules which code for specific proteins. These nuclear products are distributed around the cytoplasm by a process called cytoplasmic streaming and are translated into proteins at sites such as the growing tip.

When the growing alga reaches a length of about 1–1.5 cm, a **whorl** of small hair-like projections called **verticils** is produced. These grow and branch into delicate structures. Tip growth continues from the centre of the whorl and, after several days, another whorl appears. This is repeated a number of times until a cap primordium is produced, which grows into the characteristic structure about 0.5 cm in diameter that is the clearest morphological identifier of the species. Growth then ceases, after a developmental period of about 3 months. The cytoplasm in the hairs flows back into the stalk and the empty hairs drop off, the plant retaining nothing but their scars.

A mature *Acetabularia* can live for several years without changing its form. However, under normal conditions the alga begins after a few months to change its state in a dramatic manner. The nucleus in the rhizoid undergoes a series of mitotic divisions that result in thousands of nuclei which, with the cytoplasm, stream up into the cap. Rhizoid and stalk are left empty, while within the cap cysts are produced, each containing hundreds of haploid nuclei which are the consequence of a meiotic division of the diploid nuclei. Nuclei and cytoplasm differentiate in the cysts, producing flagellated gametes. The cap wall then dissolves, releasing the cysts, which subsequently open and release the gametes which conjugate in pairs, producing zygotes, and the cycle begins again.

1.1.1 Polarity and gradients

These events in the life cycle of *Acetabularia* are all remarkable processes, involving an intricate interplay among a number of basic cell properties of which biologists have only the most rudimentary understanding. The very large size and the simple form of the plant allow experiments that would be difficult with other organisms. The plant has great regenerative capacities. If the cap is cut off, a new one is produced by the same type of process as in normal growth: a tip is produced which generates first whorls and then a cap. What is more remarkable is that if the rhizoid is cut off as well as the cap, leaving only a non-nucleated stalk, a cap will regenerate, most frequently from the same end where it was before. The other end usually does nothing, though occasionally a rhizoid-like structure is produced, but without a nucleus.

◇ What do you infer from the regeneration of a cap on a non-nucleated stalk about the distribution of nuclear products in the alga?

◆ The cytoplasm in the stalk must contain the nuclear products required to make a cap.

If the regenerated cap is removed from this alga with no nucleus, it is unable to produce another. Evidently the store of nuclear products in the cytoplasm gets used up, and in the absence of a nucleus no more can be made.

Some non-nucleated stalk segments regenerate caps from the 'wrong' end, spontaneously reversing their normal **polarity** (i.e. the end that was nearest the rhizoid and would not normally produce a cap did so, while the other end did nothing). What causes this? It was found that one way of controlling the regenerative polarity of these stalk segments was by imposing an electrical potential difference between the ends during the early stages of regeneration: the segments then produced caps towards the electrically positive pole. When electrical currents were measured around growing and regenerating plants and the ions responsible for carrying the electrical charge were identified, it turned out that the algae actively pump chloride (Cl^-) ions into the stalk and they leak out of the rhizoid, thus maintaining a flow of current through and around the plant (see Figure 1.3). Current densities up to 380 $\mu A\ cm^{-2}$ have been recorded, which is the record so far for a developing organism. (Current density is measured as flow of current, in this case in microamperes, across unit area, in this case a square centimetre.)

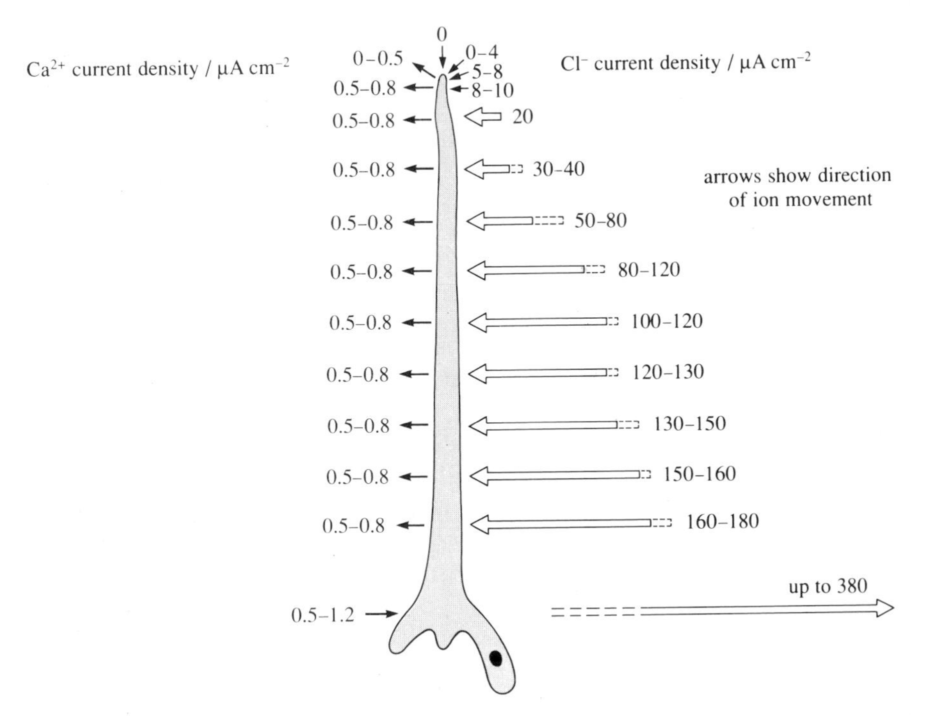

Figure 1.3 Electrical activity in *Acetabularia*.

Ions will flow only if there is a net force driving them in a particular direction. So we can infer from the observed entry of ions into the stalk and out of the rhizoid that there must be a gradient of some kind. This is an electrochemical gradient, defined by the differences of electrical potential along the alga, that drives the negatively charged chloride ions down the stalk and out of the rhizoid. Later it will become evident that gradients are involved in the spatial organization of all developing organisms. It is not known whether the chloride electrochemical gradient in *Acetabularia* is involved in generating the morphological polarity of the cell (the difference of form between the growing tip and the rhizoid), but it certainly plays a significant role in the electrical polarity. Evidence will be presented shortly that another ion is much more directly involved in the morphological changes of the cell during growth.

The results shown in Figure 1.3 indicate that electrical currents are somehow involved in normal growth and regeneration in *Acetabularia*. But are they actually necessary for these processes? When an electrical circuit was arranged so that no current could flow between the ends of an alga, there was no growth. So it appears that ionic fluxes are indeed a necessary part of the growth process.

Many different species show this phenomenon of electrical currents being a necessary accompaniment of growth and regeneration, and some developmental biologists believe that this is a universal component of development. Humans, for example, can regenerate the tips of their fingers if part of the last joint is lost. However, this happens only if the wound is left open so that a special wound epithelium is formed, across which an electrical current has been shown to flow. If the wound is closed by sewing a flap of mature skin over it, no current flows and no regeneration occurs. In regenerating fingertips the current is carried primarily by sodium (Na^+) ions, which leave through the wound epithelium and are pumped into the old epidermis further back from the tip. So different ions are involved in regenerative processes in different species.

In other species, such as newts and salamanders, if such 'currents of injury' are prevented from flowing in regenerating digits or limbs, the regeneration process stops. On the other hand, it is not possible to induce limb regeneration in species unable to do this (e.g. frogs or rats) simply by passing a current across the wound epidermis. So although electrical currents do appear to be involved in growth and regeneration, they are only one component of the process. In general, there are no simple causes in development; rather, there is a cooperative interaction of many component events, failure of any one disturbing the overall process.

1.1.2 Morphogenesis

Although Cl^- ions are the main carriers of the electrical current associated with the growth axis in *Acetabularia*, another ion is much more directly involved in the detailed shape-determining or morphogenetic events at the tip. Simply as a result of reducing the concentration of calcium ions in the seawater in which plants grow, from its normal value of $10\,mmol\,l^{-1}$ to a value of $3\,mmol\,l^{-1}$, the algae lose the capacity to produce caps though they still grow actively, producing many sets of whorls. A further reduction to $2\,mmol\,l^{-1}$ and they continue to grow but can no longer make whorls either; while at $1\,mmol\,l^{-1}$, the tip becomes bulbous and all growth stops.

◇ What do you expect to be the form of a cell growing in $2\,mmol\,l^{-1}$ calcium?

◆ Since the tip forms and the cell grows but no whorls or caps are produced, the alga must be simply a stalk that continues to elongate.

These effects on morphogenesis result also from the use of ions such as cobalt (Co^{2+}) and lanthanum (La^{3+}) that block the movement of calcium across the membrane into the cell, so it is not simply the effect of calcium on the cell wall that is involved. Why should a substance as simple as calcium play such an important role?

Any change of shape in organisms requires mechanical work, and this usually involves microfilaments which are present in the cytoplasm of all eukaryotic cells, plant and animal. These form a network throughout the cytoplasm called the **cytoskeleton**, a highly dynamic structure. Microfilaments are polymers of the protein actin. These polymers grow longer or shorter by addition or loss of actin molecules in a manner that is highly sensitive to conditions in the cytoplasm, one of the major controlling substances being calcium. Also, there are enzymes in the cytoplasm that cut up the filaments, enzymes that are activated by calcium. Associated with the actin filaments is another major protein, myosin, very similar to that which occurs in muscle. The actomyosin polymers are capable of contraction, another process that is under the control of calcium. There are other structural elements of the cytoskeleton known as microtubules, also sensitive in their state of polymerization to local concentrations of calcium. Changes of shape in cells always involve processes of depolymerization and repolymerization of microtubules and microfilaments, which give the cytoplasm its qualities of structural plasticity. So if the tip of *Acetabularia* is going to change its shape, developing first whorls and then a cap, it is not surprising that calcium is intimately involved.

However in algae, and also in plants, no significant change of shape can occur in a cell without the wall changing. Calcium is itself a major ionic constituent of the wall, and changes in its concentration can significantly alter wall elasticity. Exactly how the state of the cytoskeleton is linked to that of the cell wall, and both of these to changes in ions such as calcium, are active research problems. Since reducing calcium in the seawater interferes with normal morphogenesis, which occurs at the growing tip, it might be expected that calcium levels would be elevated in the growing tip and in the region where a whorl of hairs or a cap is produced compared with the parts of the stalk in which there is no active morphogenesis occurring.

In 1988, research by Lionel Harrison in Canada confirmed the role of calcium. He used a chemical called chlorotetracycline (CTC) which fluoresces in response to bound calcium (e.g. calcium attached to membranes or structures within the cell, such as components of the cytoskeleton). The concentration of this bound calcium will reflect the concentration of free calcium, although indirectly. What is observed is that the amount of CTC fluorescence is greatest at the growing tip of a cell, though there is also some fluorescence elsewhere (the wall, for example, also contains bound calcium, so it fluoresces). However, what is particularly interesting is the pattern of change in CTC fluorescence during whorl formation. There is a characteristic change in the shape of the tip during this process. The growing tip has a conical form (Figure 1.2). Just before a ring of whorls is produced, the tip flattens and the ring of verticils then arises at the periphery of the flattened region. If calcium is directly linked to these morphological changes, then the CTC study should reveal something interesting. Harrison observed that maximum CTC fluorescence in these flattened tips was present in an annulus corresponding to the region where the verticils are initiated. As they form and grow, the intensity increases. This work confirms the intimate relationship between calcium and morphogenesis in this cell, but it does not tell us how it acts. Similar studies have been made on other cells with tip growth such as lily pollen tubes and root hairs, which also show strongest CTC fluorescence at the tip.

The observations presented so far suggest that the detailed shape changes that occur during the development and regeneration of *Acetabularia* depend upon an integrated set of activities involving electrical currents, differential ion distributions, cell wall and cytoskeletal changes, as well as specific mRNAs transported from the nucleus to the tip and translated locally into the proteins required for tip, whorl, and cap formation. These processes, coordinated in time and over space to give characteristic shapes during development, are collectively known as a **morphogenetic field**. In *Acetabularia* this field has certain properties that are particularly easy to study, such as the strong electrical currents and the effects of calcium on morphogenesis; but similar morphogenetic fields exist in all unicellular and multicellular organisms, since they must all generate specific forms during their life histories.

Other species of marine alga, particularly the intertidal seaweeds *Fucus* and *Pelvetia*, have been extremely useful in studying electrical currents, polarity, and the role of calcium ions in growth and morphogenesis. These multicellular algae shed haploid spherical eggs with no sign of a polar axis, which are fertilized by sperm. By 12 hours after fertilization the cell begins to bulge in one region, establishing the rhizoid–stalk axis, and shortly thereafter the cell divides with a cleavage plane perpendicular to the axis (Figure 1.4). The cell with the elongating tip develops into the rhizoid, while the other cell gives rise to the thallus, both undergoing extensive growth and cell division. Experiments demonstrated that the first indication of polarity in the eggs, well before there is any sign of a shape change in the cell, is the occurrence of an electrical current that enters the future rhizoid region and leaves the future thallus. Furthermore, in these cells calcium is a major carrier of the current, and if the concentration of free calcium inside the cell is kept at a very low level by the injection of a nonlethal calcium-binding substance, the egg fails to form an axis.

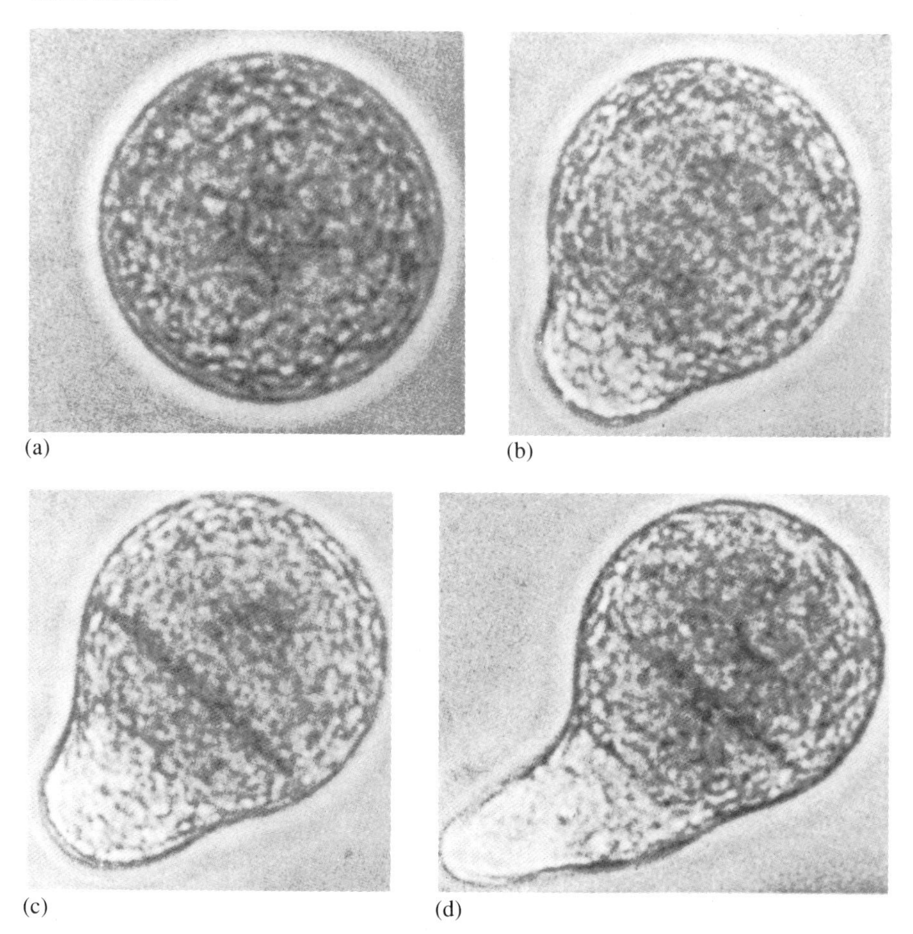

Figure 1.4 Development of *Fucus* egg.

1.1.3 Making shapes

Why does *Acetabularia* make the sequence of shapes shown in Figure 1.2? In particular, why does it make whorls of verticils that simply drop off and play no role in the adult? There are two types of answer that can be given. One type of answer is to say that in the evolution of the giant unicellular green algae there were ancestors of *Acetabularia acetabulum* in which whorls did play a role in the adult. This is perfectly possible because there are species of

Chlorophyceae that do not make caps and the whorls function as gametangia (the structures where the gametes are produced). The whorls in *Acetabularia* are then to be understood as relics; like the human appendix they were useful once but now they persist because of a kind of developmental inertia. This leaves us with two problems. Why were whorls not eliminated from this species, since it is widely believed that natural selection has remarkable powers of moulding organisms to almost any form that is useful, and getting rid of useless parts? The second problem is to account for the production of whorls in the first place. What makes these structures possible? Recognising their usefulness in certain species doesn't explain how they are generated.

The other type of answer solves both of these problems together. It is the form of explanation that has always appealed to some developmental biologists. This depends upon the idea that organisms are physical systems of a particular type, organized according to principles that make particular shapes easy to generate, while others are not. If this is the case, then it should be possible to show that a process organized according to the physical principles operating in *Acetabularia* should form a tip during its growth, that this should flatten, and that a ring of hairs should then be produced, all as a natural and spontaneous sequence. How could this idea be tested? As in other sciences, it is necessary to produce a mathematical model of the processes involved in the development of the alga, to 'grow' the model, and to see if structures like whorls arise naturally. This was done by Brian Goodwin and his colleagues at the Open University in collaboration with Christian Brière in France. The model was based on the properties of the cytoskeleton in interaction with calcium, as described in Section 1.1.2, and on the principles according to which plant cell walls are believed to grow. The model is mathematically complex, but biologically very simple — a kind of universal plant cell.

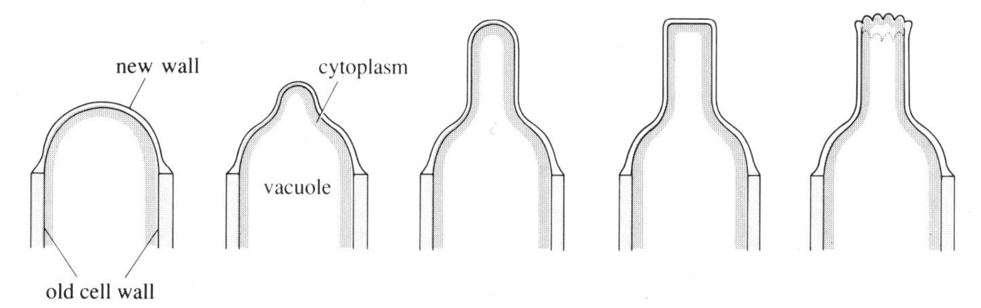

Figure 1.5 Diagram of *Acetabularia* tip regeneration.

The sequence of shapes that the model sought to explain is shown in Figure 1.5. This is what happens during the initial stages of regeneration, as described in Section 1.1.1. The sequence produced by the model is shown in Figure 1.6. First there is a dome (Figure 1.6a), representing the initial stage of regeneration. The model then predicts that calcium concentration should increase at the pole and, when it reaches a sufficient value, a bulge appears — the initiation of a tip (Figure 1.6b). This is followed by an interesting sequence. The tip extends as growth occurs (Figure 1.6c) and then suddenly, instead of the calcium concentration remaining high at the tip, it spontaneously transforms into an annulus, producing a ring of elevated calcium around the tip (Figure 1.6d). At the same time, the tip flatten. The annulus increases in concentration, and then it breaks up into a ring of peaks, each peak initiating the growth of a hair (Figure 1.6e). The model was not specifically programmed to produce this result: it happens simply as a result of the calcium–cytoskeleton–cell wall interactions and changes of shape during growth. Once a tip is initiated and growth occurs, calcium annulus formation, tip flattening and the whorl pattern develop naturally. The patterns of calcium

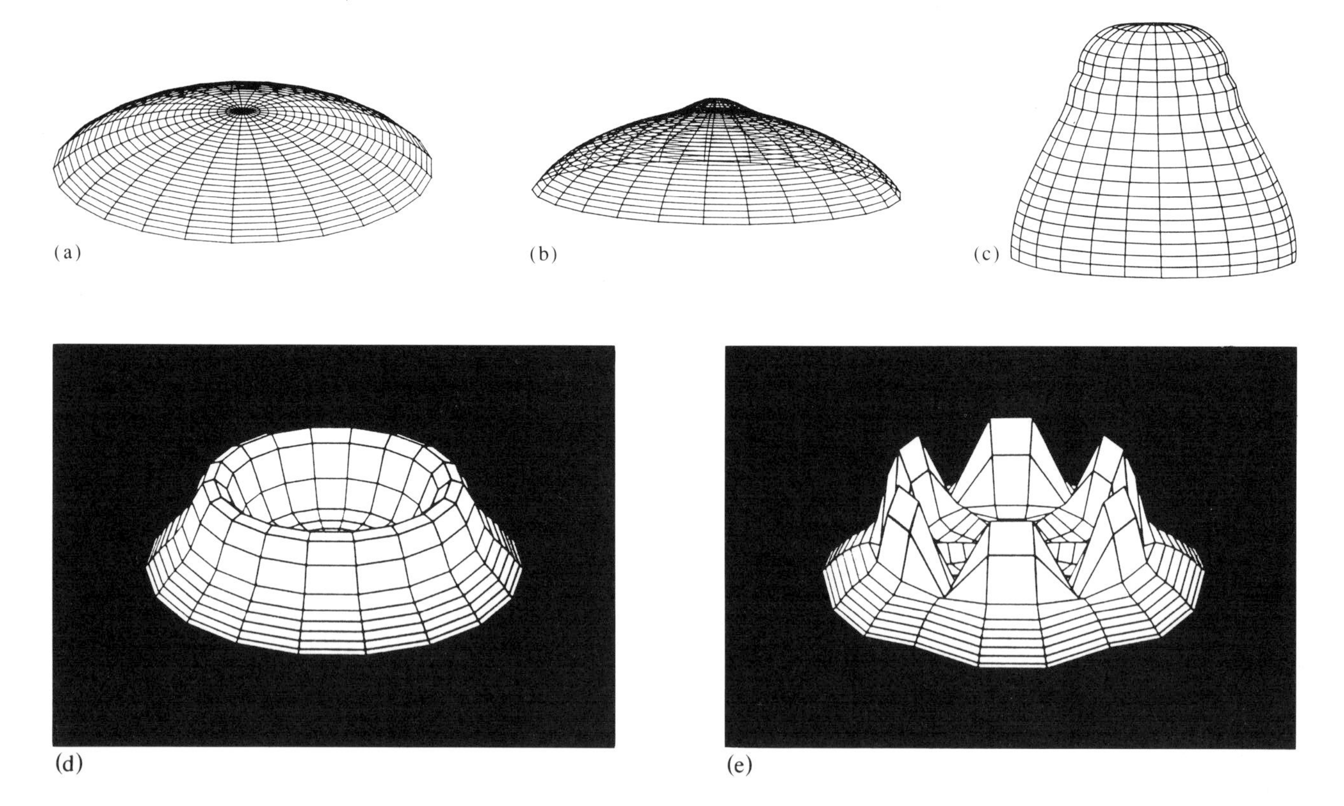

Figure 1.6 Results from mathematical model of *Acetabularia* tip regeneration. (a)–(c) Changes in tip shape. (d)–(e) Calcium concentrations following tip flattening.

change match closely the results reported by Harrison (Section 1.1.2). However, it must be remembered that his observations were on bound calcium, whereas the model describes free cytoplasmic calcium. This remains to be measured, to see if the predictions are borne out.

What the simulation clearly shows is the possibility that *Acetabularia* produces whorls because these structures are a natural consequence of the way the morphogenetic field is organized, involving calcium interacting with the cytoskeleton, and with the cell wall during growth. This could explain both how whorls are possible in such a system, and why *Acetabularia* produces them even though they serve no function in the adult. The alga is simply doing what comes naturally.

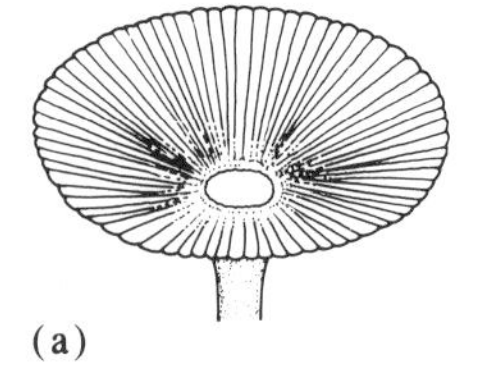

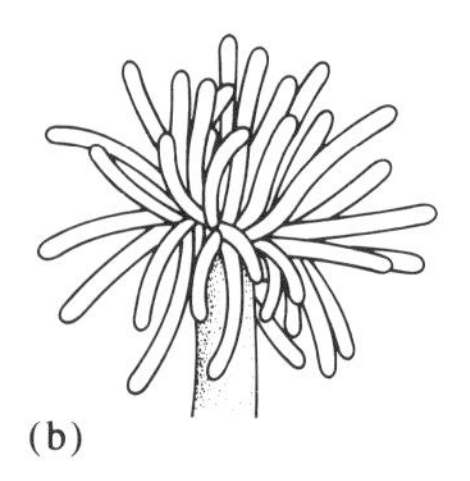

Figure 1.7 (a) *Acetabularia mediterranea* cap. (b) *Acetabularia crenulata* cap.

1.1.4 Nucleus and cytoplasm

Electrical currents, ion gradients, calcium, the cytoskeleton and the cell wall all appear to play significant roles in growth and morphogenesis. However, these are components of the universal pattern generators in eukaryotic cells with walls, and so they cannot account for the distinctive character of the morphological structures produced in different species. *Acetabularia crenulata* is closely related to *Acetabularia acetabulum*, but has the distinctly different cap structure shown in Figure 1.7. The contribution of genetic information in the nucleus to cytoplasmic morphogenesis can be simply and directly investigated in these species by transplanting nuclei. This is achieved by cutting off the rhizoids of plants and grafting them to stalks of other species as illustrated in Figure 1.8. The caps can then be cut off and regeneration allowed to occur. Will the regenerated caps have the species character of the nucleus or of the cytoplasm? It turns out that both reciprocal hybrids produce caps of mixed morphological character.

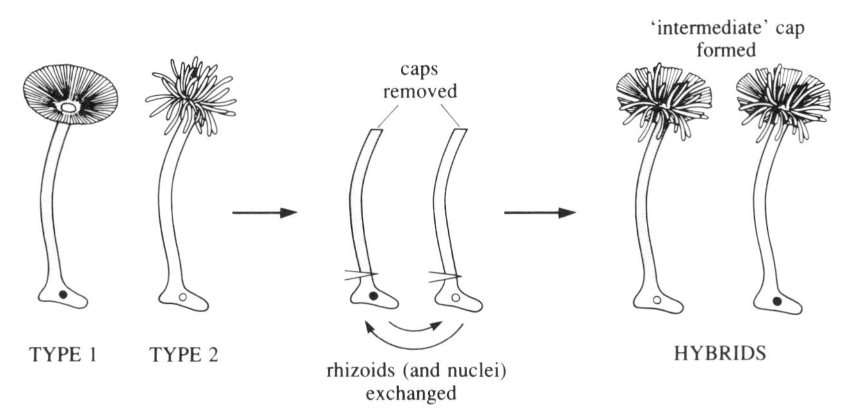

Figure 1.8 Cap grafting experiments on *Acetabularia*.

◇ Suppose that these caps of mixed species morphology are cut off and the cells allowed to regenerate new ones. What would you predict to be the form of the new caps?

◆ They are expected to have a morphology determined by nuclear type, since the nuclear products in the original cytoplasm of the hybrid will be used up during the first tip regeneration and will be replaced by new products from the nucleus.

Cap formation involves the synthesis of new proteins. Thus, specific mRNA molecules must be present in the cytoplasm. Because a cap can be formed without the nucleus, these RNA molecules must have been present in the cytoplasm before the nucleus was removed. This mRNA must be particularly stable because cap formation can occur weeks after the removal of the nucleus.

So for this organism we could suggest a hypothesis that the control of differentiation is by cytoplasmic control over the translation of stable mRNA into the specific proteins required for cap formation, rather than control at the level of transcription of mRNA in the nucleus. What evidence is there that mRNA is long-lasting? Consider the two experiments shown in Figure 1.9.

Experiment 1 — A growing *Acetabularia* is cut into two halves. The two halves of the cell are treated with ribonuclease, an enzyme that breaks down RNA (the enzyme is added to the medium around the cells). Neither half generates a stalk or cap because both the existing and newly synthesized mRNA are broken down. When the ribonuclease is removed from the medium, regeneration occurs only in the half with a nucleus.

Experiment 2 — The *Acetabularia* cell is cut close to the nucleus and the two parts treated with actinomycin D, an antibiotic that inhibits the transcription of DNA. The regeneration of the fragment with a nucleus is stopped, although the other fragment regenerates. There does not appear to be sufficient mRNA in the fragment containing the nucleus to allow regeneration.

From these experiments you could deduce that the regeneration of *Acetabularia* is dependent on the transcription of DNA into stable, mRNA. These experiments, though, are fairly crude; the addition of an enzyme that destroys all the RNA in a cell may affect the cell in other ways.

A more elegant investigation shows conclusively that the control of the production of different proteins is organized in the cytoplasm rather than in the nucleus; that is, control is exerted at the level of **translation** rather than **transcription**. *Acetabularia* produces three different phosphatase enzymes,

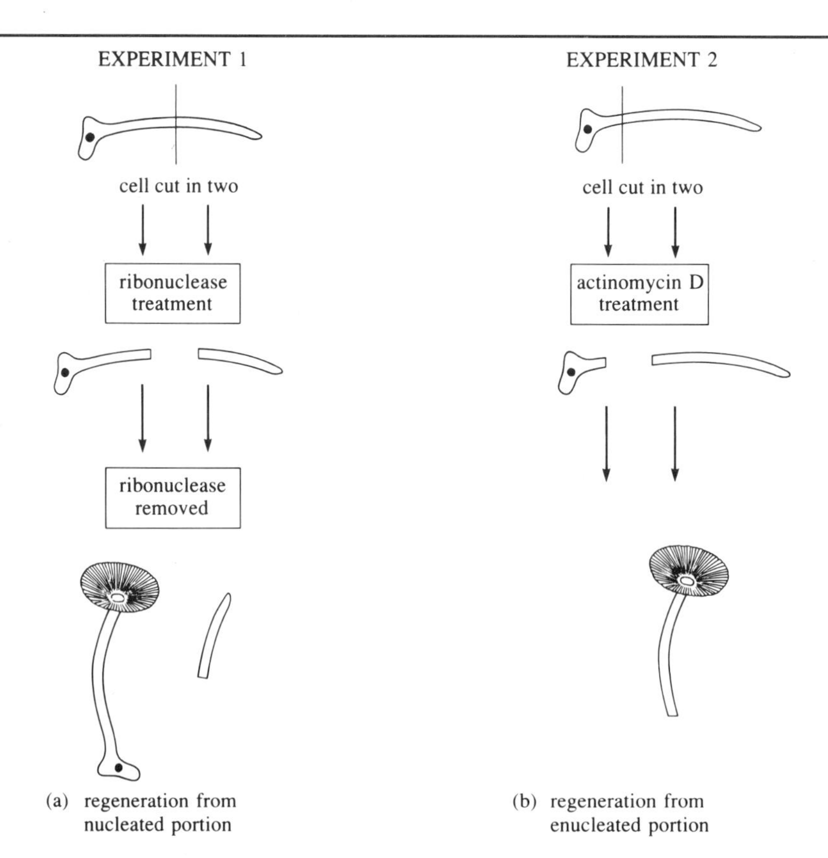

Figure 1.9 Ribonuclease and actinomycin D experiments.

which can be distinguished by measuring the pH values at which they give maximum activity. In the normal, nucleated cell these are produced in a precise sequence before cap formation. This sequence is also seen in cells that have their nuclei removed. Thus there must be specific translation in an ordered manner from the mRNA in the cytoplasm. We shall see in the next section how phosphatases may be implicated in the morphogenesis of a quite different organism, *Hydra*.

This work on *Acetabularia* was done before more direct techniques became available for observing specific mRNA and protein production during cell differentiation in developing organisms. These will be discussed in Chapters 3 and 4. What the *Acetabularia* studies reveal is that transcription of mRNA in the nucleus and its presence in the cytoplasm is necessary for the development of the alga, and that the cytoplasm controls the timing of mRNA translation into specific proteins associated with cell differentiation.

Summary of Section 1.1

Acetabularia acetabulum is a giant unicellular marine alga whose life cycle involves the growth and morphogenesis of the zygote, product of the fusion of two isogametes, into a distinctive adult form via a well defined sequence of transformations. The life cycle is completed by the formation of haploid isogametes in the cap, which are released and fuse in pairs to produce new zygotes.

Acetabularia has great regenerative powers. During regeneration electrical currents due mainly to a flow of chloride ions are a necessary component of growth and morphogenesis. Electrical currents, often involving other ions, also occur during regeneration in other species.

Calcium is an important variable in the growth and morphogenesis of *Acetabularia*. It has important effects on the state of the cytoskeleton, the network of filaments that give structure to the cytoplasm, and on the elasticity of the cell wall. The distribution of calcium changes as a whorl is produced in *Acetabularia*.

Natural selection does not explain why whorls are produced in *Acetabularia* since they serve no function in the adult. But a model that describes cell shape in terms of calcium interacting with the cytoskeleton and with the cell wall shows how whorls can be generated naturally, simply as a consequence of the dynamic interaction of cellular constituents.

Morphological differences between *Acetabularia* species are associated with genetic differences which can be revealed by nuclear transplantation experiments. Gene expression via mRNA production is necessary for development, as shown by ribonuclease and actinomycin D studies; while cell differentiation (cap formation) involves control over the translation of mRNA as shown by sequential phosphatase synthesis.

Question 1 (*Objective 1.2*). Give a description of *Acetabularia* development in terms of increasing structural complexity.

Question 2 (*Objectives 1.2, 1.3 and 1.4*). Describe how the polarity and the morphological complexity of *Acetabularia* can be controlled by environmental influences. Do these results imply that polarity and the emergence of morphological complexity are controlled during normal development by the same means? Give reasons for your answer.

Question 3 (*Objective 1.5*). Which of the following statements are *true* and which are *false*?

(a) The fact that a stalk segment without a nucleus can regenerate a cap shows that the cytoplasm can synthesize mRNA.

(b) If the cap of *Acetabularia* species 1 is removed and the stalk is then transplanted to a rhizoid of species 2, it is expected that a cap similar to that of species 2 will be regenerated. This is because the nucleus is the main factor determining cap morphology.

(c) Treatment of *Acetabularia* with the enzyme ribonuclease before a cap is formed stops cap formation. This shows that cap formation is due to translational rather than transcriptional control.

(d) The sequential production of specific proteins in cells with and without nuclei provides good evidence against the idea that only the control of transcription is involved in cellular development and differentiation.

1.2 *HYDRA*: THE REMARKABLE REGENERATOR

Hydra is a member of a large phylum of simple multicellular animals, the *Cnidaria*. Its basic body design is essentially a hollow column containing the enteron with a mouth surrounded by a domed **hypostome** and a ring of **tentacles** at one end, and an adhesive foot or **basal disc** at the other (Figure 1.10a). There are about fifteen distinguishable cell types, and some 100 000 cells in a mature individual, organized into distinctive structures. The body, up to 15 mm long, consists of two epithelial layers, the outer epidermis and the inner gastrodermis, separated by a thin extracellular matrix, the mesogloea. Some of the major cell types are shown in Figure 1.10b, which represents a section of the wall of the enteron in the gastric region as shown by the box in Figure 1.10a. The outer layer (epidermis) has interstitial cells, nematoblasts, **nematocytes**, nerve cells and the epitheliomuscular cells which predominate in this layer. They contain at their base longitudinally oriented muscle threads, the myonemes, closely associated with the mesogloea. Contractile movements of the body result from their action. These movements are controlled by electrical activity in a diffuse nerve net or plexus made of many nerve cells which are distributed throughout the body in both tissue layers.

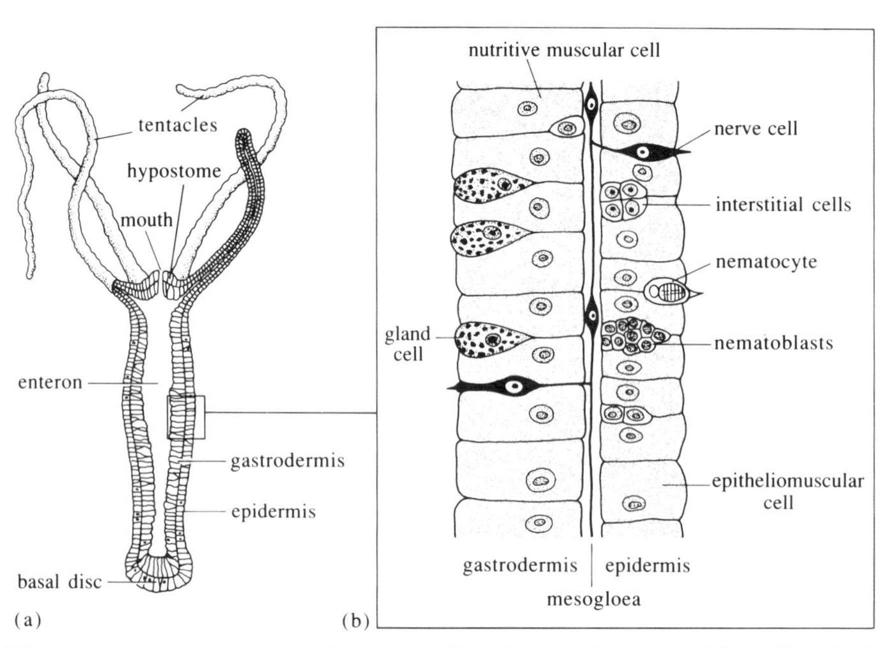

Figure 1.10 Hydra (a) whole body (b) enteron wall section.

Nematocysts are large, complex organelles that are formed within cells called nematocytes, of which there are four kinds. They are located at the surface of the body and the nematocysts can be discharged explosively on stimulation, releasing a thread-like process that wraps itself around any object such as a prey organism, thus immobilizing it. The tentacles contain many of these nematocysts. They arise from the small **interstitial cells** that are packed in at the base of the epitheliomuscular cells. In the early phase of differentiation these cells are called nematoblasts. Interstitial cells also produce nerve cells, and gametes when the animal enters a sexual phase.

Large nutritive muscular cells, with many fluid-filled sacs (vacuoles) where food is digested, predominate in the gastrodermis. Their muscle fibres, or myonemes, lie at right angles to those in the epidermis, so that their contraction narrows the enteron and causes elongation of the animal. Mucous gland cells are abundant near the mouth and enzymatic gland cells secrete enzymes for the preliminary extracellular digestion of food. Interstitial cells also differentiate into these cell types. The basal region has a relatively inactive gastrodermis and epidermal gland cells that secrete mucopolysaccharides. The cell composition varies systematically throughout the different regions of the organism.

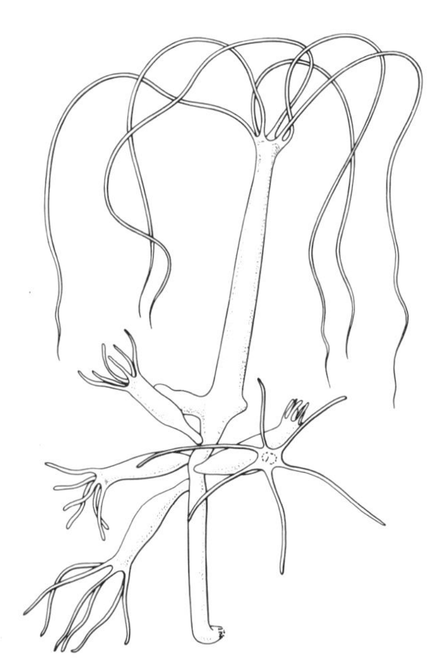

Figure 1.11 Hydra with buds (asexual reproduction).

The organism has two patterns of reproduction, asexual and sexual, the former being the predominant mode. In this process a new individual arises as a bud towards the bottom of the gut and develops into a fully formed but miniature *Hydra* which detaches from the parent and grows to adult size, when it too starts to bud (Figure 1.11). This is a type of vegetative reproduction, with a continuity of tissue between parent and offspring. It is a very widespread pattern of reproduction throughout plant and animal phyla.

Sexual reproduction is induced by a combination of starvation and reduced temperature which results in a cessation of growth and the initiation of **gametogenesis**. Cells in the outer epidermal layer that would normally contribute to the growth and development of the organism or a bud divide and form cell masses that lead to the production of testes or ovaries, both of these forming within the same individuals in some species. The mature, haploid eggs are fertilized by sperm, resulting in the diploid zygote which is then shed into the fresh water in which *Hydra* lives. This then undergoes a series of cell divisions, resulting in a mass of cells which develop into a small *Hydra* which feeds and grows into an adult. Because the organism is so small during its development from zygote to adult form, it is difficult to study. The real value of *Hydra* in the investigation of developmental processes comes from its remarkable regenerative capacity and the budding process in mature animals, which are much more accessible to experimental manipulation and observation than the embryo.

1.2.1 Regeneration in *Hydra*

The extraordinary ability of parts of *Hydra* to transform into a whole organism have been known since the mid-18th century when the Abbé Trembley, a French tutor to a wealthy Amsterdam family, collected specimens from the local canals and observed the regeneration of complete animals from fragments. Any part of the main body column of *Hydra* has this ability (Figure 1.12), though there is a lower size limit to a fragment that can regenerate, amounting to about a fiftieth of the adult size. Even the inner gastrodermal layer has the capacity, when isolated, of giving rise to a whole organism, with all the different cell types. These transformations occur without a significant amount of growth or cell division so the regenerate is a

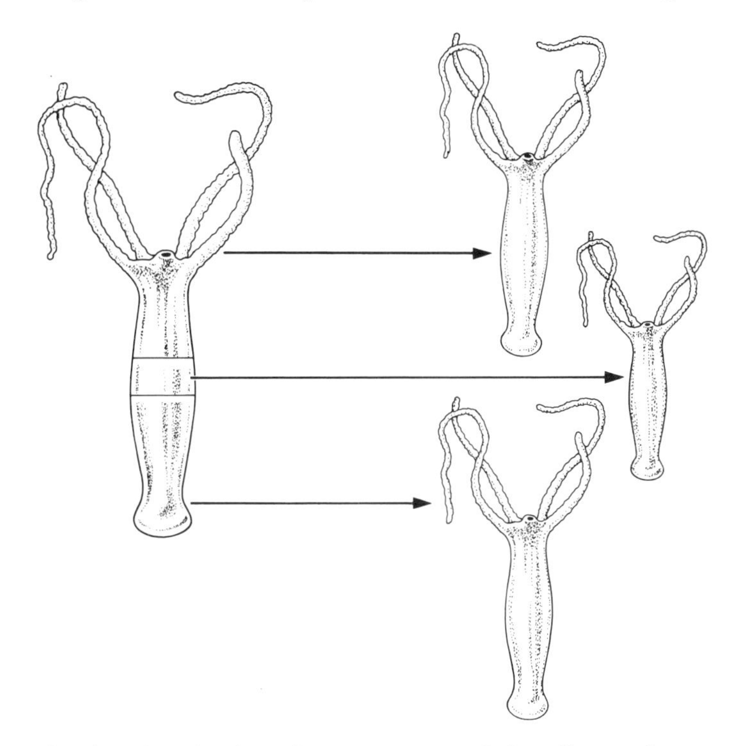

Figure 1.12 Regeneration of *Hydra* from body fragments.

small *Hydra* that then feeds and grows to normal size. Since a fragment from the mid-gastric region, for example, must form both a head with tentacles and a foot, and these involve different types of cell from those in the fragment, it is clear that during regeneration cells change their states. They also change their relative positions. However, the original polarity of the fragment is retained: a head regenerates from the end closest to the old head, while the foot forms at the other end. This is reminiscent of the same phenomenon in *Acetabularia*.

◇ Does the fact that *Hydra* is a multicellular animal and *Acetabularia* is a unicellular plant necessarily mean that the processes involved in maintaining polarity are different?

◆ No. Electrical currents, ion gradients and ion flows might well be involved in the determination of polarity in *Hydra*, just as in *Acetabularia* but with ions flowing between cells.

There is evidence from studies with certain species of *Cnidaria* that electrical currents can influence the polarity of a regenerate. Overall, the process of producing a whole organism from a part during regeneration results from a profound reorganization, involving changes of cell state and spatial relations so that a small but complete organism is produced. Different types of cell have differing abilities to undergo these changes of state, some being fixed in their differentiated states while others can form many of the fifteen or so different cell types that make up the complete organism. The interstitial cells have the potential of differentiating into a number of different cell types, as we have seen. They give rise to nerve cells, of which there are three kinds, and nematocytes, which come in four varieties, as well as gastrodermal gland and mucus cells and, in the sexual cycle, oocytes and sperm cells. They are accordingly called **multipotent** cells. An egg cell is capable of producing all the different cell types in the body, and so is called **totipotent**. Nematocytes, on the other hand, are unable to change their state and so are said to be **irreversibly differentiated**. These terms describe very basic properties of cells in developing organisms. The challenge facing the developmental biologist presented with the remarkable phenomenon of regeneration in *Hydra* is to explain how cells know what to do and where to go to produce a complete organism from a fragment.

1.2.2 The dynamics of development

One important observation puts this challenge into a context which makes the phenomenon easier to understand. It comes from a study of cell movement in an adult *Hydra* that is maintaining a stable form. In this study, small patches of cells are taken from one organism and labelled with a vital dye (a dye, such as neutral red, methylene blue or Nile blue sulphate, which enters and remains in cells but does not damage them in any way). The dyed cells are grafted into specific regions of mature animals, where they rapidly heal into the surrounding tissue. The movement of these marked cells can then be followed.

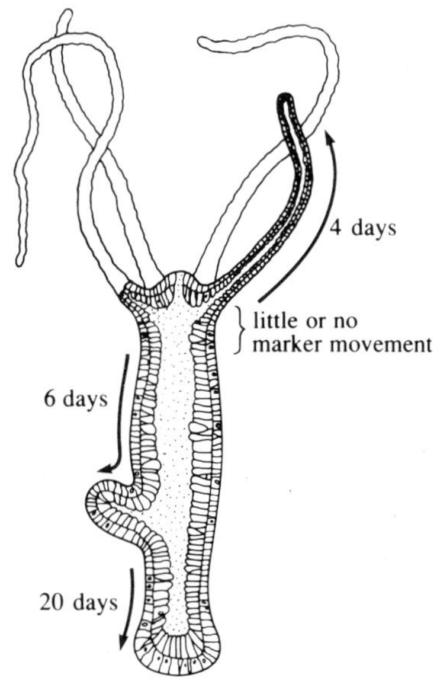

Figure 1.13 Cell movement pattern in *Hydra*.

The cells move in characteristic ways, as shown in Figure 1.13, so that the form of the mature organism is maintained like that of a fountain, involving a continuous flow of its substance while the form is relatively constant. Nematocytes, which account for 20–25% of the total cells in an organism, continually release their nematocysts in feeding animals, about 25% of the total number being lost every day. Cells are also lost from the sticky foot; and there is a continuous flow of cells into growing buds. This loss and movement is balanced by the division and differentition of cells, such as the epithelio-muscular, nutritive muscular and interstitial cells, which are distributed throughout the main body column but not the tentacles or the foot. Therefore the maintenance of adult form involves continuous movements of cells and changes of their state in a manner appropriate to their position. This is true in varying degrees of all organisms.

◇ A regenerating fragment from the mid-gastric region clearly cannot feed, having no tentacles or mouth, and so cell divisions soon cease. Nevertheless, the fragment regenerates as a complete, though small, *Hydra*. How is this possible?

◆ The interstitial cells in the fragment migrate and differentiate into nerve cells, nematocytes, and the other cell types making up the regenerating tentacles and hypostome, while other interstitial cells, epitheliomuscular and nutritive muscular cells reorganize spatially to form a new basal disc, differentiating into cells characteristic of this region.

The process of regeneration is not, then, fundamentally different from what goes on in a normal organism as far as cell movement is concerned. The spatial reorganization is simply more extensive. What needs to be understood is the nature and distribution of the influences that induce cells to change their positions and their states in a manner that results finally in a complete, functional organism.

Since cells have different properties depending on their position in the animal, one basic method of studying the pattern of influences that maintains the whole integrated structure is to move parts around relative to one another. If the head is removed from an animal by a cut such as that shown by the line marked 1 in Figure 1.14a, then a head regenerates, largely from cells in the sub-hypostomal region between lines 1 and 2. So this tissue has the capacity to form a head. This capacity or potential is referred to as competence for head formation. Suppose now that the head is removed, as with cut 1, and then the sub-hypostomal tissue is removed by cut 2 and this tissue is grafted into the mid-gastric region of another animal (the host, Figure 1.14b). Such grafts are readily carried out by cutting a slit in the host and inserting the graft, which rapidly heals into place. Will this grafted tissue now produce a head? Because of the intrinsic variability of these experiments, due to small differences in the size of the transplanted tissue and of age between donor and host animals, the results have to be treated statistically.

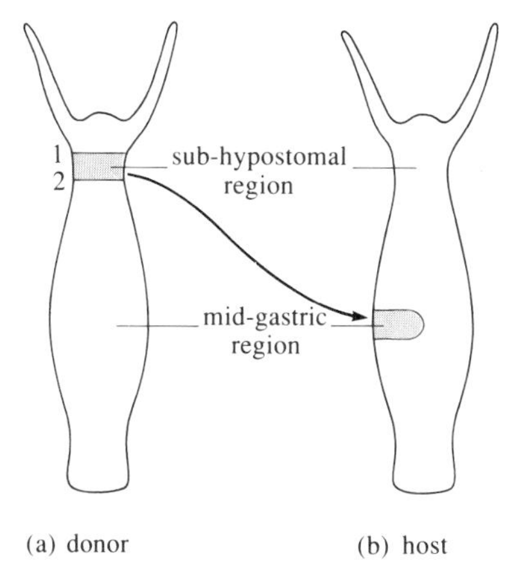

Figure 1.14 *Hydra* grafting experiment.

Some grafts result in the formation of a second head on the host while others do not. If 50% or more of the animals that receive a graft produce a second head, then the graft is said to have the capacity to induce a head. This convention is used to compare the statistics of results from large numbers of different organisms. In the majority of the experiments described above, the grafted tissue simply heals into the new position and is assimilated by the host, failing to produce a head. After a few days, one would never know that tissue had ever been added.

1.2.3 Determination

Suppose the sub-hypostomal tissue of a *Hydra* is left for varying periods of time in freshwater before being grafted into the mid-gastric region of a host. Will this tissue develop the capacity to make a head when grafted to a host before a head actually appears in the isolated tissue? This does indeed happen after 4.5 hours. By this time, isolated tissue has developed the property that when grafted to a mid-gastric region, a secondary axis is produced on the host in over 50% of the cases, complete with tentacles, hypostome, and gastric region that joins up with that of the host at the junction (Figure 1.15). Such a secondary axis is distinct from a bud in that it does not detach from the

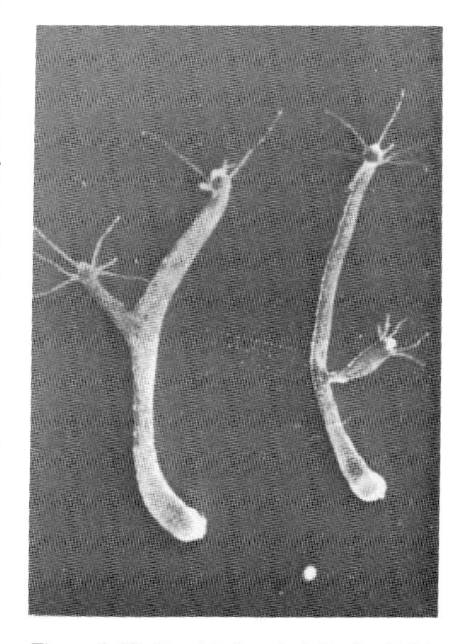

Figure 1.15 Double-headed *Hydra* (left) and normal *Hydra* with advanced bud (right).

primary axis. A head (identified by first tentacle formation) does not appear in isolated sub-hypostomal tissue until about 48 hours, so the grafted tissue develops head-producing capacity long before head structures start to appear. Sub-hypostomal tissue is then said to have become **determined** to form a head after 4.5 hours in isolation when tested in the mid-gastric region of a host.

This shows that the behaviour of grafts is determined by a balance of influences that depend both on the effect of neighbouring tissues, encouraging conformity with their state, and on the state of the grafted tissue itself. Regeneration of whole organisms from tissue fragments shows that all parts of *Hydra* (provided they are above a critical size) have the potential to make a head (and other parts). This potential is expressed only under specific circumstances, such as cutting and separating the parts from one another. Once part of the isolated tissue has changed sufficiently in the direction of becoming a head, determination occurs and the influences from the host that normally inhibit head formation are too weak to stop the determined tissue from developing into a head when grafted into a region of a host where a head would not normally form.

◇Since an isolated piece of mid-gastric tissue can produce a whole *Hydra*, with a foot as well as a head (see Figure 1.12), what do you deduce from the result that only the head end of the animal was induced by the graft (Figure 1.15), and not a foot also?

◆This indicates that the tissue changes more rapidly in the direction of head formation than towards foot formation, head-forming processes becoming determined before those for foot.

An interesting variant on this experiment reveals a very important aspect of the head influence. If sub-hypostomal tissue, newly removed from the donor, is grafted into the mid-gastric region and at the same time the host's head is removed, then the organism is left with sub-hypostomal tissue in two different positions. Will the host axis assimilate the graft and regenerate a new head where the old one was? Or will both tissues, with equal potential to form heads, proceed to do so? The second alternative is what happens, so in this circumstance the graft is determined to make head. This shows that determination is a relative concept: the same tissue (e.g. sub-hypostomal tissue) can be determined when placed in one context (the mid-gastric region of a host without a head) and not determined in another (the same region of a host with a head). Evidently the inhibitory influence that prevented head formation in the graft described in Figure 1.14 depended on the presence of a head. It seems that the head is sending a signal to other parts, effectively telling them that a head is already present and not to make another even though they have this potential. This is a familiar strategy in biological control systems and is called **negative feedback**: the rate at which something is produced decreases as it accumulates, due to an inhibitory effect on its own production.

◇Can you recall any examples of negative feedback that operate in a biological system?

◆You might have recalled blood sugar regulation, thermoregulation, water balance, or many others. For example, the negative feedback that acts in blood sugar regulation is as follows:
As the glucose concentration rises in the blood it stimulates release of insulin from cells in the pancreas. This reduces glucose concentration by stimulating its removal from the blood and conversion into stored form (glycogen) in cells. Hence glucose inhibits its own accumulation in the blood.

What accumulates in the *Hydra* example is a structure, a head. Assuming that this specific structure has associated with it specific substances, it seems perfectly reasonable to suppose that these can diffuse through the organism and inhibit the accumulation of similar substances, hence similar structures, elsewhere.

Evidently this requires that the concentration of the head inhibitor should be greatest at the head itself since this is its main source, as deduced from the observation that removal of a head allows a sub-hypostomal graft to develop into a head in the mid-gastric region of the animal.

◇ *Hydra* buds, emerging towards the basal end of the gastric region, develop heads (see Figure 1.11), whereas a sub-hypostomal graft into the mid-gastric region does not. What do you deduce about the strength of the head-inhibiting signal as a function of distance from the head?

◆ It suggests that the strength of the inhibiting influence decreases as a function of distance from the head.

This can be confirmed by grafting sub-hypostomal regions to positions of the body further from the head than that shown in Figure 1.14. The frequency with which heads are induced increases with increasing distance from the head.

We are thus led to a hypothesis about the general nature of the influences that parts of developing organisms have on each other: they inhibit one another from developing into similar structures. This is a good principle for the stability of various structures in the adult organism. But it clearly is not sufficient to describe the active process whereby the structures are generated in the first place, whether from an egg, a bud, or a fragment of the body. As a minimum, it is also necessary to have self-activating processes such as those involved in generating a bud where no bud existed before; or a head from a region where no head existed. If self-activation is also interpreted in terms of substances, then we are using the second of the control processes involved in dynamic biochemical and physiological regulation: **positive feedback**. In this case, the rate at which something is produced increases as it accumulates, due to a positive effect on its own production.

1.2.4 Morphogens

What are these postulated activators and inhibitors involved in positive and negative feedback likely to be in *Hydra*, and how could one go about identifying them? This defines one of the great treasure hunts in developmental biology, which has so far yielded tantalizingly few results. However, studies on *Hydra* have been among the few to give some success. Others will be described in later chapters.

Identifying HA

By definition, a head activator is a substance that encourages the formation of a head. A biological test (bioassay) for such a substance is to take two similar groups of animals, expose only one group to the substance and then to measure the increase in the number of animals with head structures such as tentacles after exposure to the substance, compared with the number in control animals not exposed to it.

Since *Hydra* is a very small animal, and since it is likely to contain only trace quantities of activators and inhibitors, it takes some heroic biochemistry on tiny amounts of extracted substances to track these down. However, this was achieved in the 1970s by a group in Germany led by Chica Schaller.

In general, it is reasonable to assume that basic control substances are likely to be the same in related species, and often this is true even for distant relatives. Schaller and her colleagues therefore assumed that the sea anemone *Anthopleura elegantissima*, a close relative of *Hydra* but much larger, would contain head activator (HA). Their procedure for identifying the HA was as follows.

A crude aqueous extract of the anemone was found to stimulate head formation. The crude extract was fractioned on an ion-exchange column (apparatus which retains different substances in different positions according to their charge and molecular size). The resulting fractions were each tested for their head-activating properties by bioassay. Those fractions which proved active were further processed by chromatographic and molecular sieve procedures and the resulting sub-fractions were in turn separately tested for head activating properties. This is, as you can imagine, an extremely laborious and time-consuming activity. From the 200 kg of sea anemone, Schaller and her colleagues obtained 20 μg of pure HA. This was sequenced and identified as a peptide of composition glutamine-proline-glycine-glycine-serine-lysine-valine-isoleucine-leucine-phenylalanine.

Distribution of HA and HI

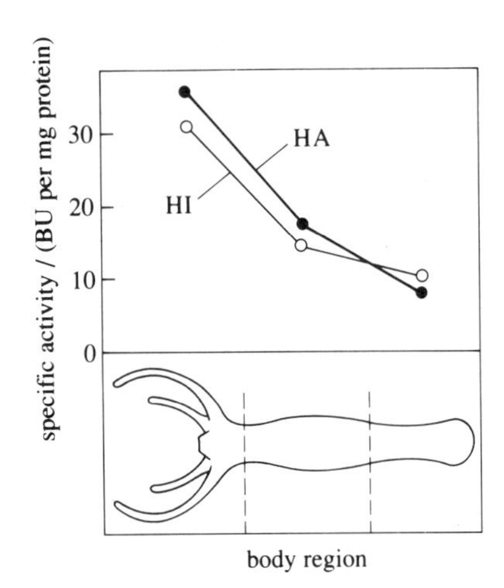

Figure 1.16 Bioassay of head activator and head inhibitor in *Hydra* from three different sections of the adult.

Extracts from three different sections of adult *Hydra* were tested for their head activator properties, using the bioassay procedure described above. This revealed the anticipated gradient of distribution of head activator, with a maximum in the head region and a minimum towards the foot.

Figure 1.16 shows a graph of the results. The numbers on the ordinate of the graph are in biological units (BU), relating to the bioassay. One biological unit of HA is defined as the amount of activator required to produce a 6% increase in the number of tentacles in regenerating *Hydra* 48 hours after the heads were removed, compared with control animals. This is an arbitrary measure, but it provides a consistent standard of comparison for the activity of different extracts.

On the same graph is shown the distribution of head inhibitor (HI). The bioassay involves exposing *Hydra* to extracts from different regions of the body and measuring their effect on the rate of bud formation, which is found to give a more accurate measure than head regeneration. The head is the first part of the bud to differentiate and the inhibitor delays this. One biological unit of HI is defined as the amount of inhibitor required to produce a 50% reduction in the number of buds after 8 hours of exposure, compared with controls.

As expected, the HI concentration, like the HA concentration, increases towards the head of the animal. This shows the correlation between state of differentiation and concentration of control substance. The inhibitor is a non-peptide, hydrophilic substance with low molecular mass. It can diffuse rapidly, but it has proved difficult to purify and identify chemically.

How does HA act? According to Schaller and her colleagues it stimulates interstitial cells to divide and to differentiate into head-specific nerve cells. Head inhibitor antagonizes this differentiation. It is then to be expected that near the head, where HA levels are high, many interstitial cells will become determined to differentiate into nerve cells with head specificity. Further away, in the gastric region, fewer cells will be so determined and the head inhibitor will tend to prevent this. Evidently the way cells behave in any region of the animal depends upon the relative influence of the two control substances, hence on their concentrations and affinities for target sites in the cells that regulate pathways of cell differentiation. Significant changes in these, or in the rates of production of HA and HI, would be expected to alter the normal morphology of the organism in some way.

◇ How could such changes in the internal characteristics of the control processes be investigated?

◆ By looking for mutants with altered morphology.

An interesting mutant called *maxi* has a body length two to three times greater than normal, so that the distance between head and foot is greatly increased (Figure 1.17). This mutant was found to have high HI contents. How does this result in a change of overall size? Remember that HI inhibits bud formation. So an elevated level along the body column will result in a failure of budding until the inhibitor level falls to a value that allows buds to form. This point will be further from the head than in a normal animal. So such a mutant will go on growing until its length and the inhibitor level are such that buds can be produced. Thus HI appears to act like a size regulator. As expected, the converse situation arises in a mutant called *mini*. In these the HI gradient is significantly lowered relative to normal, with the consequence that the budding zone is much closer to the head than usual, and the animals remain very small (Figure 1.17).

Figure 1.17 *Hydra* mutant *maxi* (left) and *mini* (right).

◇ What would you expect to be the consequence of raising HA relative to normal levels?

◆ Since HA encourages head formation, a larger head region would be expected, resulting either in an enlarged single head with many tentacles, or multiple heads.

Another very interesting morphological mutant is one in which an excessive amount of HA is produced and instead of a single head, a multiheaded *Hydra* is produced (Figure 1.18). These examples illustrate the remarkable capacity of such regulators to alter morphology. Because this influence is on the processes that generate the form of the organism, i.e. on morphogenesis, these substances are called **morphogens**. It is evident from the above examples that a powerful method of identifying them and their effects is to use a combination of grafting, biochemical and genetic techniques. More examples of these procedures and the dramatic results obtained by their means will be discussed in subsequent chapters.

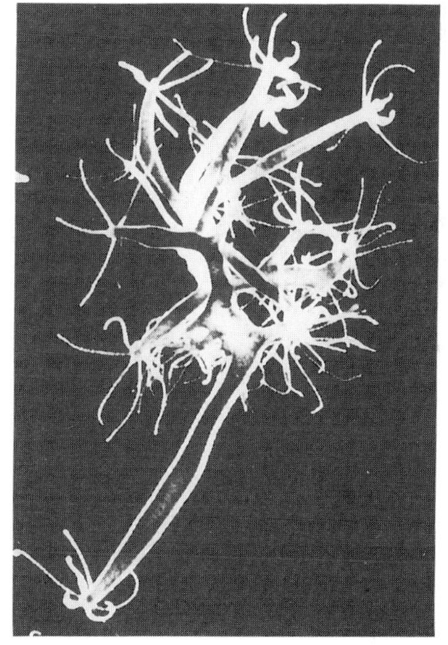

Figure 1.18 Multiheaded *Hydra*.

HA in other species

A further intriguing feature of this molecular analysis comes from comparative biochemical studies. HA has an amino acid composition that identifies it as a neuropeptide, a class of low molecular mass peptides which act as neurotransmitters and as regulators of neurosecretion in the brains of mammals. In fact, HA itself has now been isolated from mammalian brains, where it is most concentrated in the hypothalamus. It is also found in the gastro-intestinal tract, where it may play a role in pancreatic secretion. In addition, it has been identified in both mammalian embryos (human and rat) and in tumour tissue, where it appears to play a mitogenic role (it stimulates mitosis). So a substance discovered as a result of a hunt for a morphogenetic control molecule (a morphogen) in a simple organism turns out to play a diversity of roles in other species.

1.2.5 Phenocopies and metabolites

One rather puzzling aspect of the studies on these morphogens in *Hydra* is that head activator fails to produce a response in normal or wild-type animals that is anything like the effect of the mutation shown in Figure 1.18. If all that is required to produce such a multiheaded monster is an increase in HA concentration in the animal, why is it that exposure of regenerating animals to HA at high concentration in the medium does not induce multiple heads? All that happens is a modest increase in the rate of normal head regeneration, but never the formation of extra or 'ectopic' heads. Perhaps it is not sufficient to accelerate the recruitment of interstitial cells into head-specific pathways of cell differentiation to induce head formation, and a more general stimulus is required.

It is possible that the increased level of HA observed in the multiheaded *Hydra* mutant is playing a secondary stabilizing role in maintaining cells in the state of differentiation and morphology corresponding to head structures, and that the primary head morphogen remains to be identified. Such a morphogen would be something that can actually induce extra heads to form. Recently such a substance has been discovered, and the implications of its effects are very intriguing.

There is a group of metabolites that have been found to play a central role in the reception and processing of signals by cells. Since the process of development in a multicellular organism such as *Hydra* involves the passage of signals such as morphogens between cells, resulting in cell movements and changes of state as a function of relative position in the animal, it is to be expected that such molecules might play a role in morphogenesis.

One of these metabolites is called diacylglycerol (DAG), a lipid-soluble compound found in cell membranes (see Figure 1.22). More will be said about its metabolic activity later. Because of its lipid solubility DAG cannot be dissolved in water and it is necessary to prepare an emulsion with a known nominal concentration for testing its effects. In 1988 Werner Müller of Heidelberg, Germany reported the results of experiments involving the exposure of well fed, healthy *Hydra* to the emulsion for a fixed time (2–3 h) every day over a period of one to several days. Batches of animals were taken after specific periods of treatment and the gastric region was isolated by cutting off the head and the basal part including the budding zone (see Figure 1.19a). These isolated gastric sections were then allowed to regenerate in normal medium without further DAG exposure.

Untreated control animals regenerated normally as expected. With increasing number of days of exposure to the DAG emulsion, more and more ectopic tentacles and heads appeared along the body column of regenerating animals. After 4 days of treatment 33 out of 48 animals had a second head where a foot would normally be produced (Figure 1.19b), while after 8 days of treatment 40 out of 48 animals had ectopic tentacles and hypostomes at intervals all along the body column (Figure 1.19c).

Figure 1.20 shows what happens to these bipolar *Hydra* when they are fed and allowed to grow. The body column elongates and the first overt sign of proximal (towards the foot) structures is the appearance of buds (Figure 1.20a) in a position equidistant from the two heads. More buds then arise in this region and the zone elongates, the buds separating (Figure 1.20b). The region between the buds gradually assumes the characteristics of the basal stalk (Figure 1.20c) and eventually the two animals separate, having reached their normal form.

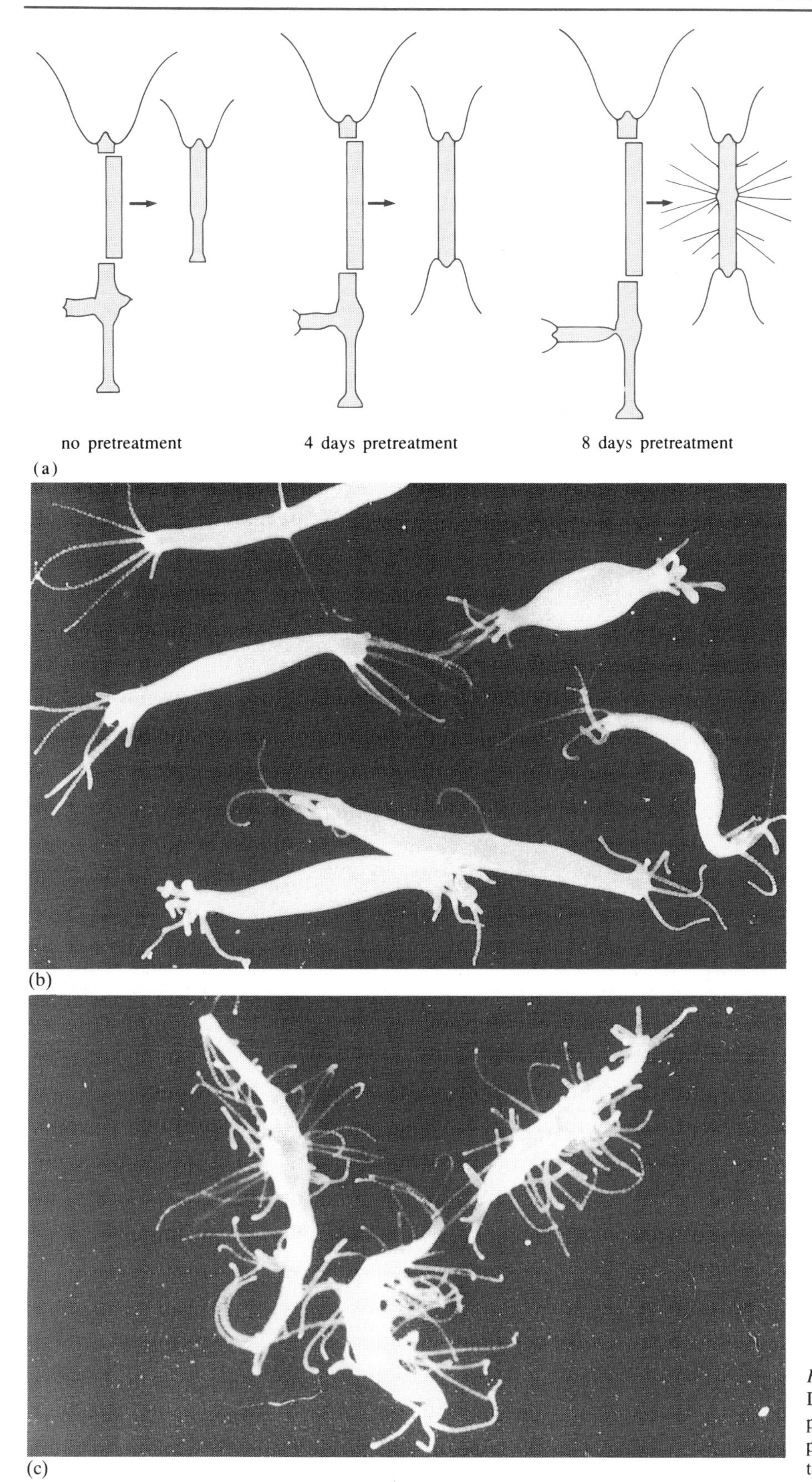

Figure 1.19 Monitoring the effect of DAG on regeneration in *Hydra*. (a) Experimental design. (b) *Hydra* after four pretreatments. (c) *Hydra* after eight pretreatments.

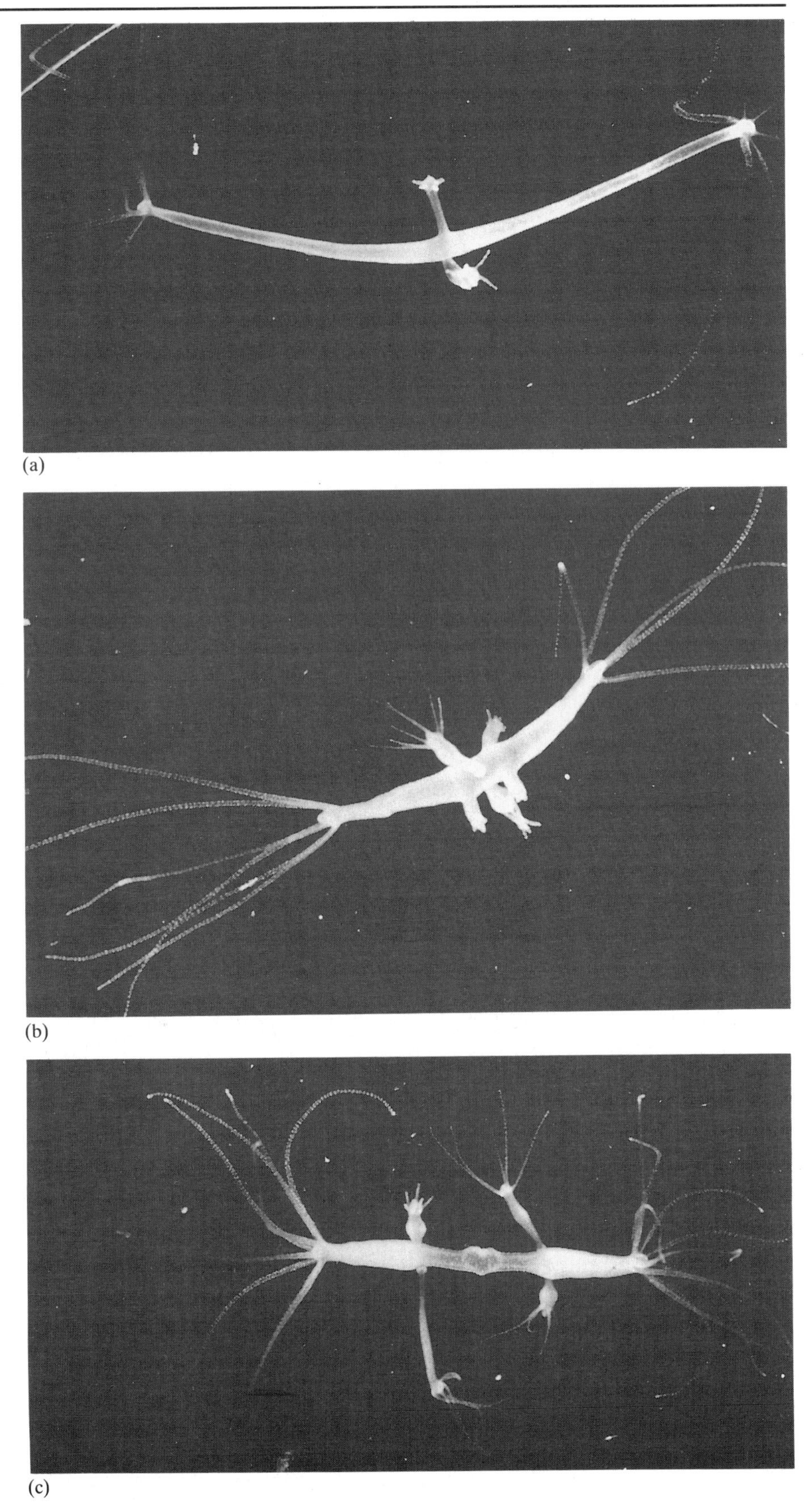

Figure 1.20 Normalization of bipolar *Hydra*.

◇ What would you expect to happen if a normal *Hydra* is exposed for long periods of time to DAG, without cutting and regeneration?

◆ Since DAG appears to have the capacity to induce new heads to form, it is to be expected that it would cause a gradual transformation of a normal animal into a multiheaded form of the type shown in Figure 1.18.

When intact, continually fed *Hydra* were given daily exposures to DAG they gradually elongated and, after 6–10 days of treatment, a disordered array of tentacles appeared in the middle of the gastric zone (Figure 1.21a). With continued daily exposure to DAG, hypostomes developed in the middle of the tentacle clusters and grew out into secondary axes, like those induced by sub-hypostomal grafts in Figure 1.15. At the end of two weeks, as many as nine such heads were observed on one animal (Figure 1.21b), spaced at fairly

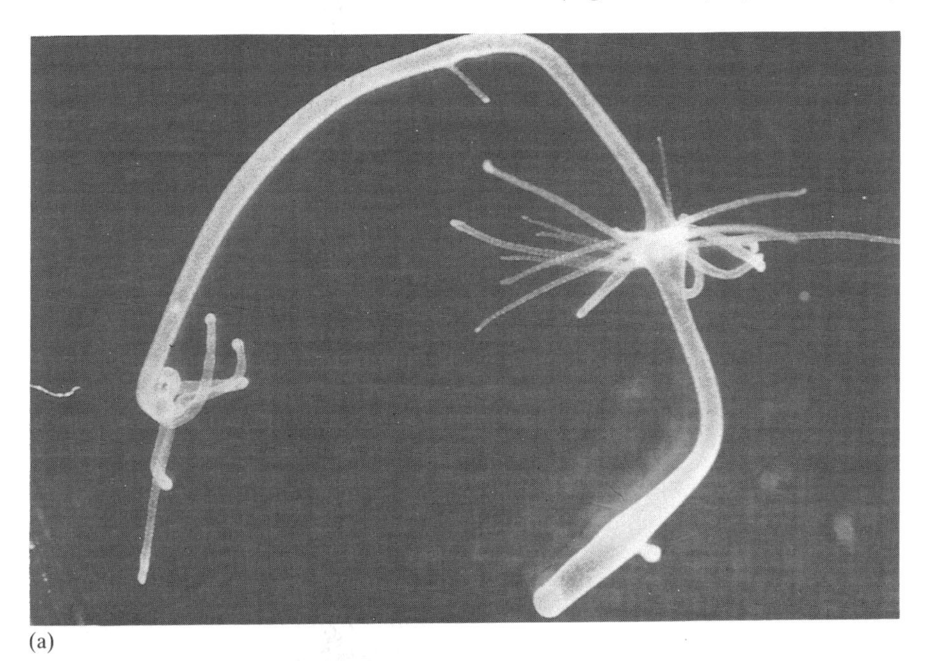
(a)

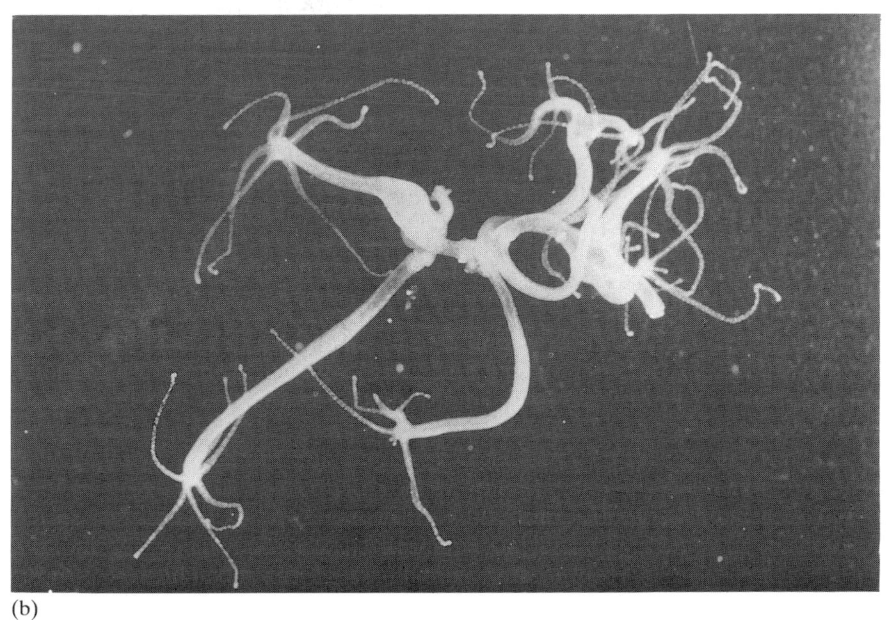
(b)

Figure 1.21 Development of a multiheaded *Hydra* exposed to daily doses of DAG. (a) After eight days. (b) After fifteen days.

regular intervals along the elongated body axis. By three weeks, some individuals had a total of eighteen heads, involving branching of the secondary axes. These multiheaded monsters were like those observed in the mutant form of Figure 1.18. Copies of morphological mutants, produced by exposure of genetically normal organisms to environmental stimuli, are known as **phenocopies** ('phenotypic' or morphological copies of mutants). They show that the mutant gene is not necessary to produce the altered form; organisms with a normal genome can transform to the same morphology with appropriate stimuli.

◇ Does this mean that the mutant gene involved in generating the form shown in Figure 1.18 must be acting by increasing the level of DAG in the organism?

◆ This is a possibility, but it is not a necessary conclusion. In any metabolic process the same result can be produced by different agents because of the many components involved. DAG is one component in a chain of metabolites and signalling substances, any of which could be affected by the mutant gene so as to give multiple branching heads.

These are very striking results that could provide significant insight into the processes involved in morphogenesis in *Hydra.* The intracellular signalling system into which DAG is coupled is summarized in Figure 1.22, and is described in some detail in Book 3 (Chapter 3) of this series*. Figure 1.22 summarizes the main components:

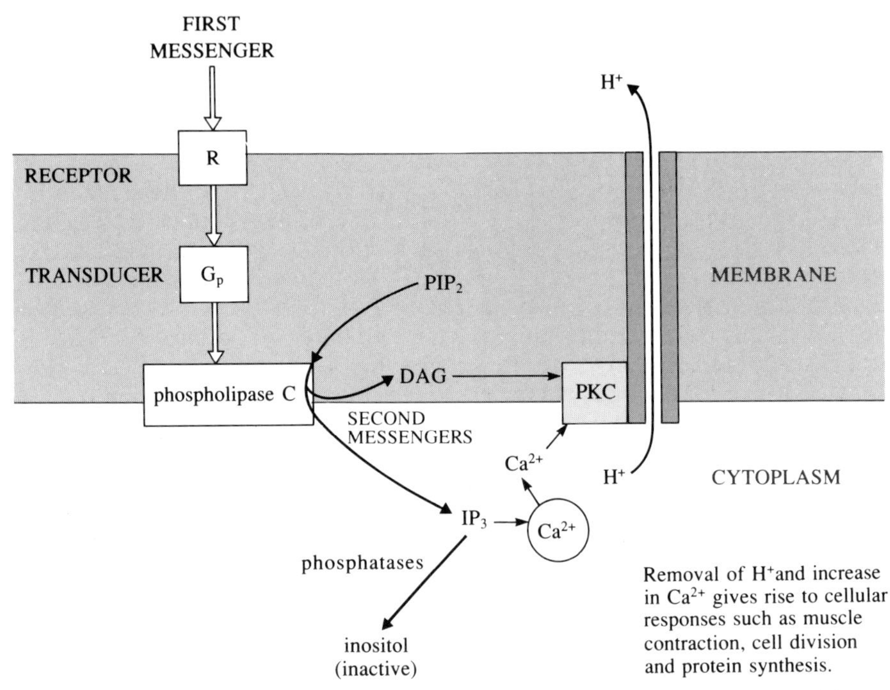

Figure 1.22 Intracellular signalling involving DAG.

Receptors (R) on the outer surface of the cell membrane are coupled by a 'G-protein' (G_p) to an enzyme (phospholipase C) on the inner surface of the membrane. In the cell membrane is a phospholipid known as PIP_2. When a chemical signal or 'first messenger' activates the receptor, the G_p acts as a transducer, activating the phospholipase C to cleave the PIP_2 into two molecules. The cleavage products are a sugar called inositol triphosphate (IP_3) and DAG. These are molecular signals known as 'second messengers'

*Michael Stewart (ed.) (1991) *Animal Physiology*, Hodder and Stoughton Ltd, in association with the Open University (S203, *Biology: Form and Function*).

which are then broken down further by another set of enzymes found in the cell. This is the process that controls the release of calcium ions in the cell, and causes a diversity of further effects depending on cell type. For example, an interstitial cell might be stimulated to divide, an epitheliomuscular cell to contract, a nerve cell could fire, and membrane pumps and channels can be activated or inhibited.

The enzymes which break down the IP_3 are phosphatases which, as you will recall from Section 1.1.4, are involved in the control of cap formation in *Acetabularia*. The IP_3 is a water soluble molecule and causes the release of Ca^{2+} from intracellular stores. So the phosphatase activity influences the Ca^{2+} levels which in turn affect pathways of cell differentation.

DAG is a lipid and so it does not enter the cytoplasm but remains in the membrane where it activates a specific enzyme called protein kinase C (PKC). It does so by increasing the enzymic affinity for Ca^{2+}, which is required for activity. Protein kinase C stimulates a membrane-bound pump which removes protons (H^+) from the cell. The resultant rise in cytoplasmic pH in conjunction with the IP_3-induced increase in intracellular Ca^{2+} concentration is believed to stimulate increased synthesis of RNA in cells, but it has many metabolic effects as well.

DAG treatment of *Hydra* is thus likely to have a fairly complex set of metabolic consequences whose overall effect is to alter the pathway of differentiation of cells in the body column from their normal gastrodermal and epidermal state towards the formation of tentacles and hypostomes. It is interesting to note that head induction results in a number of heads of normal size distributed at more or less evenly spaced intervals along the body column, rather than producing a single monstrous head with many tentacles.

Nothing has been said about control of foot formation. This process is modulated in the same way as head formation by a foot activator (FA) and an inhibitor (FI) that have been isolated but not fully characterized biochemically. The FA is a small polypeptide, while the FI is a non-peptide hydrophilic molecule. So both are similar to the head-controlling substances. Their action and distribution mirror precisely that of HA and HI, with gradients that have a maximum towards the foot of the animal. However, no substance such as DAG has been discovered that induces multiple feet. The same conclusions apply to these regulators as to those affecting head formation; their action may be to modulate the *rate* at which tissue changes in the direction of foot formation, rather than acting as a primary initiator of this change.

This research illustrates how processes in developmental biology tend to get resolved into substances, because these are what can be most readily and accurately identified and measured. These substances act upon cells in interesting and significant ways, as in the proposed action of HA in inducing cells to enter head-specific pathways of differentiation. However, an understanding of morphogenesis itself, the generation of tentacles, a hypostome, buds and a foot, all spatially coordinated, requires that the action of morphogens and the responses of cells be integrated into the tissue level of organization and form. This is the level of the morphogenetic field, whose properties underlie the capacity of parts to become wholes of characteristic form during regeneration and reproduction.

Summary of Section 1.2

Hydra is a relatively simple multicellular organism whose body is made up of about 15 different types of cell that constitute the body. It has both sexual and asexual modes of reproduction.

Hydra has great regenerative abilities, small fragments of the organism being able to develop into small, whole organisms which then grow to normal size. Regeneration involves changes of cell state and of cell position.

The adult form of *Hydra* is maintained by a continuous production, differentiation, movement, and loss of cells, so morphological constancy depends upon cellular flow and transformation. This flow can be observed by marking groups of cells with vital dyes and recording their movements. The capacity of tissue to change state as a function of position can be examined by moving parts from one region of the organism to another and observing the result.

Tissue can be examined for its state of determination to produce particular structures by transplantation to test sites in the host organism. A study of this kind, on head determination in sub-hypostomal tissue, led to the proposal that substances exerting negative and positive feedback effects are associated with the formation of particular structures in the organism.

Morphological control substances can be identified by their effect on the rate of formation of a structure such as head, i.e., by bioassay. A head activator and a head inhibitor in *Hydra* were identified by this means, and pure head activator was obtained by an extraction procedure on 200 kg of sea anemone, a close relative of *Hydra* but much larger. Morphological mutants of *Hydra* have been shown to have altered levels of head activator and inhibitor.

Diacylglycerol (DAG), a lipid-soluble component of cell membranes, induces extra heads in *Hydra*, producing the same phenotype as a multiheaded mutant. DAG is a component of a signal transmission system that involves changes in Ca^{2+} levels and key metabolites involved in many regulatory processes. It may be exerting its effect on morphogenesis via the influence of these substances on the differentiation of cells and head formation.

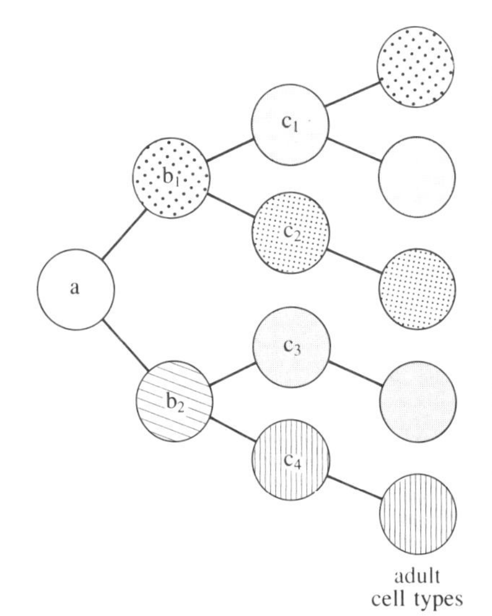

Figure 1.23 Cell differentiation (schematic).

Question 4 (*Objectives 1.6, 1.7 and 1.8*) What processes are involved in the regeneration of a whole *Hydra* from a part, and what cellular properties make this possible?

Question 5 (*Objective 1.7*) Assuming that an organism is made of the 5 cell types shown at the right in Figure 1.23, identify cells that are (a) totipotent; (b) multipotent. Which of the latter category is most multipotent?

Question 6 (*Objective 1.9*) Sub-hypostomal tissue from the region between lines 1 and 2 in Figure 1.14a is isolated in freshwater for 4.5 hours and then grafted into the sub-hypostomal region of another animal. Would you expect this sub-hypostomal tissue in a sub-hypostomal site to show (a) the same, (b) a higher or (c) a lower frequency of head induction than that observed in the mid-gastric graft shown in Figure 1.14b. Give the reasoning behind your answer.

Question 7 (*Objectives 1.10 and 1.11*) As seen in Figure 1.21, the effect of DAG treatment is to produce heads at intervals along the body axis. In terms of the properties of morphogens described, how would you account for the result rather than the production of a single gigantic head?

1.3 AMPHIBIANS: THE DEVELOPMENT OF COMPLEXITY

The life cycle of a newt is shown in Figure 1.24. Once again we see the basic characteristics of continuity between the generations, parts of the parents developing into whole adult organisms which themselves produce parts capable of generating new adults and so on. But this cycle, despite similarities, has properties that distinguish it significantly from those of *Acetabularia* and *Hydra*. First, the commitment to sexual reproduction is complete, there being no other cycle, unlike the process in *Hydra*; while sexual differentiation is also complete, reproduction normally requiring the participation of male and female gametes, which are produced by sexually differentiated individuals. The fertilized egg develops into a larva before undergoing metamorphosis to the adult organism. Both the larva and the adult, have a complex morphology, sharing the same basic body plan with the rest of the vertebrates, which include humans. Amphibian eggs are large and readily available, and the embryo is easily studied with the use of a microscope and simple methods of manipulation. So amphibians have long been favourite animals for embryological studies, from which have come many of the basic concepts of development.

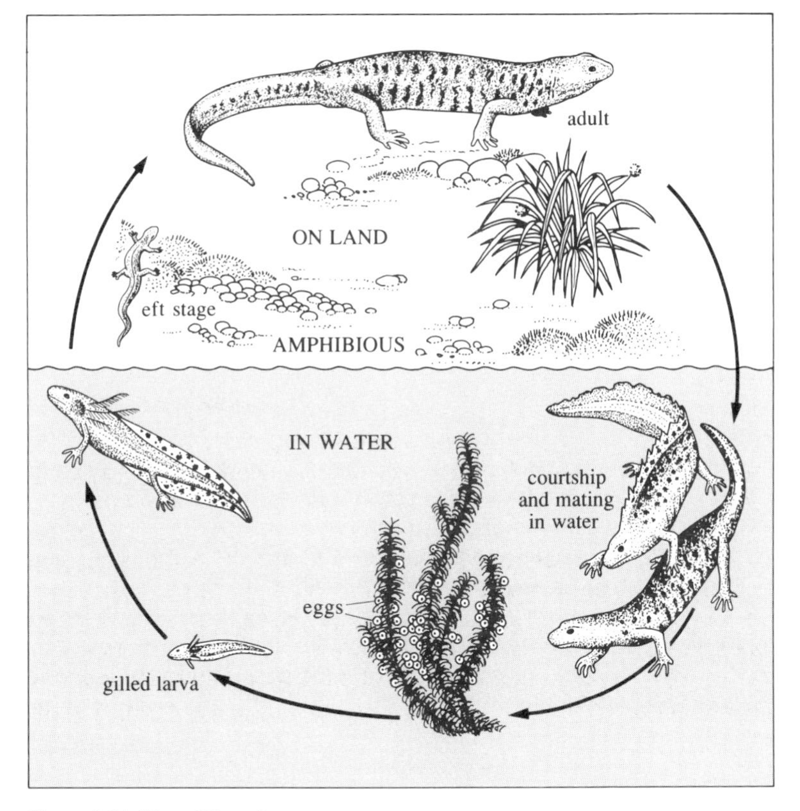

Figure 1.24 Newt life cycle.

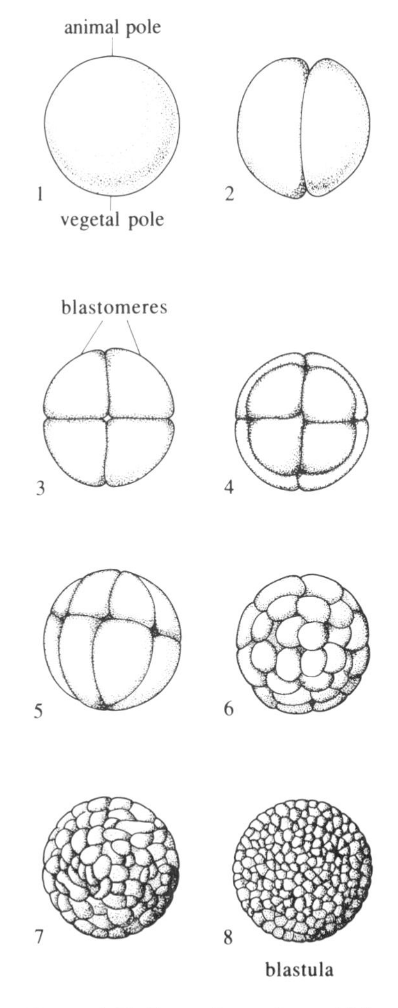

Figure 1.25 Formation of amphibian blastula.

1.3.1 Early amphibian development

The first stage of amphibian development after fertilization is a series of cleavage divisions in which the giant egg cell, usually 1–2 mm in diameter, divides in an ordered pattern to produce a ball of much smaller cells (Figure 1.25), called a **blastula**.

Many different multicellular lifecycles start in this way but the details of the pattern vary among them. The species shown undergoes **holoblastic cleavage** (division is complete) which is characteristic of amphibians and results in separate cells (called **blastomeres**) that stick together and form intercellular junctions.

Cleavage starts first at the animal pole, and ends at the vegetal pole (see Figure 1.25). Food reserves for the embryo are present as yolk, which has a graded distribution in the egg with a maximum towards the vegetal pole. Such animal–vegetal polarity, present in the egg, is a feature of many animal species. This polarity is established during the growth of the egg in the maternal ovary.

The next phase of development after the blastula involves a remarkable process in which the ball of cells turns itself into a multilayered structure. The process, known as **invagination**, is rather like what happens if a beach ball is punctured and then kicked: the ball collapses and the inner surface on one side makes contact with that on the other, making a large dimple. The amphibian embryo achieves this in a much more elegant manner, as shown in Figure 1.26. Developmental biologists describe the distinguishable steps in morphogenetic processes as **stages**. Stage 9 differs from stage 8 of Figure 1.25 only in that there are more cells, whose outlines are no longer shown. At stage 10, cells in a particular region of the embryo have started to move inside, forming a slit or groove on the surface which spreads as more cells engage in this inward movement (stage 11). These moving cells migrate over the inner surface of the ball, spreading inside until they form a complete surface and the groove closes into a small circle, the **blastopore**. The result is the stage 12 gastrula. (These processes will be described in more detail in Chapter 3, when the behaviour of cells underlying these gastrulation movements will be considered in more detail).

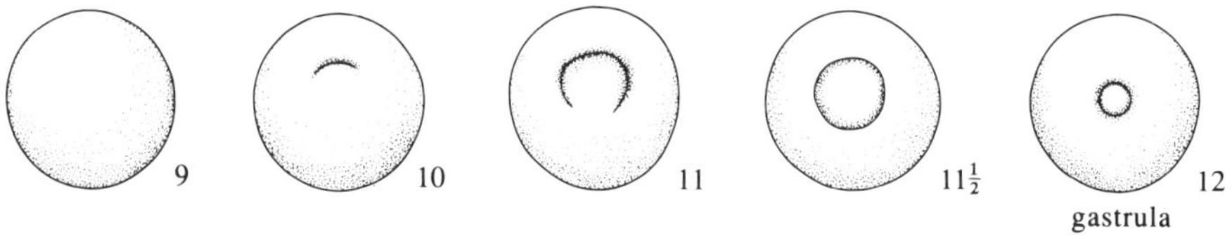

9 10 11 $11\frac{1}{2}$ 12 gastrula

Figure 1.26 Formation of amphibian gastrula.

So far everything looks symmetrical, and there is no obvious sign of a head to tail axis. If, however, the gastrula is cut in half, there is clear evidence that a rather complex pattern of relationships between cells in different parts of the developing embryo is beginning to emerge (Figure 1.27). The cells that have moved inwards from the dorsal lip of the blastopore have formed a distinct layer of cells (the **mesoderm**) beneath the outermost cells (the **ectoderm**). And a third layer (the **endoderm**) forms the lower surface of the space beneath the mesoderm. These are the three primary germ layers of the embryo, out of which all the tissues of the mature organism will be formed. In

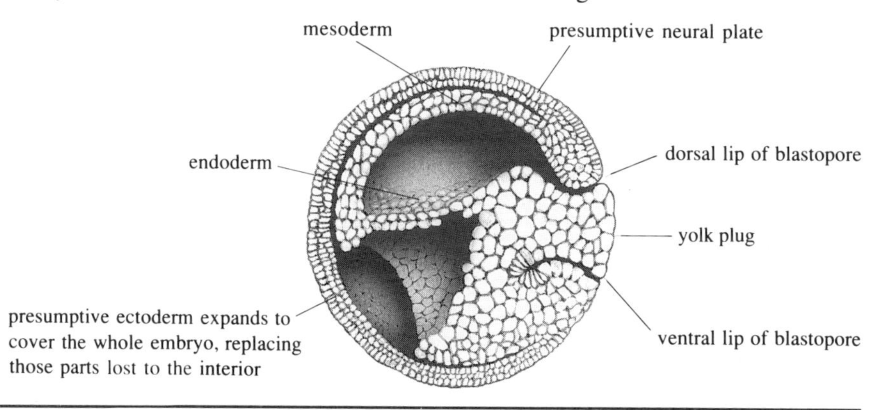

Figure 1.27 Gastrula structure.

the gastrula the asymmetries that result in the bilateral form of the adult animal are beginning to emerge. In relation to Figure 1.27 these presumptive relationships are: dorsal = top, ventral = bottom, posterior = right (blastopore), anterior = left.

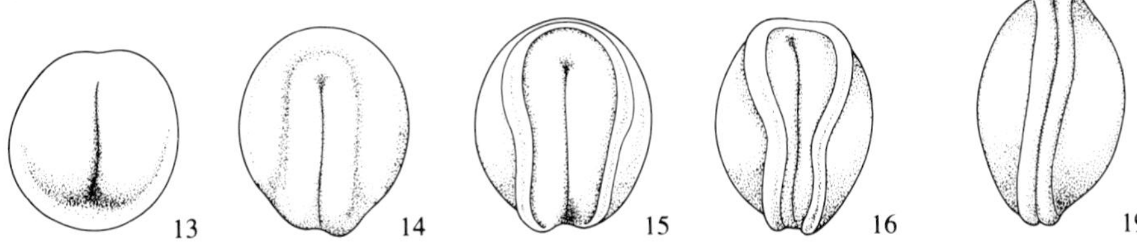

Figure 1.28 Formation of amphibian nervous system and neurula.

The first clear sign of a longitudinal axis in the developing embryo is the formation of a groove on the surface (stage 13, Figure 1.28) which deepens and then results in the formation of **neural folds** on the surface that gradually come together to initiate the nervous system as shown in stages 15, 16 and 19. (These steps are separated by a few hours in the newt, but rates of development vary greatly between amphibian species). The upper part will form the head and the lower part the tail. The outcome of these foldings and movements is the **neurula** (stage 21), shown in side view, head to the right, tail left, back up and belly down. This is clearly a bilaterally symmetrical organism. The neurula develops further as shown in stages 24, 26, 29, 31, 33, and 34 to the hatched larva of stage 35 (Figure 1.29). This then grows into a gilled larva of the type shown in the life cycle of Figure 1.24.

Acetabularia, *Hydra*, and the newt all have a primary axis, distinguishing one end from the other. However the newt, and indeed all vertebrates, also develop right and left halves. How does this bilateral symmetry arise?

◇ Give examples of bilateral symmetry in the parts of higher plants.

◆ Leaves are often bilaterally symmetrical, and a number of species produce flowers with such a form (e.g. pea, snapdragon, several species of orchid), but plants as a whole are never bilaterally symmetrical overall.

The bilateral symmetry of the amphibian is usually initiated at the moment of fertilization, when the sperm enters the egg at some position in the upper animal half of the egg, displaced from the pole. The three points on the egg surface defined by the animal pole, the vegetal pole, and the point of sperm entry define a plane through the egg, and this is normally the plane of the first cleavage division (stage 2, Figure 1.25). It is also the plane of bilateral symmetry of the future embryo so that the first two blastomeres correspond to the presumptive (future) right and left halves of the organism. However, this is by no means a mechanical sorting of the egg into predetermined halves, for if the two cells resulting from first cleavage are separated from one another, each will develop into a complete though half-sized larva. These twins then feed and grow into normal sized larvae which subsequently metamorphose into completely normal newts.

The first such separation of the initial pair of blastomeres was achieved by the German embryologist Hans Driesch towards the end of the last century using sea urchin embryos, not amphibians, but the result was the same: half-sized but otherwise fully normal larvae were produced that grew into perfectly normal adults. Because they come from the same egg, these are identical twins. This established the totipotent property of the first two blastomeres: their capacity to generate a complete organism, with no parts or cell types missing. Embryos can also be separated into two multicellular parts at later stages and still develop into complete organisms.

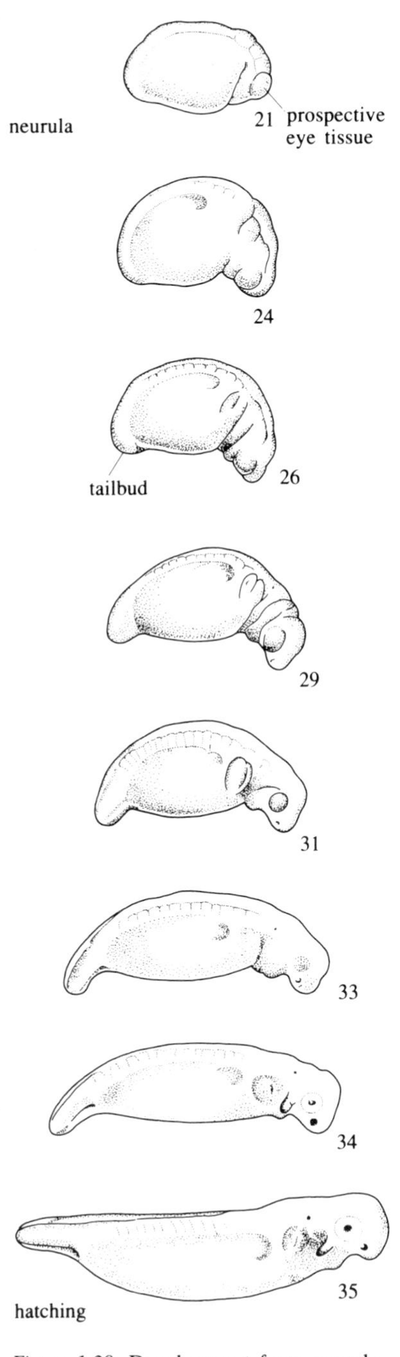

Figure 1.29 Development from neurula to hatched newt larva.

◇ What is the evidence that human embryos can be separated into two parts each of which develops into a complete human being?

◆ The occurrence of identical twins.

1.3.2 Regulation

The capacity of parts of embryos to produce whole organisms is referred to as **regulation**. It implies that cells which would normally have developed into a particular part of the organism have the capacity to develop into others, and so it involves the properties of totipotency and multipotency already described. As we shall now see, two factors are important in determining the capacity of embryonic parts to regulate: which part is isolated, and at what stage in development.

The position of the plane of first cleavage in the amphibian egg is variable (Figure 1.30), and it sometimes happens that, while still passing through animal and vegetal poles, it is at right angles to the plane containing the sperm entry point. If the two blastomeres resulting from such a cleavage are separated, the one containing the sperm entry point fails to regulate; it produces a symmetrical form with structures characteristic of the ventral part of a normal embryo, the belly. The other blastomere which would normally have given rise to only the dorsal part of the organism regulates to give a normal larva. So even at this early stage of development, not all parts of the embryo have equivalent regulative capacities.

There is evidently some spatial organization set up between the moment of sperm entry and the time of first cleavage that results in a difference in the properties of the future ventral half of the embryo as compared with the dorsal half. In a number of species of amphibian with considerable pigment in the egg, which increases towards the animal pole, visible evidence of this difference can be seen soon after fertilization in the form of a **grey crescent**. This is a pale crescent-shaped region that develops on the side of the egg opposite to the sperm entry point, and identifies the future dorsal region of the larva. So the sperm not only activates the egg; it also initiates the process of symmetry-breaking that creates the second major axis in the amphibian embryo and results in a bilaterally symmetrical organism. The relationships of grey crescent and blastomeres to regulative and non-regulative parts is shown in Figure 1.30.

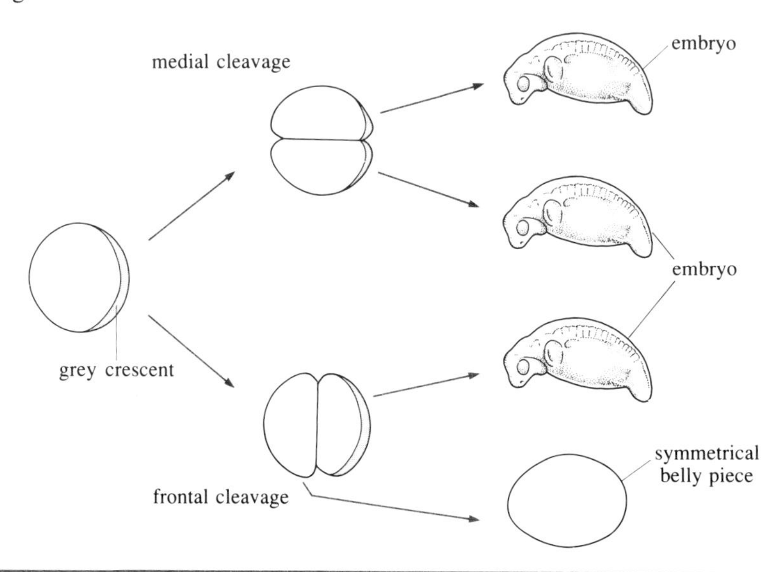

Figure 1.30 Variations in regulative capacity in different halves of the embryo after the first cleavage.

As mentioned earlier, amphibian embryos can be divided into two halves along the plane of bilateral symmetry at stages of development later than the first two blastomeres and each half can regulate to give a whole larva. However, this capacity decreases sharply during gastrula formation, and by the time this process is complete, grossly asymmetrical larvae which fail to survive result from such a separation. In general the regulative capacities of embryos decrease as developmental age increases and they become progressively more fixed or determined with respect to their normal fates, producing only the structures that they would have in the undisturbed embryo.

1.3.3 Morphogenetic fields

When Driesch made his discovery about the remarkable regulative capacities of separated sea urchin blastomeres, he realized that he had uncovered one of the deepest and most intriguing properties of developing organisms. He compared the result with what happens when a bar magnet is separated into two parts. The original magnet has a pair of poles between which there is a magnetic field, detectable by its influence on test particles such as iron filings. If the magnet is cut into two parts so that the poles are separated, the result is not two isolated poles but two complete magnets, each with a pair of poles and a magnetic field between them. Driesch used this metaphor to describe the properties of morphogenetic fields, which we have already encountered in *Acetabularia* and *Hydra*. These are the processes in developing organisms, distributed in space and organized in time, that generate coordinated wholes. The separated blastomeres, each of which regulates to produce a whole organism, like the whole magnetic field of the half magnet, provide a classic example of the field property in organisms. The field concept is essential in science for the analysis of processes that result in spatially organized phenomena, whatever their nature may be. The movement of planets in elliptical orbits (gravitational field), the spiral flow of water down a drain (hydrodynamic field) and the tornado (aerodynamic field) are all familiar examples, each of which involves a field with specific properties, resulting in particular types of behaviour. If you disturb the spiral flow of water down a drain by putting an obstacle in its path, the flow pattern is altered, but the original form is reconstituted when the obstacle is removed. Similarly for the tornado, which removes most obstacles itself. The capacity to reconstitute a whole from separated parts, or after other types of disturbance, is characteristic of fields. As described in the previous section this type of behaviour in developing organisms is referred to as regulation in morphogenetic fields. The properties of these fields lie at the foundation of organized patterns of development and hence are at the basis of reproduction, life-cycles, and evolution; truly a fundamental concept in biology.

Several examples illustrating the properties of morphogenetic fields have already been presented, such as regeneration in *Acetabularia* and in *Hydra*, where the coordinating influences in space and time that result in a restoration of normal form were described. An equivalent term that is often used is **developmental field**.

1.3.4 Studying the properties of morphogenetic fields

Just as physicists use test particles to determine the presence and intensity of magnetic fields, so developmental biologists use test systems to investigate the presence and intensity of morphogenetic fields. A procedure that has resulted in a great deal of insight into the properties of these fields is to take cells from a particular region of one embryo (the donor) and put them into a different region of another embryo (the host). Of course this test material is not a passive recorder of field state; it is living tissue that interacts with its new environment. There are basically two possible outcomes, as observed in the sub-hypostomal *Hydra* grafts onto the mid-gastric zone of hosts used to test the head-forming capacities of sub-hypostomal tissue. The graft can be assimilated into the new environment, changing its state so as to conform to the neighbouring host tissue; or it can develop along some path to which it was committed, generating a particular structure and influencing its new neighbours in the process.

Induction, competence and evocation

Consider the following experiments. Suppose some cells are dissected from a region far from the blastopore where cells are moving inside a stage 10 embryo (see Figure 1.26) and are transplanted to a different region, also far from the developing blastopore of another embryo, at the same stage. These cells heal rapidly into their new position and the embryo develops into a perfectly normal larva: the cells are assimilated into their new environment and develop into structures unlike those they would have made, consistent with their new neighbours. But suppose that the transplanted cells come from the region of the blastopore itself and are transplanted to another stage 10 host at a position other than its blastopore region (Figure 1.31). This embryo now develops quite differently. The host blastopore develops normally, gastrulation proceeds, and the neural folds emerge, resulting in the neural plate (Figure 1.31b). But a second blastopore is formed by the grafted tissue, and this results in the formation of a second neural plate (Figure 1.31c,d) leading to a secondary embryonic axis. The embryo ends up as a pair of Siamese twins (Figure 1.31e) though usually the second embryo is smaller and less well developed than the primary. The grafted tissue in this case **induced** a new axis to form, as did the sub-hypostomal *Hydra* graft after 4.5 hours of development in isolation.

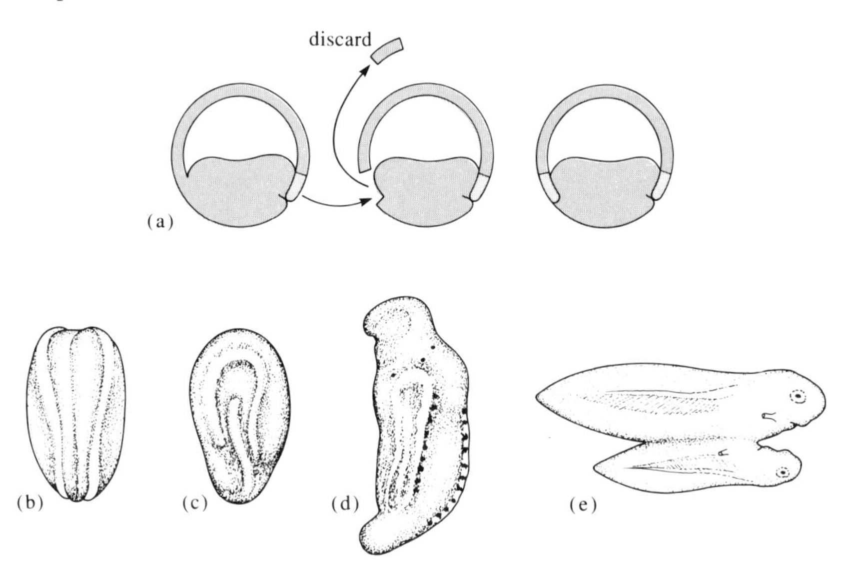

Figure 1.31 Induction of a secondary axis in an amphibian.

The amphibian induction experiment was carried out in 1924 by two German developmental biologists, Hilde Mangold and Hans Spemann, and it is one of the most celebrated of all embryological results. More details of this process will be discussed in Chapter 3. Here we note a few of its characteristics. First, there is the stability of the dorsal lip tissue in maintaining its commitment to the invagination process despite the new environment in which it finds itself in the embryo. Instead of being assimilated into the host, the tissue has persisted in the organization of the cell movements that result in a new blastopore: the tissue is determined in this state relative to the perturbation of changing location. Second, not only does the tissue persist in its previous state, it induces neighbouring cells to participate in forming the new axis. This can be seen by using differently pigmented or stained host and graft tissue: the secondary axis consists of cells from both host and graft. This shows also that the host cells opposite the original blastopore, which would not normally form a neural plate, are **competent** to do so; i.e. they have this potential, which is expressed when they are stimulated in the appropriate manner by an inductive stimulus.

Another classic example of embryonic induction reveals a very intriguing aspect of the process. The experiment involves grafting tissue from the neurula of one species into another to see if the reason for the production of particular structures in one species but not in another is due to lack of inductive stimulus or lack of competence to respond. Two such experiments can be done with newt larvae and frog tadpoles, which have distinctly different mouth parts and associated organs (Figure 1.32a), but have similar belly epidermis. The frog tadpole has a small mouth with no teeth, and has mucus-secreting suckers. The newt larva has a wide mouth, teeth and balancers.

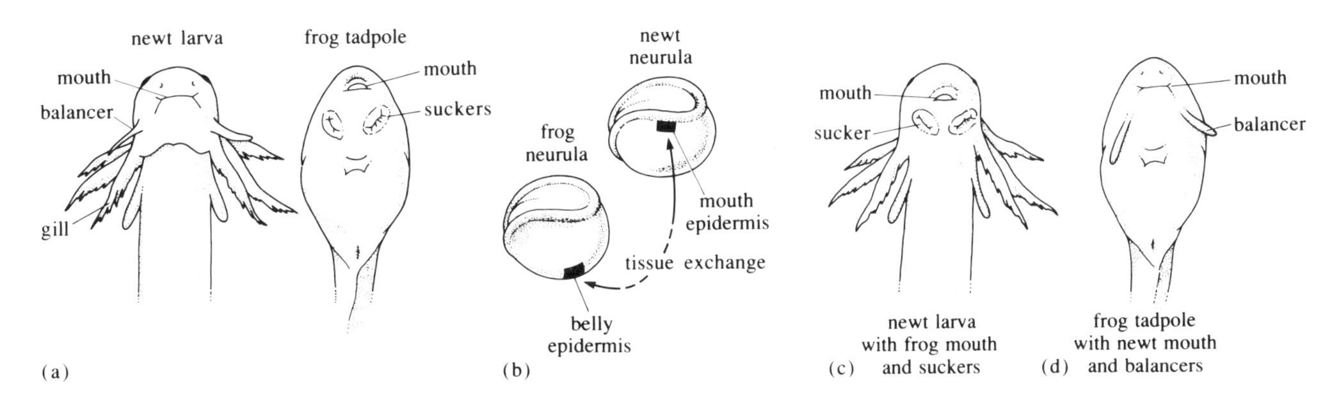

Figure 1.32 Interspecific induction experiments.

In one experiment, a bit of neurula ectoderm that would normally form part of the belly epidermis was taken from a frog neurula and transplanted into a newt neurula in a position which gives rise to mouth parts (Figure 1.32b). At the same time the prospective mouth tissue removed from the newt neurula was grafted into the 'belly' position in the frog neurula. The newt larva which subsequently developed had a frog mouth and suckers (Figure 1.32c), while the frog tadpole developed normally.

In the second experiment, a bit of newt neurula ectoderm from the prospective belly region was exchanged with the mouth region of the frog neurula. In this case the frog tadpole developed with a newt mouth and balancers (Figure 1.32d), while the newt larva developed normally.

This shows that the newt can induce the formation of frog mouth parts in frog ectoderm, and the frog can induce newt mouth parts in newt ectoderm. So the reason for the production of the species-specific structures lies not in the inducing stimulus, but in the ectodermal response. It is believed that the inducer is a relatively nonspecific stimulus in both species, and it evokes a response in the ectoderm that is characteristic of the species. This **evocation** involves the organization of processes that give rise to particular spatial structures in which gene activity specific to the species is presumably involved. This is known as permissive induction. There are also examples in which the inducing stimulus carries with it specific information so that the inducer instructs the responding tissue to form a particular structure, as will be described in Chapter 3.

1.3.5 The hierarchical nature of development

In the life cycle of *Acetabularia* the adult form arises from the spherical zygote through a sequence of increasingly complex forms. In *Hydra*, a bud is initiated as a bump on the body axis and develops a progressively more complex form until a miniature but complete *Hydra* is generated, which then detaches from the adult. In the development of the newt, the spherical egg undergoes transformation to the larva through a well defined sequence of changes of body form.

We can distinguish two basically different ways in which a complex form can be generated. Consider first the painting of a picture. This could be done by starting at some point (say the top left-hand corner of the canvas) and painting in the full detail of the final picture in a sequence of adjacent regions so that when the last region is painted in, the picture is complete. This procedure would be sequential. On the other hand, the painting could be sketched initially in the most general manner, delineating sky from land, grass from trees from lake, but with nothing recognizable at first. The clouds could then be sketched into the sky, some colour tones added to the landscape, and more detail added progressively within the initial spatial sketch. Modifications could be introduced as the painting proceeds, to balance form and colour. This procedure would be sequential and hierarchical: within any region of space, the progressive sequence of detail can be described as a division of an initially undifferentiated region into one with progressively more spatial detail, until the final degree of spatial complexity is achieved.

Simply looking at its external form, it is evident that the developing embryo follows the latter strategy. For example, in Figure 1.29 the neurula has a 'sketch' of the regions in which more spatial detail will appear later. The somites (body segments) arise as subdivisions of the dorsal fold — the tail takes shape out of a bud, the eye develops spatial detail within an initial mound of undifferentiated tissue.

If we look inside the embryo at the different tissues, organs and cell types that are produced while these external changes of form are occurring, the picture is considerably more complex, as described in Figure 1.33. Some of the terms used will be familiar and some will not. Chapter 3 will clarify a number of the details. The point to notice is that as the embryo changes through cleavage to the blastula, then transforms into a gastrula, to a neurula and then to the tailbud stage (Figures 1.25–1.29), there is a progressive increase in the number of distinguishable tissues whose cells then differentiate further into the cell types of the larva. The hierarchical nature of this process is again obvious: complexity arises progressively within spatial domains laid down earlier. It is not like printing a picture from a die, in which process all the details of the picture are produced at once, nor is it like a painting produced by a purely sequential process, as described earlier. Furthermore in the vertebrates each of the major 'decisions' taken by tissue with respect to the alternative pathways available to it involves an irreversible commitment at some stage in its development: the tissue becomes fully determined with respect to a particular set of options, so that no environment that allows development to continue is able to divert the tissue into another pathway with a different set of options. For example, once ectoderm is on the pathway to neural plate, with the options of forming the brain or the spinal cord, it

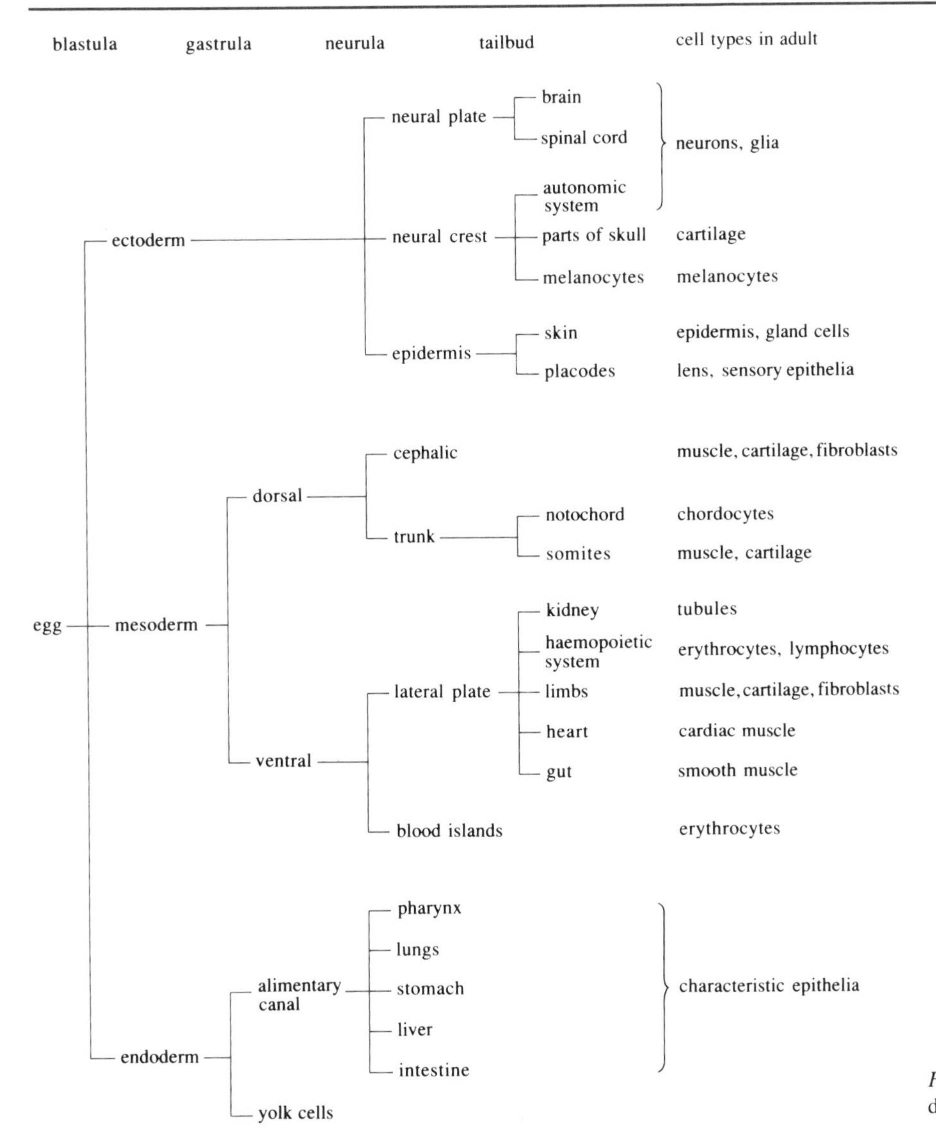

Figure 1.33 Hierarchical processes of development in vertebrates.

cannot be diverted to form epidermis with the options of generating skin or placodes (from which parts of the eyes, ears and other sense organs are formed). At the moment the technical terms are not important; the significant point to recognize is that with each decision, embryonic tissue loses some of its multipotential, becoming more restricted in the range of adult structures and differentiated cell types that it can produce.

◇ Referring to Figure 1.33, which of the three germ layers in the blastula (ectoderm, mesoderm, endoderm) is most multipotent (produces the greatest diversity of cell types in the adult)?

◆ Mesoderm.

There comes a time in development, therefore, when the embryo is essentially a mosaic of specified parts, which have no options left and are all determined to develop into specific differentiated structures. Different species of embryo reach this mosaic state at different stages of development, some arriving there very early so that their regulative phase lasts for only a brief period. Other species such as *Hydra* retain full regulative properties in the adult. The extent to which regulative capacity persists in the adult determines its regenerative powers.

◇ Name some parts of the human body that retain regulative properties (regenerative or healing capacity) after birth.

◆ Fingertips; skin, as in wound healing; liver and kidney; blood-producing tissues.

1.3.6 Preformation and epigenesis

The above descriptions do not tell us how this hierarchically organized process, in which finer and finer spatial details gradually emerge, is actually generated. From the beginning of the systematic study of development, which goes back to Aristotle, there have always been two different ways of explaining this process:

1 All the details of the adult structure are laid down or coded in the egg in some form, the progressive emergence of detailed shape being the result of an ordered reading of a pre-existing plan. The task of developmental biology is then to discover and decipher this plan.

2 The adult structure emerges by a process of interaction between parts which are not prespecified but arise from the intrinsic dynamic properties of developing organisms as self-organizing systems of a particular type. The task of developmental biology is to describe the nature of this dynamic process.

◇ On the basis of what has been described so far about development, assess the validity of these two positions.

◆ Since separation of the first two blastomeres in the amphibian and the sea urchin results in two complete larvae, if there is a plan in the egg then it is clearly not a rigid one that is read out mechanically, as a blueprint is used for the construction of a house. So the first alternative must involve some means of reorganizing the reading of the plan according to conditions.

Since different species develop in a repeatable and regular manner, development cannot arise simply from the general dynamic properties of a process that involves interaction between spontaneously generated parts to produce progressive complexity. Some aspect of inheritance must stabilize the specific sequences of events that distinguish the development of different species.

These two different viewpoints result in somewhat different research programmes for the study of development, both converging on the same goal but from complementary perspectives. The first one has traditionally been described as a theory of **preformation**, since the adult form of the organism is in some sense preformed in the egg. It results in research whose objective is to discover this preformed structure or information in the egg and its expression in the developing embryo, which in contemporary biology has its focus on genes and the macromolecules they produce. Examples of this will be considered in detail in later chapters. The second viewpoint is identified with the theory of **epigenesis**, which emphasizes the capacity of developing organisms to generate increasing complexity spontaneously without any pre-existing plan. This results in a study of the ways in which developing organisms respond to various types of disturbance, resulting in either normal or abnormal forms as a result of interactions following the disturbance.

Examples of this type of experiment have already been considered, which gave rise to the concept of the morphogenetic field. Physics is founded on both the concepts of particle and field, and neither is more fundamental than the other; they are complementary. Equivalently, particles (molecules in this case) and fields play complementary roles in the understanding of developmental processes, neither being more significant than the other.

Summary of Section 1.3

Amphibian development starts with a series of cleavage divisions of the fertilized egg and proceeds through a well defined sequence of stages involving morphogenetic movements that result in a three-layered, bilaterally symmetrical neurula. Bilateral symmetry is initiated at the moment of sperm entry into the egg, the two blastomeres resulting from first cleavage corresponding normally to left and right halves of the future organisms. If separated, each blastomere develops into a half-sized but otherwise fully normal larva that then grows into a normal adult.

Regulation is the capacity of parts of embryos to give rise to wholes. The extent to which parts can regulate varies with the position of the part in the embryo, and with developmental age.

The field concept is fundamental for the description of organized processes in space and time, and is widely used in physics. Morphogenetic fields describe the coordinated space–time activities that underlie morphogenesis and regulation in embryos.

The properties of morphogenetic fields can be studied by moving bits of tissue from one part of a developing embryo to another. This has revealed the basic phenomena of induction, competence, and evocation which describe the stimuli and responses characteristic of tissues in morphogenetic fields.

Development is a hierarchical process in which detailed structure emerges gradually and progressively, starting with general axial order and proceeding to increasingly finer spatial structure within pre-established order. As parts emerge, cells differentiate to particular types, and there is a general tendency for the regulative properties of the embryo to diminish. However, some regulative capacity remains in the adult, which varies greatly between species and determines their regenerative capacity.

Two different interpretations of the causes of development, known as preformation and epigenesis, emphasize the role of pre-existing genetic information in generating species-specific structure, and of dynamic interactions that spontaneously give rise to spatial order, respectively. An understanding of both types of activity in the generation of patterns of development is required for a satisfactory theory of how organisms of specific form are generated.

Question 8 (*Objectives 1.2 and 1.12*). Compare the processes of axis formation in *Acetabularia* with that in the amphibian embryo.

Question 9 (*Objective 1.13*). Tissue taken from the blastopore region of a stage 10 gastrula (Figure 1.26) is transplanted to the diametrically opposite position of the *same* embryo. The result is an embryo with two equal axes that develop into Siamese twins of equal size and degree of development, rather than the unequal embryos of Figure 1.31. Give an interpretation of this result.

Question 10 (*Objective 1.14*). You are given fertilized eggs of a species of marine worm and asked to find out if the embryo has regulative properties.

You separate the first two blastomeres and observe that they develop into half, not whole, embryos. Give two possible interpretations of the observation, and design further experiments to distinguish between them.

Question 11 (*Objectives 1.15, 1.16 and 1.17*). Which of the following are true and which are false?

(a) The first evidence of bilateral symmetry in an amphibian embryo is when gastrulation begins.

(b) Morphogenetic fields describe the dynamic properties of developing organisms that give them the capacity to generate integrated whole organisms despite damage and disturbance during development.

(c) The theory that development is coded in the genes is an example of a theory of preformation.

(d) The formation of different structures in different species is a result of the differences in the inducing stimulus that evokes the formation of the structures.

(e) The concept of hierarchical development is that the sequence of morphogenetic events proceeds from the genes to cells to tissues to the whole organism.

OBJECTIVES FOR CHAPTER 1

Now that you have completed Chapter 1, you should be able to:

1.1 Define and use, or recognize definitions and applications of, each of the terms printed in **bold** in the text.

1.2 Describe morphogenesis in *Acetabularia* in terms of the emergence of complexity. (*Questions 1, 2 and 8*)

1.3 Outline the evidence for a relationship between morphological and electrical polarity in *Acetabularia*, and suggest how it is acting. (*Question 2*)

1.4 Present evidence that calcium is involved in the morphogenesis of *Acetabularia*. (*Question 2*)

1.5 Describe how the influence of genes on morphogenesis and the mechanisms of gene control can be studied in *Acetabularia*. (*Question 3*)

1.6 Describe the regenerative potential of *Hydra*. (*Question 4*)

1.7 Explain the importance of multipotent cells in *Hydra* regeneration. (*Questions 4 and 5*)

1.8 Describe the processes involved in the maintenance of the adult form in *Hydra*, and how these relate to regeneration. (*Question 4*)

1.9 Illustrate how the inducing properties of parts of *Hydra* are studied experimentally. (*Question 6*)

1.10 Describe the evidence for the basic principles of control used to explain the stability of the adult form of *Hydra* and its regeneration from parts. Predict the expected distribution of control substances based on these principles. (*Question 7*)

1.11 Present genetic evidence for the existence of morphogens in *Hydra*, and describe a phenocopy of a morphological mutant. (*Question 7*)

1.12 Describe how bilateral symmetry arises in the amphibian embryo. (*Question 8*)

1.13 Explain the concept of regulation in amphibian embryos, and describe how it is studied experimentally. (*Question 9*)

1.14 Distinguish between regulative and mosaic properties of embryos. (*Question 10*)

1.15 Present evidence for the hierarchical nature of development. (*Question 11*)

1.16 Describe how to distinguish experimentally between induction and competence as the source of species-specific differences underlying the emergence of distinct morphological structures in different species. (*Question 11*)

1.17 Describe the difference of developmental emphasis between the theories of preformation and epigenesis, and the types of research programme that they encourage. (*Question 11*)

GAMETES AND FERTILIZATION ◆ CHAPTER 2 ◆

2.1 EGGS AND EMBRYOS

The starting point for the process that leads to a complex multicellular adult organism is generally a fertilized egg. Even before the genetic material in the sperm joins with the maternal chromosomes to form a diploid nucleus, the sperm has an influence on subsequent development resulting from its contact with the egg. Since sperm–egg interactions are generally crucial for development, we will provide in this chapter a little of the background to the complex process of fertilization.

Following fertilization there is synthesis of new macromolecules which allow rapid cell growth and cell division. These processes require a source of energy and cellular 'building blocks' which have to be 'packaged' into the egg; sustaining the developing embryo until it is able to feed itself directly or, as is the case in placental mammals, be able to derive nutrition from its mother. So the egg, as well as providing half of the genetic material, provides the metabolites for early development.

2.2 GAMETES

2.2.1 Eggs

Eggs are highly specialized types of cell that develop in an ovary. They start as quite ordinary cells, **primordial germ cells**, which grow and divide by mitosis, producing a large number of **oogonia**. In mammals, this takes place in the embryonic ovaries, so that by the time a female mammal is born the mitotic phase is complete. The oogonia then go on to grow and become large **oocytes**. Mouse oocytes, for example, are about 80 μm in diameter. In birds and reptiles, the oocyte can eventually be 100 000 times the volume of one of the original germ cells. In amphibians, egg growth starts after metamorphosis and takes 3 years. In that time the diameter increases from 50 μm to about 1.5 mm, 27 000 times initial volume.

The oocytes then progress to meiosis, but most remain dormant in the first meiotic prophase. When the human female embryo is in the second to seventh months of gestation, 1000 or so oogonia divide rapidly, producing around seven million oogonia. The numbers then fall, as most oogonia die, leaving around two million at birth. The numbers continue to decline until puberty, when there will be about 400 or so oocytes which have the potential to mature over the woman's reproductive lifespan. Human oocytes can stay in this dormant phase for up to 50 years. At puberty, cyclical changes in hormones allow one or two eggs each month to progress through the remaining meiotic divisions to produce an **ovum**. This whole process, from germ cell to ovum, is known as **oogenesis**, and it is summarized in Figure 2.1.

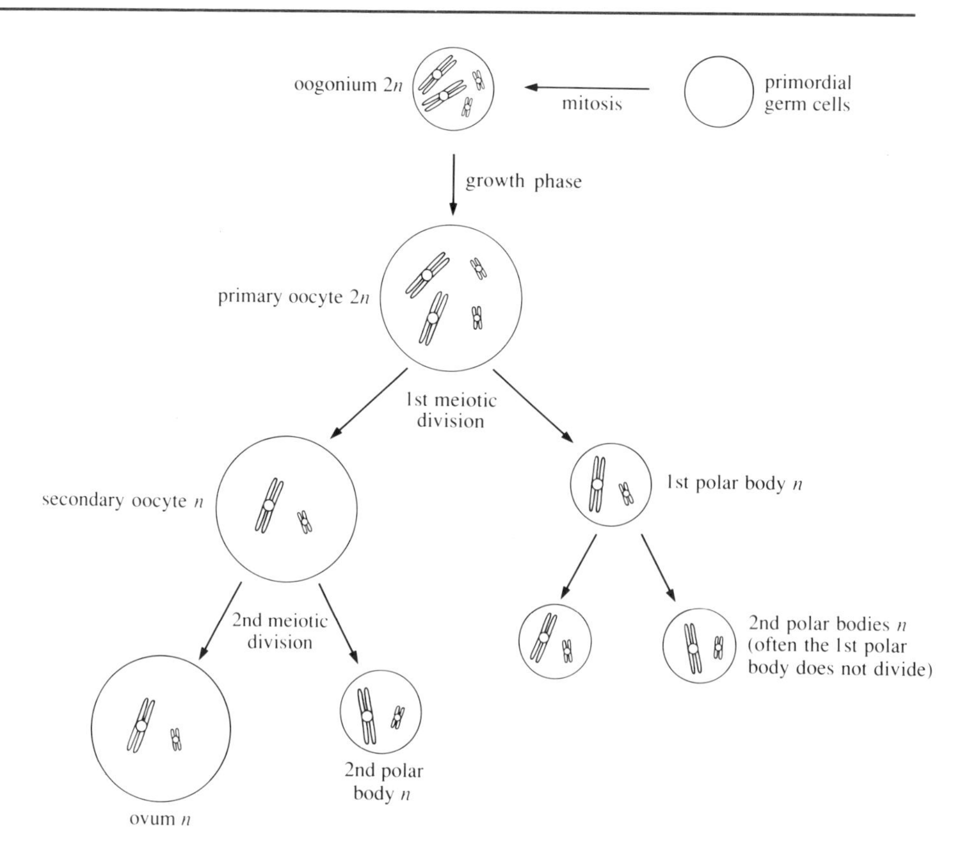

Figure 2.1 Oogenesis.

An essential feature of these meiotic divisions — and this is typical of all animals, not just mammals — is that the products of division have equal numbers of chromosomes but very different amounts of cytoplasm. Virtually all the cytoplasm is associated with the one cell that will become the ovum. The others, known as **polar bodies**, have very little cytoplasm and do not develop any further. They have no function in fertilization but, as you will see, may have a role in the subsequent development of the egg. The ovum which is formed is a very large cell with half as many chromosomes as are in the somatic cells of the organism.

The process of oogenesis can take a very long time. In humans the minimum time will be around 12 years, i.e. from the time of germ cell formation before birth to the first ovulation. In frogs an oocyte which is in the ovary at the time of metamorphosis from the tadpole stage takes three years to grow to a mature egg. In the hen, much of the main growth of the oocyte occurs in the days immediately before ovulation. The volume of the oocyte can increase by 200 times in just 7 days. The hen's egg with the yolk can weigh as much as 50 g. But the nucleus and cytoplasm which will be involved in fertilization will still be the size of the original oogonium, about 5 μm diameter.

2.2.2 Sperm

The sperm are also highly specialized cells that differentiate from precursors in the embryonic testis. These primordial cells give rise to **spermatogonia** which will eventually develop into mature **sperm**. The whole process of sperm production from spermatogonia, summarized in Figure 2.2, is known as the **spermatogenic cycle**.

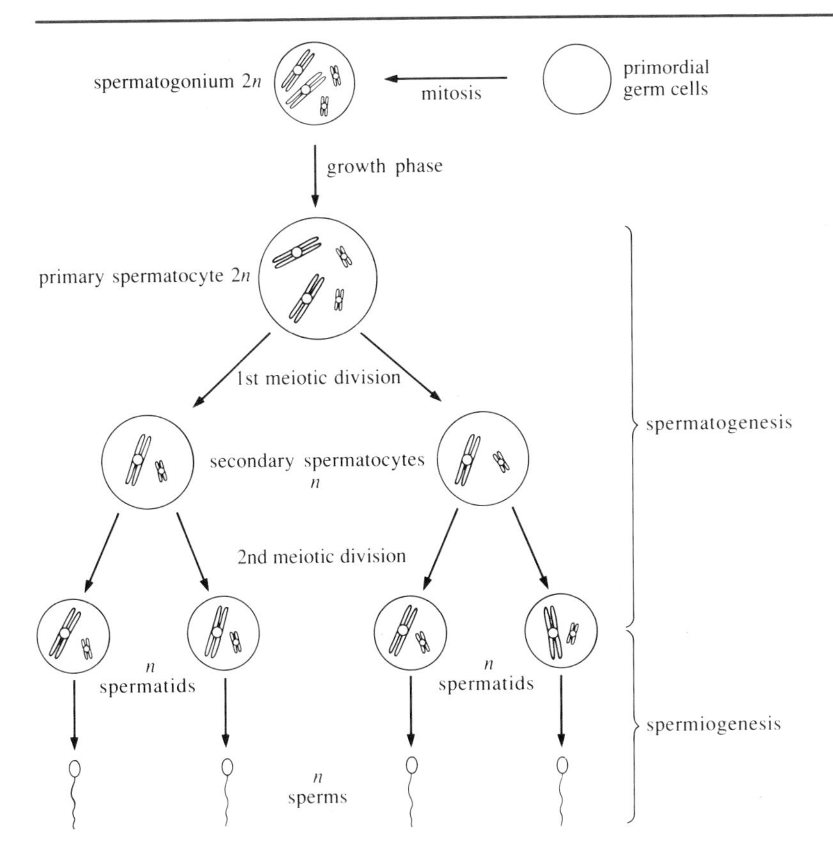

Figure 2.2 The spermatogenic cycle.

The spermatogonia divide by mitosis and at the onset of puberty some of them increase in size. These larger cells are known as **spermatocytes** and go through meiotic divisions to form **spermatids**. The meiotic division stage is called **spermatogenesis**. The spermatids then differentiate into sperm. This final stage is known as **spermiogenesis**. (Note the potential for confusion between the terms 'spermiogenesis' and 'spermatogenesis'. In general reading you may also find 'spermatogenesis' used to refer to the whole process of sperm production — a useful shorthand but not strictly correct.) The mitotic and meiotic processes in the spermatogenic cycle are very similar to those in oogenesis, differing in that (a) there are many more divisions of the primordial germ cells in the production of male gametes and (b) four spermatids, and hence four sperm, are formed from each spermatocyte.

◇ How does this differ from the events in oogenesis?

◆ In oogenesis there is only one gamete (ovum) produced from each oocyte.

The process of spermiogenesis and spermatogenesis are under hormonal control in the sense that critical levels of androgens at the appropriate time are necessary to allow the production of sperm. Hormone levels greater than these critical values do not change the rate of the process. On the other hand, absence of androgens will prevent it. The rate of sperm production is remarkably constant within a particular species. For example in humans it takes 64 days for a primordial germ cell to divide and develop into a mature sperm. The hormonal control of spermatogenesis and the role of the various cells in the testis is a complex process. It is described in Book 3 of this series* and you will meet it again in Chapter 6 of this book.

*Michael Stewart (ed.) (1991) *Animal Physiology*, Hodder and Stoughton Ltd, in association with the Open University (S203, *Biology: Form and Function*).

2.2.3 Development of the egg

We have already outlined the mitotic and meiotic processes involved in the development of an ovum, and described the vast increase in size. But how does this occur?

Given the great diversity of the animal and plant kingdoms it would be surprising if there was a universal solution to building up the food reserves in the egg.

In most vertebrates, specialized **follicle cells** surround the oocyte throughout its growth and development. Initially these form a single layer — the **primordial follicle**. The oocyte remains within the primordial follicle during its dormant state. Figure 2.3 is an electron microscope photo showing a cross-section of two oocytes with associated follicle cells.

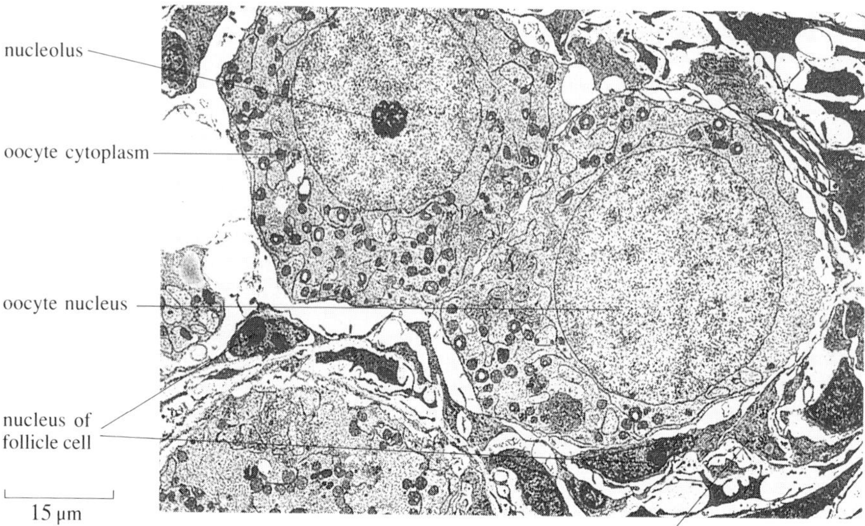

Figure 2.3 Micrograph of two young oocytes surrounded by follicle cells.

At sexual maturity the primordial follicles start to grow and can increase in size from 20 μm to 400 μm. Most of this increase is due to the oocyte itself increasing in mass. The DNA in the chromosomes is active during this growth, and is transcribed to produce large amounts of mRNA, rRNA and tRNA. It has been estimated that some amphibian oocytes will produce as many as 10^{12} ribosomes prior to ovulation.

◇ Suggest an explanation for this high level of nucleic acid synthesis.

◆ The egg is building up a supply of the components essential for protein synthesis which will be necessary for rapid cell division and growth after fertilization. The maternal mRNAs are templates for protein synthesis in the next generation.

In amphibians, the majority of yolk proteins are synthesized in the maternal liver. The follicle cells thus transfer these proteins to the egg.

In mammals, as the oocyte enlarges, the follicle cells start to grow and divide so that the oocyte is surrounded by several layers of cells. These are known as **granulosa cells**. The granulosa cells secrete a layer of glycoprotein material, the **zona pellucida** around the oocyte (Figure 2.4).

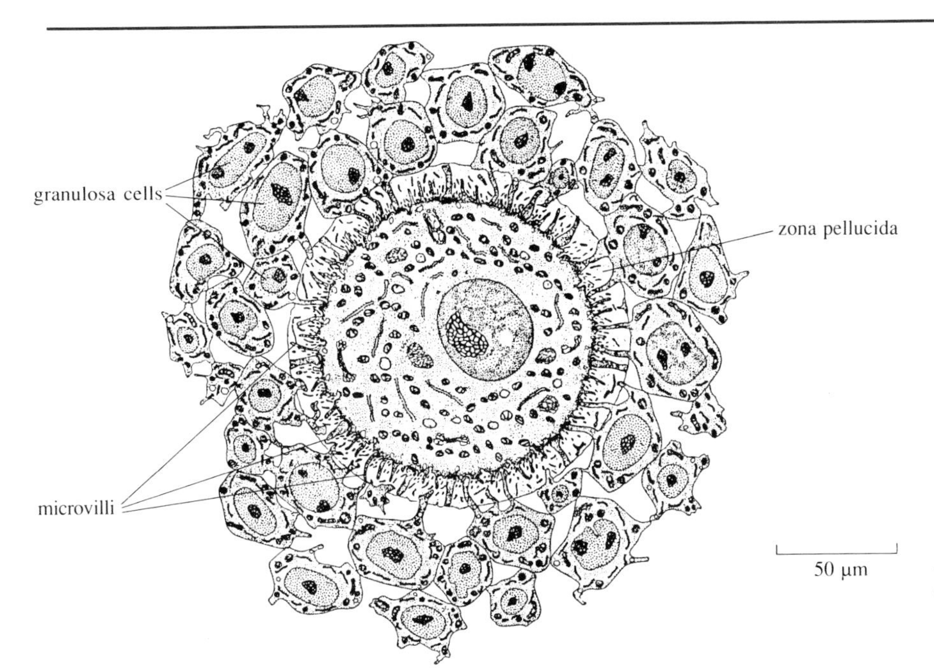

Figure 2.4 Typical young mammal oocyte with granulosa cells and zona pellucida.

As the egg approaches maturity a cavity appears within the follicle cells, and the follicle is now referred to as a **Graafian follicle**. The oocyte is not separated from the granulosa cells. Numerous 'microvilli' (tiny finger-shaped growths) develop both from the oocyte and from the follicle cells. The microvilli maintain cytoplasmic contact and greatly increase the area of contact between the oocyte and the follicle cells. Figure 2.5 shows the arrangement of the granulosa cells and mature oocyte in a typical mammal.

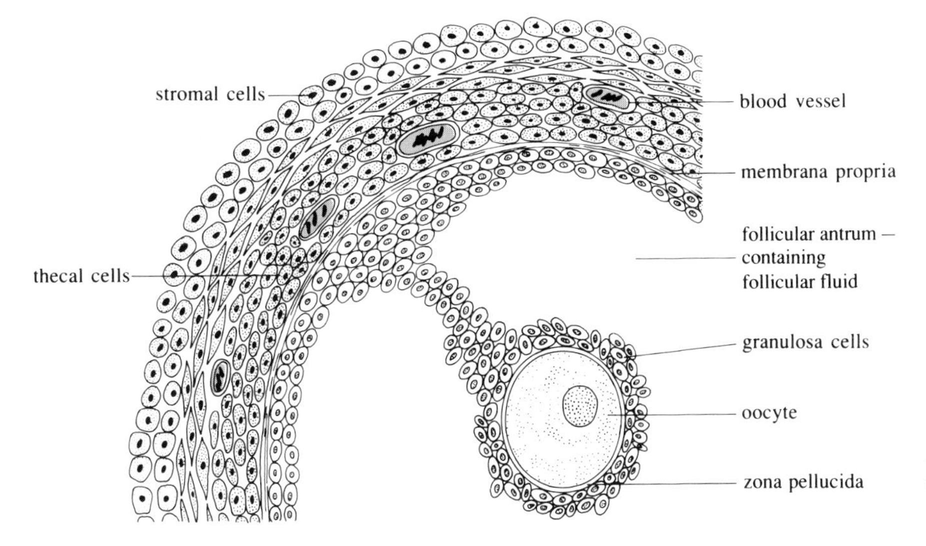

Figure 2.5 Graafian follicle with mature oocyte.

◇ Why could an increase in contact area be important for the oocyte?

◆ An increase in contact area is likely to be important in facilitating the transfer of materials from the follicle cells to the oocyte.

Microvilli are not found in the oocytes of all species. For example in sea urchins, and in some insects and molluscs, the association of oocyte and follicle cells is complicated by the presence of **nurse cells**. These nurse cells arise from divisions of the oogonia and remain connected to the oocyte.

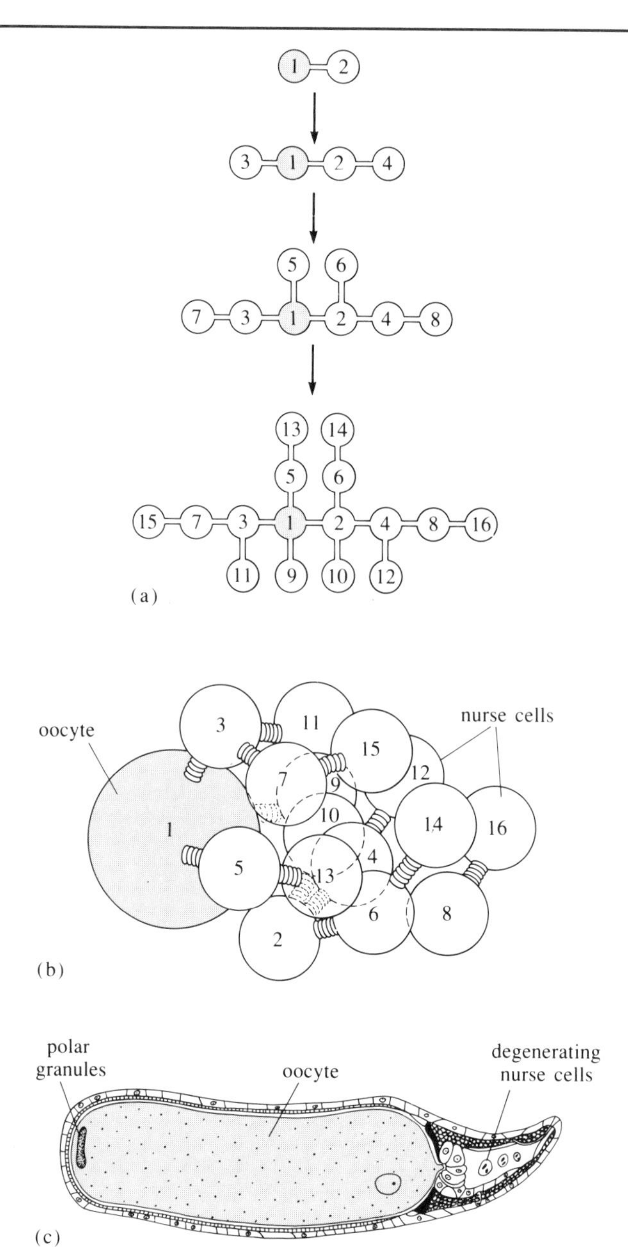

Figure 2.6 Oogenesis in the insect *Drosophila melanogaster.* (a) Cell divisions form a cluster of 15 nurse cells connected to the oocyte. (b) Three-dimensional arrangement of nurse cells and oocyte. (c) Section view of mature oocyte.

Figure 2.6 illustrates the process of oogenesis in *Drosophila*. The contents of the nurse cells flow into the oocyte. The main product of the nurse cells is RNA. Some yolk is synthesized by the follicle cells but most comes from the fat body, an organ in the insect. When the oocyte is mature the nurse cells degenerate and become separated from the oocyte by the egg case.

Studying the microscopic structure of the *Drosophila* oocyte reveals that the end which was previously connected to the nurse cells will become the anterior end of the egg, larva and the adult fly. So even as the egg is formed, there are positional effects on embryonic development.

For example, the egg cytoplasm can influence the formation of gametes in the embryo. This can be inferred from the role of **polar granules** which are seen as the egg develops. These granules are localized at the posterior region of the egg and will eventually become incorporated into the adult germ cells. Transplantation of polar granules to a different region in another egg will induce germ cell formation in the transplant region of the host egg. Thus the capacity to form germ cells must be localized in the egg — in the polar granules — during oogenesis. Since injecting polar granules into ectopic sites (other than normal) causes germ cells to form in those locations, it is vital for normal development that the granules are in their normal posterior position.

It is clear then that both the nature and position of the polar granules is crucial for subsequent germ cell development. How could the granules become localized at the posterior end of the egg?

In insects such as the silk moth *Hyalophora cecropia*, which has a similar ovary to *Drosophila*, there is evidence that a small electrical potential difference (normally +10 mV at the nurse cells relative to the oocyte) can influence the movement of cytoplasmic components. Under normal conditions this potential difference appears to cause movement of proteins from the nurse cells into the oocyte. But under experimental conditions, when the potential difference is reversed, proteins move from the oocyte into the nurse cell. So it is likely that such currents influence the movement and location of polar granules in *Drosophila*.

The effect of events during oogenesis on subsequent development is clearly seen in mutations which disrupt the relationship between the oocyte and nurse cells in *Drosophila*. Most mutations of this type will block formation of a viable egg but there is one particular mutation, known as 'dicephalic', which results in a very interesting morphology. Homozygous females carrying this mutation, as you might guess from the name, produce embryos with two heads. The abdomen is replaced by a mirror image of the head and thorax. In the eggs produced by dicephalic mutants, the nurse cells are separated into two groups, one at each end of the oocyte.

This is direct evidence that specific changes to the spatial pattern of cells in the developing oocyte may produce a specific alteration in the spatial structure of the embryo. Such effects will be looked at again in Chapter 4 when we consider the effects of particular genes on development.

2.2.4 Meiotic divisions in the egg

As we mentioned in Section 2.2.1, the oocyte remains dormant in an arrested state of meiosis. In most animals fertilization starts before meiosis is completed. Only in a few groups (anemones and sea urchins) is meiosis complete at the time of fertilization. Work on the eggs and embryos of a marine worm, the nemertine *Cerebratulus lacteus* (a proboscis worm) has provided much information on how meiotic divisions can influence development. Eggs from this organism are particularly useful for experimental study as they are released before any meiotic divisions have started.

The larva of *C. lacteus*, illustrated in Figure 2.7a, has a distinct apical tuft and a gut. An egg is taken just after release from the ovary and sperm entry but before the first meiotic division. You will recall from Chapter 1 that eggs often have a distinct pigmented animal pole and a vegetal pole where the yolk reserves are concentrated. If the egg of *C. lacteus* is then cut in two across the equator, separating the animal and vegetal regions of the egg, the two halves develop quite differently. In most cases both halves will develop a gut region. But of the animal halves, only about 28% will develop tufts (Figure 2.7b). If the egg is cut equatorially at the time the second polar body is extruded, fewer than half of the animal halves develop a gut region but nearly all develop tufts.

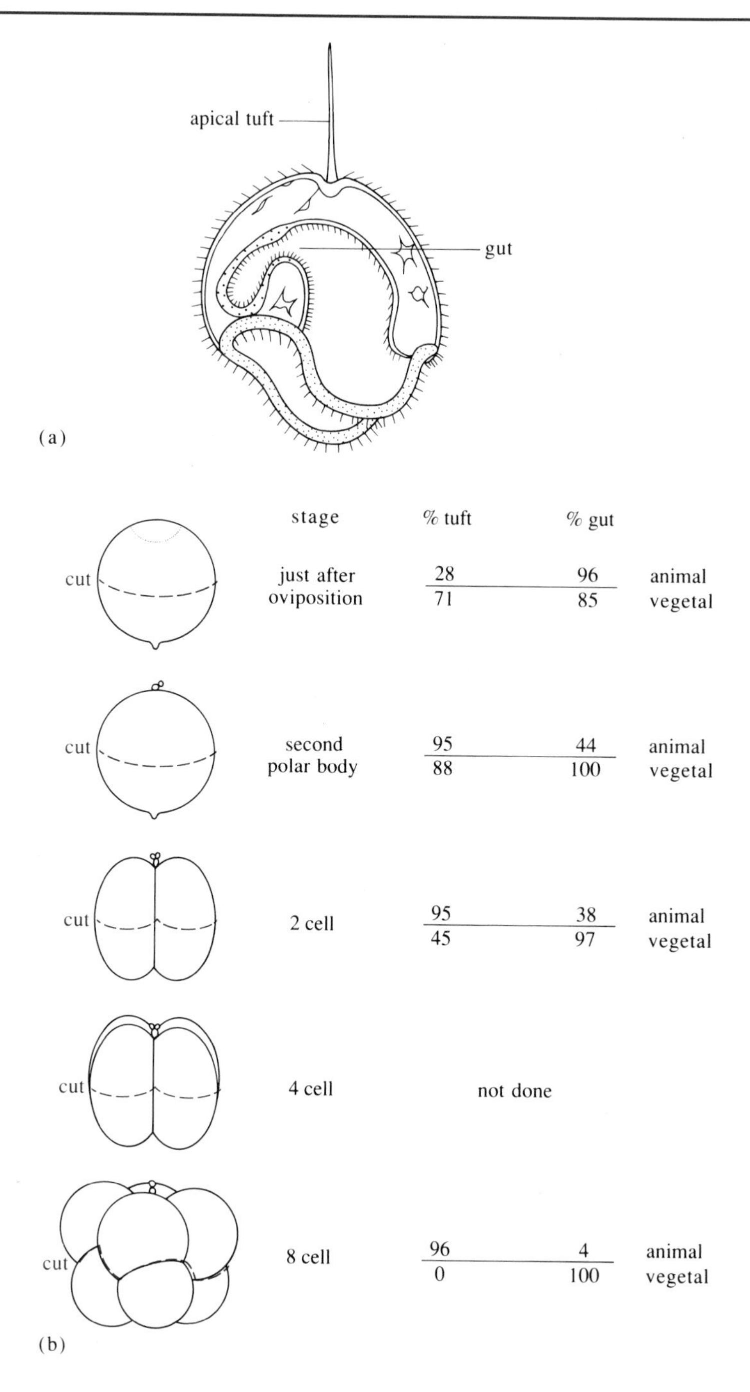

Figure 2.7 *C. lacteus.* (a) Larva. (b) Egg division experiment and results — animal above line, vegetal below line.

◇ Do these observations provide any evidence that there is redistribution of factors within the cytoplasm of the egg as a result of meiotic division?

◆ Yes. The meiotic processes must have redistributed whatever factors control apical tuft and gut development between the animal and vegetal regions of the egg.

It is thought that the meiotic spindles are involved in this redistribution. If the egg is treated with a chemical such as ethyl carbamate which dissociates the meiotic spindle then redistribution of factors is delayed. Evidently the process of meiosis can itself influence development.

Summary of Section 2.2

In oogenesis, primordial germ cells in the embryonic ovary grow and divide by mitosis to form oogonia. These grow to hundreds (or thousands) of times their original volume, forming oocytes. The oocytes undergo meiotic divisions to form the female gametes. The meiotic process generally starts early in development but is arrested until the reproductive phase of the animal's life cycle. The female gamete or ovum has one set of chromosomes, the maternal contribution to the genetic material of the organism and has a very large volume of cytoplasm.

In spermatogenesis, primordial germ cells in the testis give rise to spermatogonia. These undergo numerous mitotic divisions and some of the cells enlarge to form spermatocytes. The spermatocytes divide by meiosis to form spermatids which differentiate into sperm. Unlike egg production there are many more divisions of the primordial germ cells and four gametes are formed at each meiotic division. Although a sperm has the same amount of genetic material as an ovum, it has a very much smaller volume of cytoplasm.

The ova of different species develop in different ways. In most vertebrates the oocyte is surrounded by specialized follicle cells which are important in transferring nutrients to the oocyte. In some insects and other species, the oocytes are also accompanied by nurse cells which arise from divisions of an oogonium. Nurse cells provide the oocyte with RNA. Experiments on *Drosophila* show that the subsequent development of the embryo is affected by the position of the nurse cells and by the position of the polar granules within the oocyte cytoplasm during oogenesis.

Meiotic processes, as illustrated by experiments on *C. lacteus*, can alter the distribution of cytoplasmic factors within the egg. These can lead to subsequent changes in the processes of development.

Question 1 (*Objective 2.2*) Draw a simple diagram to show the stages in the development of an ovum from one oogonium. Your diagram should indicate the relative number of chromosomes you would expect at each stage. How long might you expect this process to take in a mammal which was sexually mature when 1 year old and ceased reproduction when 5 years old?

Question 2 (*Objective 2.3*) List the key differences in the development of sperm compared with the production of an ovum.

Question 3 (*Objective 2.4*) Given the role of nurse and follicle cells, what particular features, in terms of biochemistry, would you expect to find associated with them?

2.3 FERTILIZATION

2.3.1 Is sperm necessary?

For an embryo to develop, it is normally necessary for the genetic material from both parents to combine. This is true for most species but as in many aspects of biology there are significant exceptions.

Development of a viable embryo from an unfertilized egg is known as **parthenogenesis** and this is quite common in animals such as aphids and bees. In fact in bees, fertilized eggs produce females while unfertilized eggs produce males. Even in non-parthenogenetic species the egg can often be induced by artificial means to develop through a number of embryonic stages.

◇ What can you deduce about the relative importance of the egg or the sperm to development given the phenomenon of parthenogenesis?

◆ That the egg of many species contains all the necessary cytoplasm and genetic material to allow development to proceed.

So when reading through this section on fertilization, bear in mind that under certain conditions and/or in some species the sperm has no role in development.

2.3.2 Sperm meets egg

The first stage in **fertilization** starts with contact between the gametes. In animals which have internal fertilization (the male sheds sperm inside the reproductive tract of the female) this is relatively straightforward. However for animals which have external fertilization and release gametes into the environment, as in many marine organisms, there are obvious problems over (a) how sperm and egg can meet under such dilute conditions and (b) what prevents sperm of the wrong species fertilizing the egg. These problems are solved by two mechanisms.

A large number of aquatic species which shed gametes into the water, produce **species-specific sperm attractants** and thus overcome the low concentration of sperm relative to eggs. The egg may produce specific chemicals which attract only sperm of the same species. For example, in two species of sea urchin, small peptides 10 and 14 amino acids long have been identified which have this specific sperm attraction role.

The wrong sperm are prevented from fertilizing the egg by **species-specific sperm activation**. Before the sperm can enter the egg an **acrosome reaction** must occur. This is illustrated in Figure 2.8.

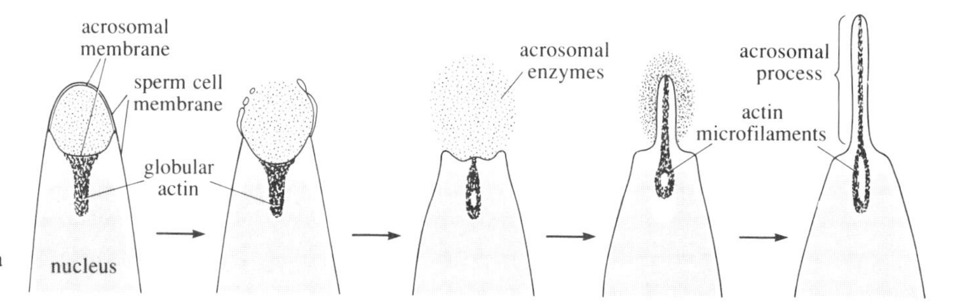

Figure 2.8 Acrosome reaction in sea urchin sperm.

When the sperm touches the egg, the sperm's acrosomal membrane breaks down and releases enzymes which will break down the egg jelly and provide a path for the sperm to the egg membrane. At the same time, actin molecules within the acrosome assemble to produce microfilaments. This is called the **acrosomal process** and it extends outwards, eventually fusing the acrosome with the egg membrane.

The acrosome reaction is generally species-specific in that the egg jelly of one species will only stimulate the reaction in sperm of that species. The major species-specific recognition is between the acrosomal process and the **vitelline envelope**, which surrounds the egg membrane. The vitelline envelope is analogous to the zona pellucida of the mammalian egg. The acrosome contains a species-specific protein known as **bindin**. On the vitelline envelope surface are specific receptor sites for this protein. Studies have shown that these receptor sites will only bind bindin from sperm of the same species.

The precise details of the acrosome reaction in mammals are not fully known and there are significant differences between species in the timing of the reaction. For instance in the guinea pig the acrosome reaction occurs in the absence of an egg. Here it is thought that the reaction starts at a certain time after ejaculation. An important feature of most mammals is that the sperm are unable to develop the acrosomal process until they have been in the female reproductive tract for some time. During this period the sperm undergo **capacitation**. Experimental work shows that during the period of capacitation there is a change in the cholesterol:phospholipid ratio in the sperm membrane. These changes are thought to assist the membrane fusion of the acrosome reaction.

2.3.3 Gamete fusion

Once the correct sperm has been recognized by the vitelline membrane or zona pellucida, the membranes of the sperm and egg fuse. Figure 2.9 shows an electron micrograph of the point of cell fusion in the sea urchin.

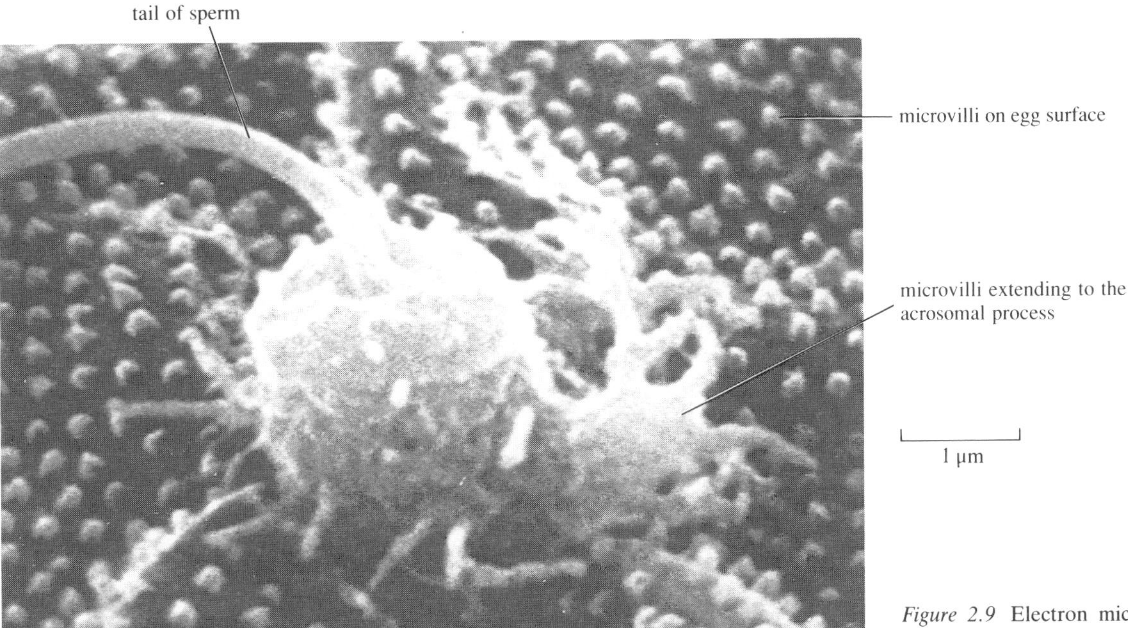

Figure 2.9 Electron micrograph of the fusion of sea urchin sperm and egg.

You can see that the egg surface is covered with small projections — microvilli — and that a number of these seem to extend and surround the acrosomal process. This forms what is known as the **fertilization cone**. The sperm nucleus and tail then pass into the egg cytoplasm.

In humans, one ejaculation can contain over 300 million spermatozoa yet only one sperm will fertilize an egg.

◇ What would be the genetic consequences if an egg were fertilized with two sperm?

◆ The egg would have three copies of each chromosome. As the cell divided, daughter cells would receive unequal numbers of chromosomes, since the mitotic apparatus can accurately partition only homologous chromosomes. This would result in an abnormal embryo, if not death.

Most organisms have mechanisms that prevent the disastrous consequences of multiple fertilization. For example in sea urchins, after the first contact with a sperm there is a rapid change (under 0.1 seconds) in the electrical potential of the egg membrane. This allows Na^+ ions to move in, raising the potential of the entire membrane from its resting value of about $-70\,mV$ to zero or above. Research work suggests that there is a specific acrosomal protein which opens the channels for sodium ions into the egg. Sperm are unable to bind to membranes with potentials more positive than $-10\,mV$, so binding of other sperm is blocked within a tenth of a second.

Figure 2.10 illustrates how penetration by a number of sperm is further prevented in sea urchin eggs. The vitelline envelope is linked to the egg cell membrane by proteins, and lying just inside the egg membrane are **cortical granules** (Figure 2.10a). Following the change in the electrical potential of the egg cell, the cortical granules fuse with the egg membrane and release enzymes. These enzymes dissolve the proteins linking the vitelline envelope to the membrane (Figure 2.10b). Then mucopolysaccharides in this region create an osmotic potential which causes water to move in, physically separating the envelope from the egg membrane (Figure 2.10c). This water-lined envelope, now called the **fertilization envelope**, forms within one minute of sperm binding and blocks penetration by other sperm.

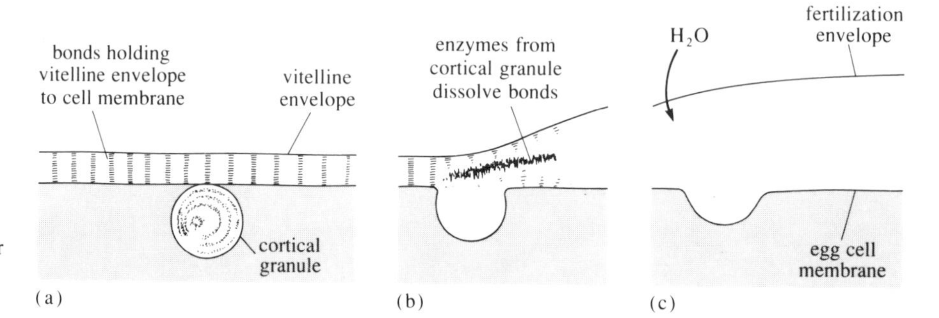

Figure 2.10 One of the mechanisms for prevention of polyspermy in sea urchins.

The systems described above are not found in all animals and there are a number of other mechanisms which prevent **polyspermy** (fertilization by more than one sperm). In mammals the cortical granules appear to modify sperm binding sites on the zona rather than creating a physical barrier as such. However in some reptiles and birds, polyspermy does occur but mechanisms exist which destroy the extra sperm.

2.3.4 Fusion of the genetic material

Fertilization is completed when the genetic material of the sperm and egg combine to form a diploid nucleus. As you might expect by now, there are the usual species differences but as an illustration, Figure 2.11 shows the sequence of events that are thought to occur in the hamster egg.

You will notice that the egg nucleus completes its meiotic divisions as the sperm enters the egg. The genetic material of the egg and sperm are known as **pronuclei** until the first mitosis begins. Normal diploid nuclei are present once the two sets of genetic material have combined and have gone through one mitotic division.

It is easy to concentrate on the genetic events here but there are also significant changes occurring in the cytoplasm. These changes are linked to the initial sperm fusion.

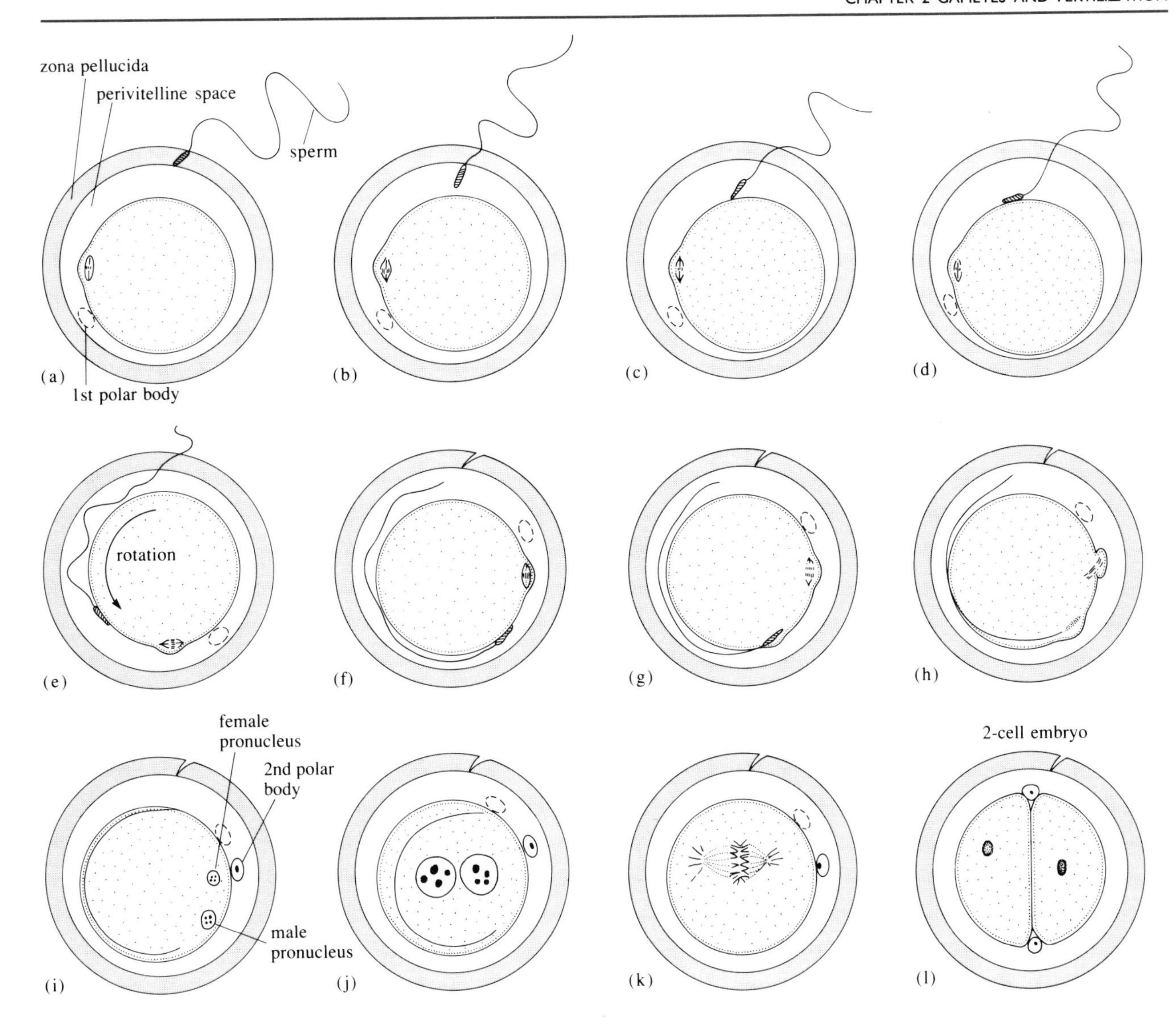

Figure 2.11 Fertilization in the hamster. (a) Sperm attaches to the zona. (b) Enzymes produced by the sperm digest the zona pellucida so the sperm can move into the perivitelline space. (c) and (d) When the sperm attaches to the egg its head is placed parallel to the egg membranes. (e) and (f) Because the sperm head is firmly fixed, movement of the sperm tail causes the egg to rotate and the entire sperm enters the perivitelline space. (g) The sperm and egg membranes fuse. (h) The genetic material is released from the sperm. (i) and (j). The male and female pronuclei swell and meet in the centre of the egg. (k) The pronuclei membranes breakdown and the first mitotic division commences. (l) The 2-cell embryo. You can see what remains of the polar bodies at this stage.

The release of the cortical granules is dependent on release of calcium ions and it appears that calcium ions are also essential in activating the series of metabolic reactions that occur in the egg. It is possible to block calcium release by injecting chemicals that bind calcium ions. This prevents an increase in egg metabolism. Normally there is a rapid increase in DNA and protein synthesis. Figure 2.12 shows the difference in the rate of protein synthesis (measured by the uptake of radioactively labelled leucine) in fertilized and unfertilized eggs.

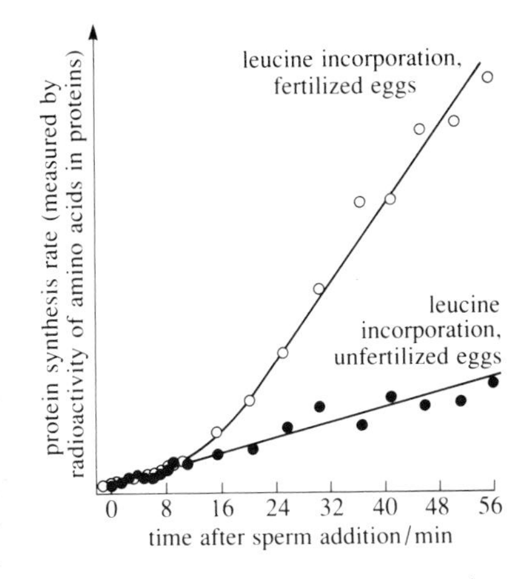

Figure 2.12 Protein synthesis rates in fertilized and unfertilized sea urchin eggs.

As well as metabolic activation, major changes in the arrangement of the egg cytoplasm occur as a result of fertilization. These changes are crucial for the embryonic events that follow genetic fusion. One of the best examples here, first observed at the beginning of this century, is from the egg of the urochordate *Styela*.

Before fertilization, the egg has a yellow cytoplasm overlaying grey yolky cytoplasm (see Figure 2.13a). When the sperm enters the egg the yellow cytoplasm streams down to the vegetal pole of the egg (Figure 2.13b). Clear cytoplasm, derived from the breakdown of the oocyte nucleus, also streams downwards (Figure 2.13c). The sperm pronucleus then migrates towards the animal pole, 'drawing' the cytoplasm with it to create the arrangement of cytoplasm seen in Figure 2.13d.

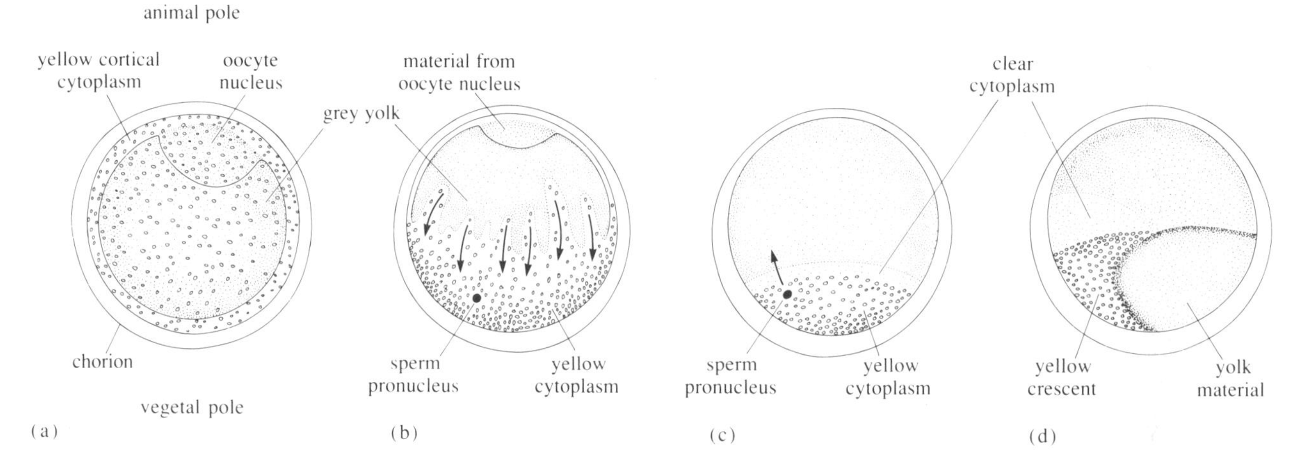

Figure 2.13 Cytoplasmic rearrangement in the urochordate *Styela*. (a) Before fertilization, yellow cortical cytoplasm surrounds grey yolky cytoplasm. (b) At the moment of sperm entry, the yellow cortical cytoplasm and the clear cytoplasm (derived from the breakdown of the oocyte nucleus) stream down to the sperm at the vegetal pole. (c) As the sperm pronucleus migrates towards the animal pole the yellow and clear cytoplasms move with it. (d) The final positions of the clear and yellow cytoplasms.

The significance of this, in terms of development, is that these cytoplasmic areas are always associated with particular tissues in the embryo. If the cytoplasmic pattern is disrupted by centrifugation then an abnormal embryo is formed with tissues in the wrong relationship to each other. Figure 2.14a shows a normal *Styela* embryo and Figure 2.14b shows one formed after centrifugation of the egg.

The importance of the cytoplasm in determining embryonic development will be emphasized later in this book.

At the end of fertilization, a process which can take about 20 hours in mammals, the genetic material from the egg and sperm has fused, proteins are being synthesized, DNA is replicating and the cytoplasm has been

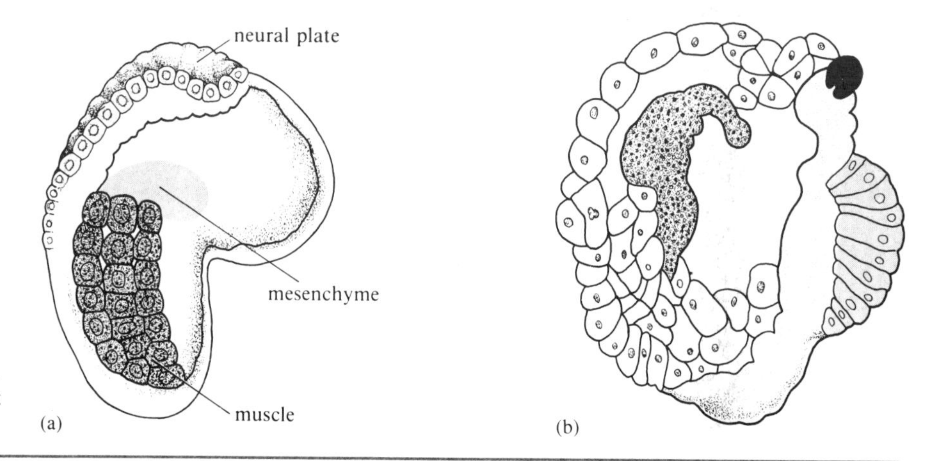

Figure 2.14 *Styela* embryo. (a) Normal. (b) After centrifugation during the first cell division.

reordered. The next stage in development is to move from a single cell to a multicellular embryo. To try to understand the processes involved in these changes we need to consider how individual cells can cooperate to build structures — the process of morphogenesis. This is where the next chapter begins.

Summary of Section 2.3

Although fertilization and gamete fusion are generally essential for normal embryonic development, some species such as bees and aphids will normally produce fertile adults from unfertilized eggs. Under experimental conditions eggs can often be stimulated to go through at least some embryonic stages by artificial means in the absence of sperm.

For fertilization to occur, the sperm of the correct species has to penetrate the egg. In species which do not have internal fertilization, there are species-specific sperm attractants (small peptides). Species-specific sperm activation also ensures that only the sperm of the correct species will enter and fertilize the egg. Only sperm of the same species as the egg will initiate the acrosome reaction which allows the sperm and egg membranes to fuse.

There are mechanisms that prevent fertilization by more than one sperm. For example, changes in the electrical potential of the egg membrane produced by contact of the first sperm prevent subsequent sperm binding. Separation of the vitelline and egg membrane will also prevent polyspermy.

Fertilization, as well as allowing fusion of the genetic material, initiates cytoplasmic changes in the egg which influence subsequent development. Experimental work on *Styela* provides a good example of this. Centrifugation of the egg after fertilization will redistribute the cytoplasm and lead to abnormal embryonic development.

Question 4 (*Objective 2.5*) You are doing experimental work on the zygote of *C. lacteus* and have some embryos at the 8-cell stage. What would be your predictions as to the ability of the animal and vegetal halves of the embryo to develop a gut, if you cut the embryo across its equator?

Question 5 (*Objectives 2.6 and 2.7*) What are the mechanisms that ensure fertilization by one sperm, of the correct species, in aquatic organisms which shed eggs and sperm into water?

Question 6 (*Objective 2.8*) Give an example showing the importance of cytoplasmic localization associated with fertilization.

OBJECTIVES FOR CHAPTER 2

Now that you have completed Chapter 2 you should be able to:

2.1 Define and use, or recognize definitions and applications of, each of the terms printed in **bold** in the text.

2.2 Briefly outline the features of oocyte production. (*Question 1*)

2.3 State the major way that sperm production differs from oocyte production. (*Question 2*)

2.4 Give examples of the role of follicle and nurse cells in oocyte growth. (*Question 3*)

2.5 Describe how meiotic division may influence development in the proboscis worm, *Cerebratulus lacteus*. (*Question 4*)

2.6 Summarize the various mechanisms that ensure that the egg and sperm of the same species fertilize each other. (*Question 5*)

2.7 Write two or three sentences describing one process which is thought to prevent more than one sperm fertilizing an egg. (*Question 5*)

2.8 Give examples of how fertilization may influence cytoplasmic organization in the egg. (*Question 6*)

Developing organisms start off simple and end up complex. How does this come about? We are used to thinking of complexity in terms of the number of different parts that make up an instrument or a machine. A clock is fairly simple by contemporary standards, though it was not considered so in the 18th century. A jumbo jet is extremely complex. However, organisms illustrate another aspect of complexity. A newt or a human being is obviously extremely complex in terms of the number of distinguishable parts or components, such as cell types, tissues, organs, etc. But these components are not produced separately and then assembled, as in a machine. They are generated during the process of embryonic development. So we have to think of an organism's complexity as something that involves time as well as space — in other words as a generative process. This distinction between organisms and mechanisms was recognised by the 18th century philosopher, I. Kant, who provided very perceptive definitions that are particularly relevant to contemporary biology, in which the concepts of organism and mechanism tend to become confused. He defined a mechanism as a functional unity in which the parts exist for one another; i.e. separable, pre-formed parts interact with one another to achieve some integrated function as in a clock. An organism, on the other hand, is both a functional and a structural unity, in which the parts exist for and by means of one another. This means that the parts are not preformed but are generated during development by coordinated interactions that maintain the functional and structural unity of the developing organism.

In this chapter we will first consider the key events in embryonic development. The single cell produced as a result of fertilization is 'cleaved' to produce a number of small cells. The resulting blastula then goes through a series of major cell movements — gastrulation — which creates the fundamental shape of the embryo. Crucial to gastrulation, and subsequent organ formation and morphogenesis, are the processes of cell movement and cell adhesion. These are discussed in some detail. We then consider how signals between different tissue layers induce particular pathways of differentiation: the complex process of embryonic induction. Inductive processes result in the emergence of new types of cell, while differential cell adhesiveness organizes cells spatially relative to one another and thus generates shapes. These interactions arise in an initially homogeneous system which maintains a coordination throughout the whole despite the development of distinct parts. The chapter concludes with the emergence of organized patterns in developing organisms.

3.1 PROPERTIES OF INDIVIDUAL CELLS

Figure 3.1 Outline of a lymphocyte in tissue culture, showing amoeboid movement.

If we take an embryo, separate the constituent cells chemically and plate the cells out in suitable medium in a Petri dish, the cells do not just remain static but move around. There are two types of movement. Some cells (called **amoebocytes**) move, as their name suggests, rather like amoebae: directional movement is associated with extension of newly formed pseudopods at the leading edge and a compensating retraction of cytoplasm from the trailing edge (Figure 3.1). Exactly how this movement occurs is not known, but cytoplasmic streaming is an obvious feature. Cells of this type are highly mobile in the adult tissue *in vivo*.

In contrast, other types of tissue cell, for example **fibroblasts**, show a characteristic gliding form of movement. Such cells adhere to a solid or semi-solid substratum, and glide across it with no obvious pseudopodial formation. During this movement the cells are typically triangular in shape, with the broad base of the triangle forming the leading edge or **leading lamella**. The leading lamella bears structures called ruffles on its edge (see Figure 3.2).

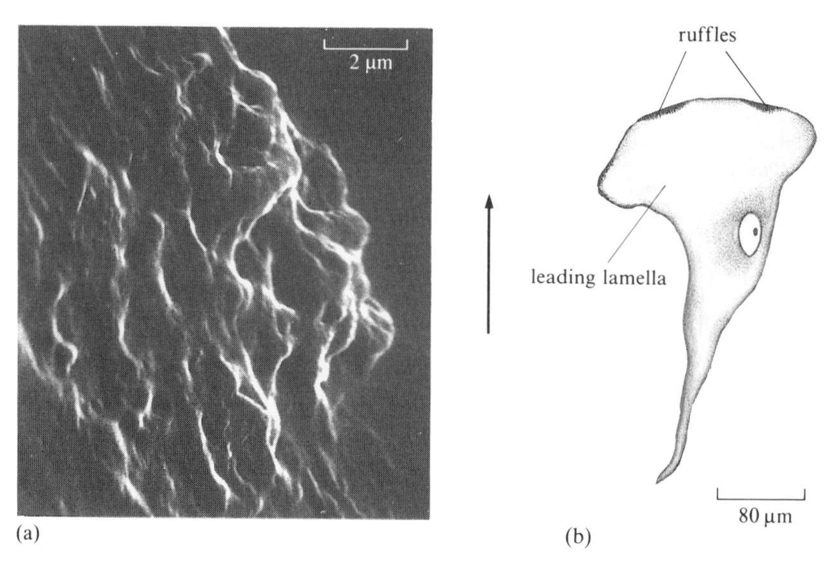

Figure 3.2 A fibroblast from a chick heart. (a) A scanning electron microscope picture of an instantly frozen cell shows the ruffled membrane structure (×5700). (b) A fibroblast moving over the substratum in the direction of the arrow.

These ruffles may be involved in cell movement, but how they work is not clear. Some workers suggest that the ruffles are evidence of intermittent contact of the cell membrane with the substratum. The continuing undulating movement of the edge of the leading lamella may produce waves of adhesive contact with the underlying substratum, thus moving the cell forward.

An alternative explanation is that cell movement involves the forward extension of the edge of the leading lamella and subsequent attachment to the substratum. If the extended edge now contracts, the whole cell is drawn forward. In this case ruffles appear when the front edge fails to attach successfully to the substratum, so causing the edge to bend upwards and backwards. So, rather than being directly involved in cell movement, ruffles might indicate a periodic failure of the cell and substratum to adhere.

Not only cell movement, but also cell alignment under culture conditions, can give information about cell behaviour. In the 1940s Paul Weiss, a pioneer in this area, discovered that fibroblast cells from embryonic chick hearts oriented themselves on fine parallel grooves cut into a glass plate. This suggested that cell movements might be oriented by special characteristics of the substratum over which they move. Weiss termed this phenomenon **contact guidance** and thought it might be explained by the secretion of macromolecules laid down on the substratum by the migrating cells. According to Weiss, the molecular network of such a mat could be oriented by the groove. Despite a later discovery that cells in culture do release macromolecular materials that coat the substratum, Weiss's theory is disputed. For instance, as more surface is available within grooves than in the flat areas between them, there may be a greater probability of contact for leading

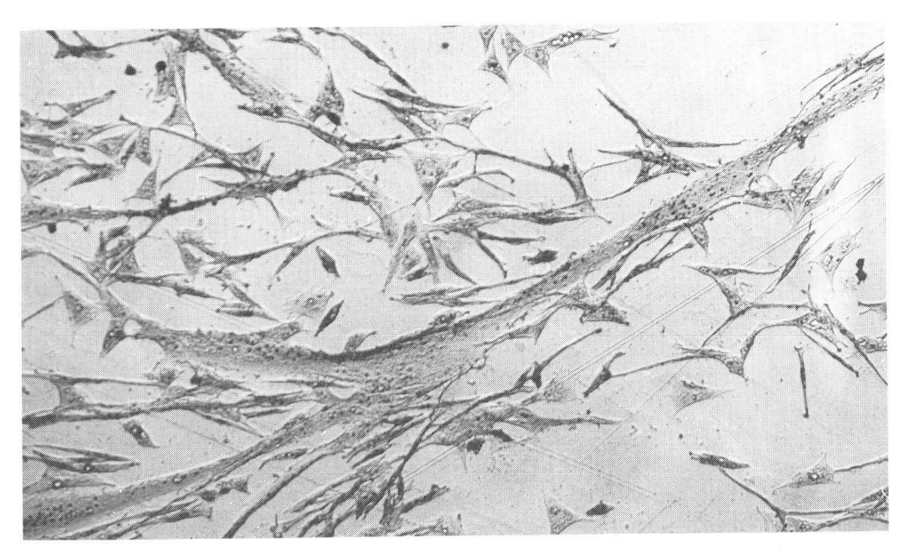

Figure 3.3 Fibroblast-like cells aggregate together to form parallel bundles and differentiation into myotubes takes place. Under a microscope, the distinctive banding pattern of the myotubes may be seen.

lamellae in the grooves. Thus contact guidance might not depend on special orienting properties or secreted materials but purely on the contact between groove and cells. Alternatively, the adhesive characteristics of the surface within the groove might differ from those of the plain surface outside. In each instance, the cell would become trapped in the grooves.

The alignment of cells in culture can also give information on the mechanism of differentiation. It was discovered, around 1960, that if a suspension of chemically disaggregated cells from chick muscle are placed in a suitable medium in a plastic Petri dish, the cells de-differentiate to form fibroblast cells that cling to the substratum. If a layer of the protein collagen is put on the Petri plate before the experiment, the cells become fused together and eventually differentiate into whole muscle tubes (Figure 3.3). Clearly, the interaction between substrate and cells is of prime importance.

Another interesting example of cell interactions in culture was studied in Edinburgh in the early 1970s by Tom Elsdale who was interested in how, under certain conditions, fibroblast cells in tissue culture can align themselves together in parallel bundles (Figure 3.4). By using time-lapse cinematography, he noticed that during bundle aggregation the cells underwent a kind of see-sawing motion together.

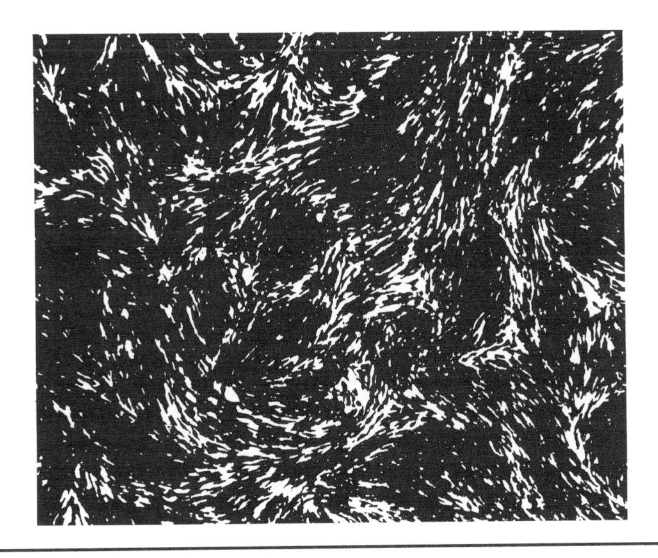

Figure 3.4 Bundles of fibroblasts in tissue culture. Each bundle consists of many individual spindle-shaped cells arranged in parallel.

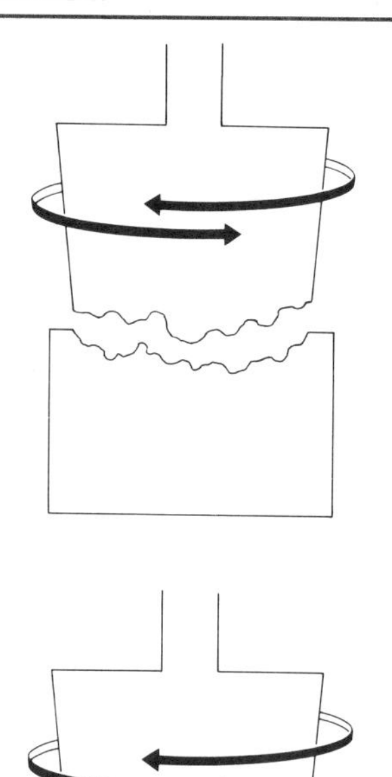

Figure 3.5 The principle of the inherently precise machine used as a metaphor by Elsdale to explain fibroblast bundle formation. The grinding of two rough blanks of glass at random gives perfectly smooth spherical lens surfaces.

Elsdale made the novel suggestion that the fibroblast bundle was behaving as an example of what in engineering is called an inherently precise machine. One such machine grinds spherical lenses by randomly rubbing two blanks over one another as shown in Figure 3.5. Similar random movements (the 'see-sawing' observed by Elsdale) could be acting in the fibroblast bundles, only instead of forming a spherical shape, the nature of the cellular material forces the cells into a parallel bundle. This idea is highly speculative, but it raises the possibility that large groups of cells, all doing identical simple things (like see-sawing) can undergo a primitive form of morphogenesis.

A final property of cells *in vitro* to be mentioned here is that of **contact inhibition**. When one fibroblast contacts another in cell culture, the advancing cell contracts somewhat, its leading lamella is paralysed and movement stops. The effect may be either physical or chemical — a chemical effect, for example, could be a localized accumulation of acid metabolites between two opposed cell surfaces. Normal cell lines from different origins tend to show mutual contact inhibition. But sometimes contact inhibition is specific to the cell type and this may be of considerable importance *in vivo*. For example, invasive cancer cells are not inhibited by contact with fibroblasts in culture.

What relevance do these various *in vitro* phenomena have for normal morphogenesis? The *in vivo* behaviour of cells in development is hard to analyse because of limitations imposed by the microscope, although technical advances such as scanning electron microscopy are making things a bit easier. For example, the role of contact guidance and contact inhibition in early embryology is not known. At most, these *in vitro* studies can indicate several general features of cell behaviour that might form a tentative basis for understanding the behaviour of cells during morphogenesis.

With this background on cell movements we can return to the events that follow fertilization — cleavage and gastrulation. These are the topics of the next two sections.

3.2 CLEAVAGE

As you know from Chapter 1, during **cleavage** the egg divides to form a number of small cells — **blastomeres**. Cleavage is characterized by cell division with little cell growth. The consequence of this is that the cytoplasm of the fertilized egg is distributed between the blastomeres.

In organisms such as echinoderms, where a small amount of yolk is evenly distributed in the egg, cleavage divisions are **holoblastic** (see Chapter 1). An orderly sequence of these cleavages, alternating between vertical (**meridional**) and horizontal (**equatorial**) result in a hollow ball of cells. This is shown in Figure 3.6.

Where the egg has significant amounts of yolk, as in amphibian eggs, the cleavage divisions in the vegetal, yolky cytoplasm progress at a slower rate, so blastomeres of different sizes are formed in the animal and vegetal regions. The cells at the animal pole are smaller because the second cleavage division starts at the animal pole before the first cleavage division reaches the vegetal pole. This is illustrated in Figure 3.7.

By the time 128 cells are formed a cavity called the **blastocoel** develops within the ball of cells. We will come back to this later.

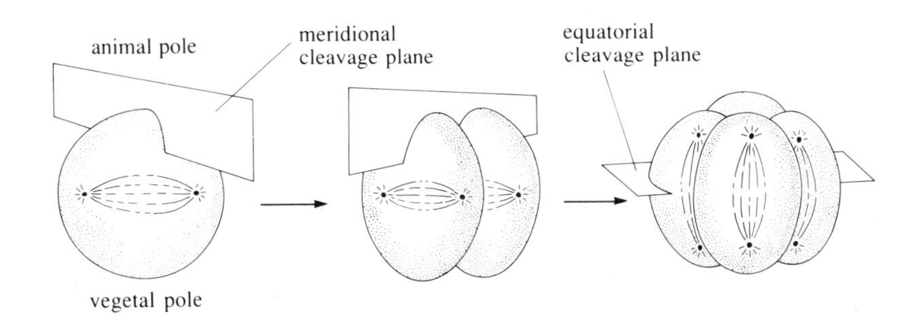

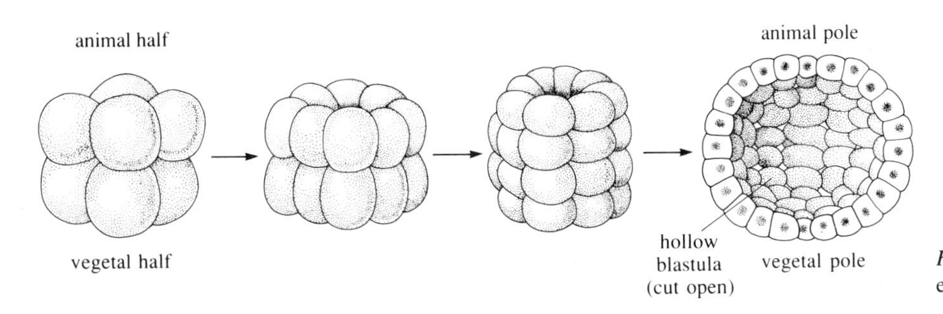

Figure 3.6 Holoblastic cleavage in the echinoderm *Synapta digita*.

Where the egg is virtually all yolk, as in birds, reptiles and some fish (the elasmobranchs) cleavage divisions are restricted to a disc-like region at the animal pole as shown in Figure 3.8.

The hollow ball of cells, the **blastula**, or the disc, the **blastodisc**, is the starting point for cell movements which progressively create the form of the embryo. Morphogenesis starts with a process known as **gastrulation**. The next section looks at this process in sea urchins and amphibians.

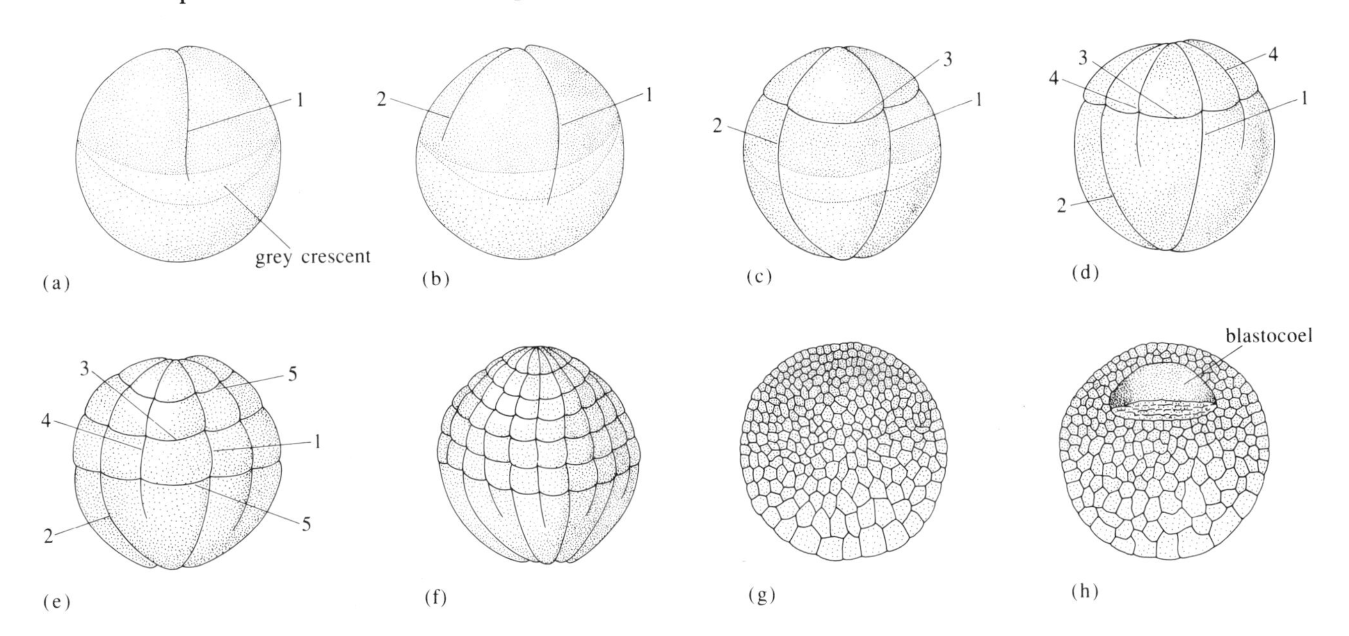

Figure 3.7 Cleavage of a frog egg. Cleavage furrows are numbered in order of appearance. (a) First division begins. (b) The vegetal yolk impedes the cleavage such that the second division begins in the animal region of the egg before the first division has divided the vegetal cytoplasm. (c) The third division is displaced toward the animal pole. (d)–(g) After continued division the vegetal hemisphere contains larger and fewer blastomeres than the animal half. (h) Cross-section of stage (g).

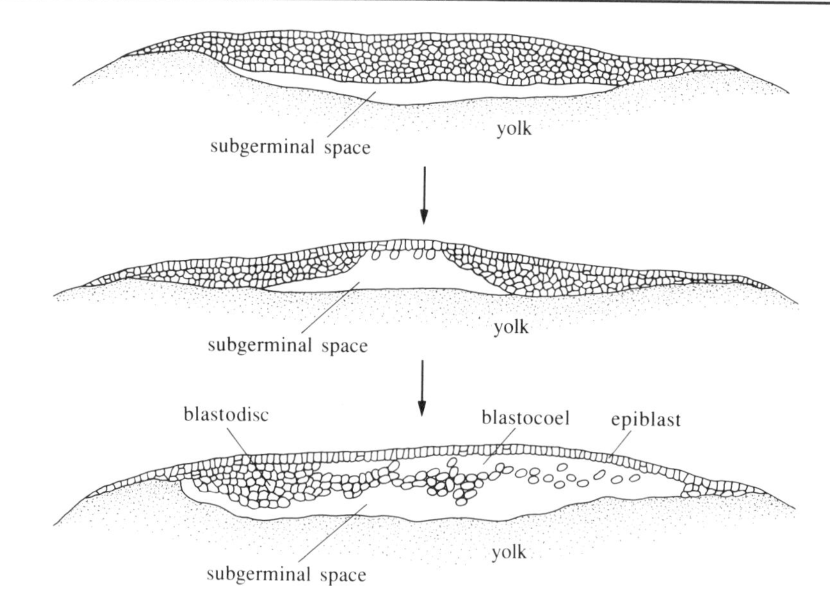

Figure 3.8 Formation of the blastodisc in the avian egg.

3.3 GASTRULATION

Development of the blastula involves movement and shape changes in layers and groups of cells and creates the embryonic stage known as the **gastrula**. The position of cells in the gastrula approximates to the position the descendants of the cells will have in the adult organism.

3.3.1 Sea urchin gastrulation

Figure 3.9 shows the general features of the process.

◇ What events in Figure 3.9 could be responsible for formation of the archenteron?

◆ Migration of cells into the blastocoel, folding inwards of cell sheets, followed later by extension of pseudopods and the pulling of the archenteron towards the animal pole.

In the 1960s, T. Gustafson and Lewis Wolpert, from an analysis of echinoderm development, proposed that the events of sea urchin gastrulation could be accounted for by two cellular properties, changes in cell adhesiveness and in cell shape.

First, let us concentrate on changes in adhesiveness. Gustafson's and Wolpert's argument was as follows. Consider an idealized cell in contact with a flat base (Figure 3.10). If there is little adhesion between the cell and the underlying base, then the cell will not have much contact with the base and (at least in the case of our idealized cell) will be rounded in shape (Figure 3.10a). If the amount of adhesion increases, then there will be more contact between the cell and the base, and the cell will become flatter by stretching (Figure 3.10b). Hence the extent of contact between a cell and its base (or perhaps between two cells) depends on a balance between the forces that tend to increase mutual contact (e.g. increased adhesiveness) and the forces that resist deformation.

Figure 3.9 Gastrulation in the sea urchin embryo. Read through the annotation in order, stages 1–8.

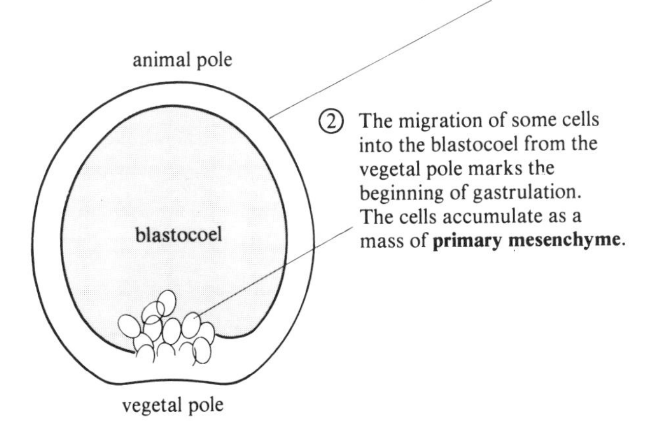

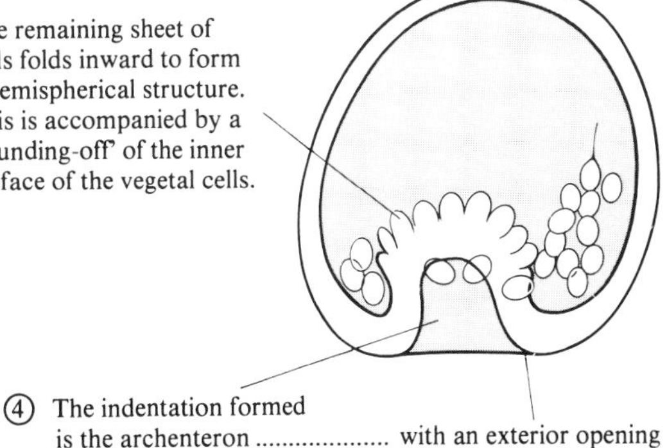

⑤ Primary invagination ceases after about 2 hours. Following a short delay, secondary invagination begins ....

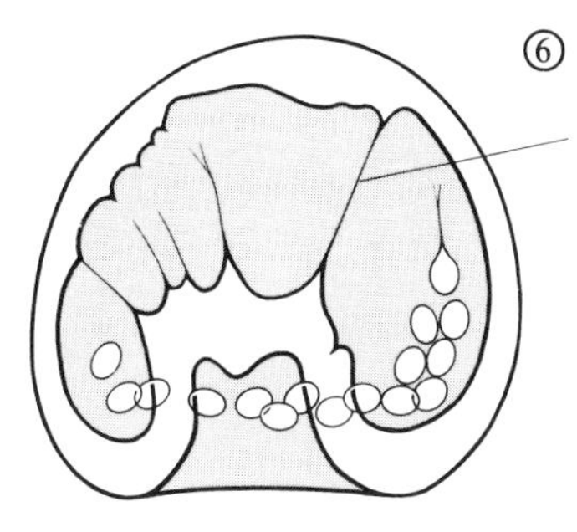

⑦ The archenteron is pulled towards the animal pole and invagination is completed.

⑧ In later development the archenteron becomes the larval gut .... and the primary mesenchyme forms the basis of the larval skeleton

If we take the argument one step up from single cells to sheets of cells, can we produce specific tissue shapes by this mechanism? In the sea urchin, the cells at the vegetal pole (where the invagination to form the embryonic gut starts) are attached to a membrane called the **hyaline layer** on their outer edge. If we consider that the adhesion between the cells is moderate, a diagrammatic representation of the cells and membrane might look like Figure 3.10c. Changes in cell adhesiveness lead to changes in shape. For example, if the cells become more adhesive to both themselves and to the membrane, then Figure 3.10d would show the resulting situation. If the cells became more adhesive only to themselves, then Figure 3.10e would be the result.

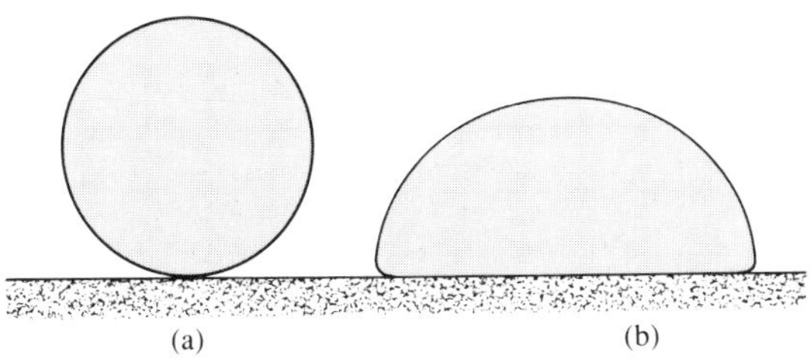

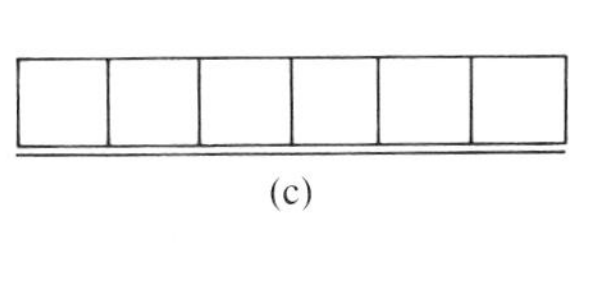

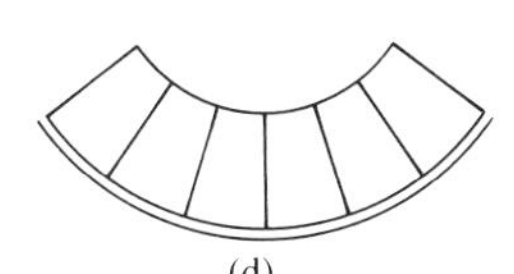

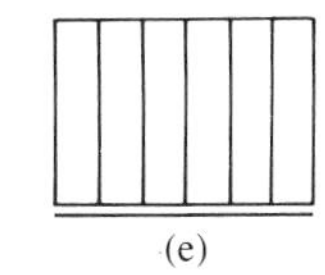

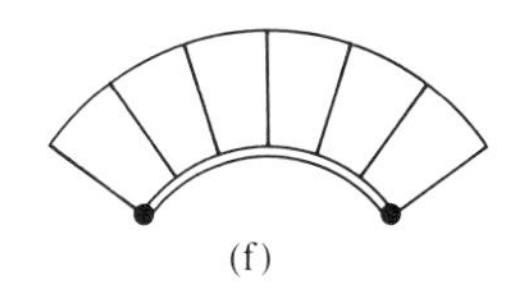

Figure 3.10 (a) and (b) Change in cell shape that might accompany a variation in the adhesiveness of the cell. (c)–(f) The effect of changes in cell adhesiveness on the form of a sheet of cells.

Gustafson and Wolpert proposed that the cells at the vegetal pole before invagination resembled those in Figure 3.10e. They suggested that contact between adjacent cells is then reduced, but contact between the cells and the hyaline layer remains the same. If the ends of the cell sheet are fixed (by some characteristic of the wall or the invaginating blastula), the sheet cannot spread so it curves. This situation is shown in Figure 3.10f. The curve this time is inwards, forming the **archenteron**, a cavity that later forms the gut. Gustafson and Wolpert's model therefore suggests that **primary invagination** could result from a change in the adhesiveness of certain cells at the vegetal pole.

Turning now to consideration of the second of the two properties, changes in cell shape, we can look at what happens when primary invagination has got underway. The second phase of gastrulation is marked by the formation of pseudopods by the cells at the archenteron tip. These cells seem to have a stretching function, the pseudopods anchoring themselves to the wall of the blastula around the animal pole (see Figure 3.11) and then contracting, and so mechanically help the elongation of the invaginating gut. This process seems to be very important, for if the blastocoel is treated with sucrose solution, producing a change in pressure inside the cells, the pseudopods break down and archenteron invagination stops.

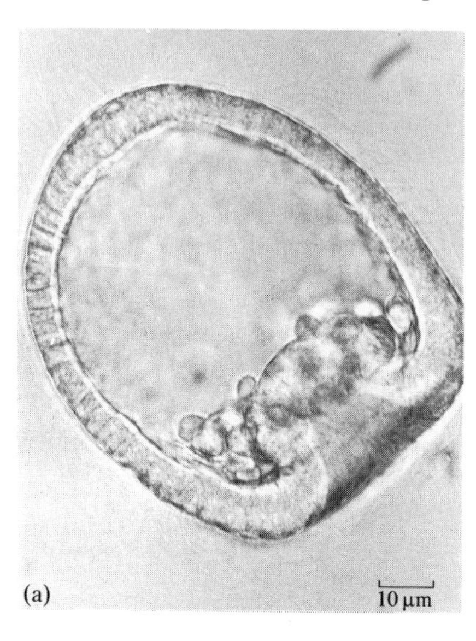

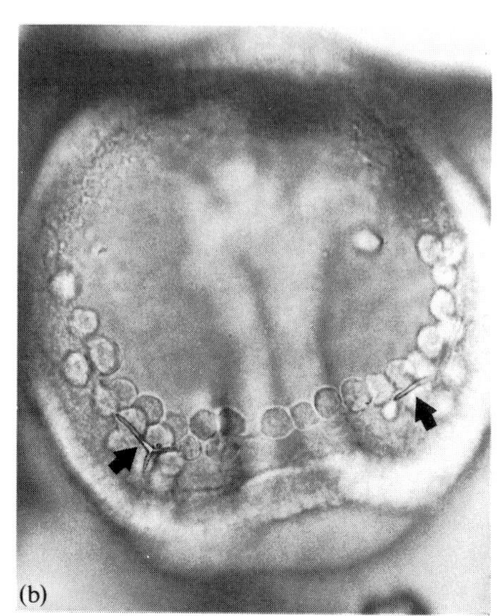

Figure 3.11 (a) An early sea urchin gastrula showing primary invagination with primary mesoderm cells as a random cluster. (b) A late sea urchin gastrula with a fully invaginated archenteron. A ring of primary mesenchyme cells is evident with two upwardly directed branches. Two skeletal rudiments are arrowed.

With the accent on changes in cell adhesiveness during the primary invagination of the gut, and cell shape changes with pseudopodial formation during **secondary invagination**, Gustafson and Wolpert underlined the importance of simple cell properties in seemingly complex morphogenetic events.

3.3.2 Amphibian gastrulation

Can simple cell properties explain morphogenesis in vertebrates as well? Most of the material to be covered in this Section deals with amphibians, and we restrict ourselves to a consideration of amphibian gastrulation, which seems more complex than gastrulation in the sea urchin.

How does vertebrate gastrulation differ from that of the sea urchin? In the USA in the 1940s, Johannes Holtfreter put forward an explantion of the role of individual cells in amphibian gastrulation. The general features of this process are shown in Figure 3.12, which you should study in some detail. Stages 1 and 4 are an early and late gastrula sectioned vertically in the median

Figure 3.12 Gastrulation in amphibians such as the frog, shown in sequence, stages 1–5.

① Gastrulation begins when a shallow groove develops at the embryo surface. This is the blastopore, formed by the inward movement (or invagination) of some surface cells. The upper rim of the blastopore is called the dorsal lip. The cavity formed is the archenteron, or primitive gut, which begins to enlarge.

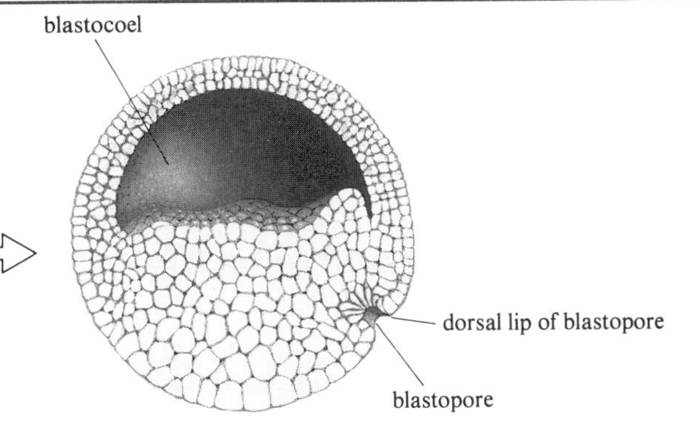

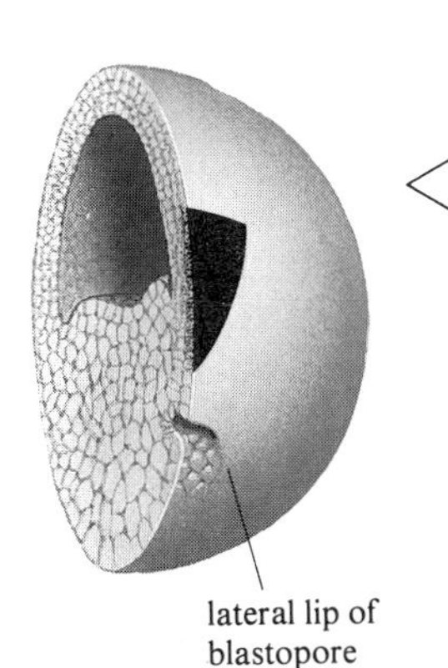

② The addition of dye to the early gastrula surface reveals subsequent cell movement. Stained areas stretch towards the blastopore, roll over its lip and disappear inside.

③ As more and more surface cells invaginate, the edges of the blastopore gradually expand laterally and a circular rim begins to develop at the embryo surface.

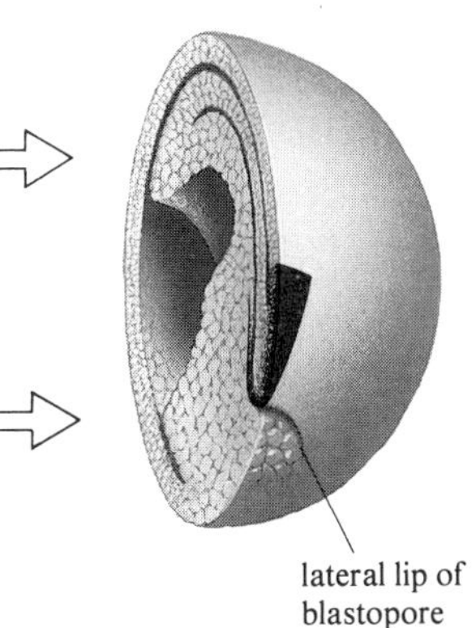

④ As a result of cell invagination via the blastopore, new internal tissues are formed.

Cells forming the roof of the archenteron are the **mesoderm** . . .

while those forming the side walls and floor are called **endoderm.**

Meanwhile, the darkly pigmented, presumptive **ectoderm** expands to cover the whole embryo, replacing those parts lost to the interior.

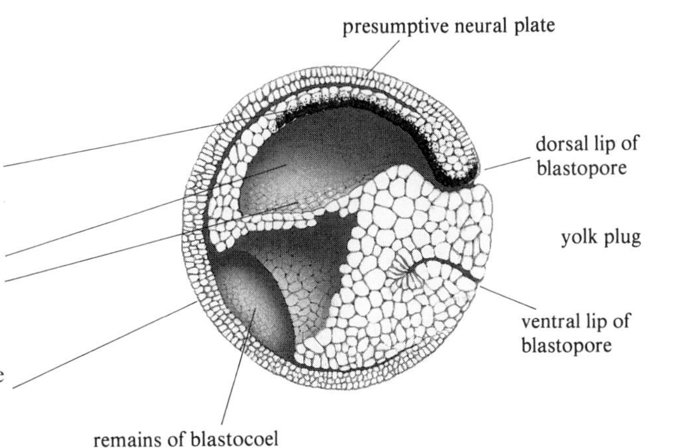

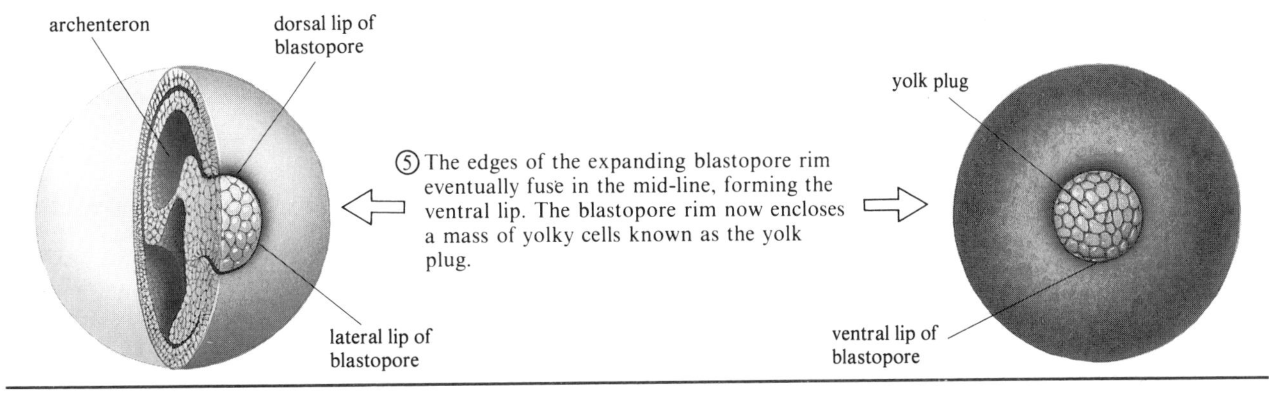

⑤ The edges of the expanding blastopore rim eventually fuse in the mid-line, forming the ventral lip. The blastopore rim now encloses a mass of yolky cells known as the yolk plug.

plane (along the axis of the animal–vegetal pole). Stages 2 and 3 are embryos cut in the median plane, but viewed at an angle. Stage 5 shows a late gastrula viewed from the side bearing the yolk plug; to the left is an angled view of an embryo cut in the median plane; to the right a view of a whole embryo, showing the darkly pigmented ectoderm surrounding the non-pigmented yolk plug cells. As gastrulation ends, the blastopore rim contracts and finally covers the yolk plug altogether. So all the material of the vegetal region eventually disappears into the interior of the embryo.

◇ What three features are shared by sea urchin and amphibian gastrulation?

◆ Formation of a blastopore; invagination of cells to give an archenteron; changes in cell shape round the blastopore.

Holtfreter stressed the morphogenetic importance of a continuous **surface coat**, which he claimed was established around the amphibian egg before fertilization. This coat was supposed to give close contact and communication between the peripheral cells of the embryo; for example, cell movements starting at one place would be communicated to adjacent parts, and so the surface coat could coordinate morphogenetic movements.

The surface layer has never been conclusively identified. The cells at the periphery of the bastula certainly seem to be closely associated with one another, and are united in a single cell sheet. Holtfreter proposed that the special properties of the peripheral cells are important in the formation of the blastopore (Figure 3.13a). Holtfreter observed through the microscope that as invagination begins, the cells connected to the invaginating gut become elongated, stretching into the blastocoel (in rather the same manner as the sea urchin pseudopods we saw in Section 3.3.1). Do the **bottle cells**, as these elongated cells are called, cause the blastopore to develop, or are they simply a consequence of the folding of the surface?

Holtfreter tested this by removing presumptive blastopore cells from an embryo: the isolated cells soon rounded up into a solid ball covered by a darkly pigmented surface layer. If this spherical mass is stained and added to a piece of endoderm tissue, it adheres to it and sinks in (see Figure 3.13b). As the cells move inwards they extend, eventually becoming bottle shaped, while still retaining their attachment to the peripheral blastopore cells, which remain in contact with the outer cells of the endoderm. The pulling force of the inwardly moving bottle cells tends to drag the surface layer into the substratum to form a groove, which Holtfreter supposed was equivalent to the blastopore. The experiment can be repeated with vegetal endodermal gastrula

bottle cells

dorsal lip of the blastopore

(a)

(b)

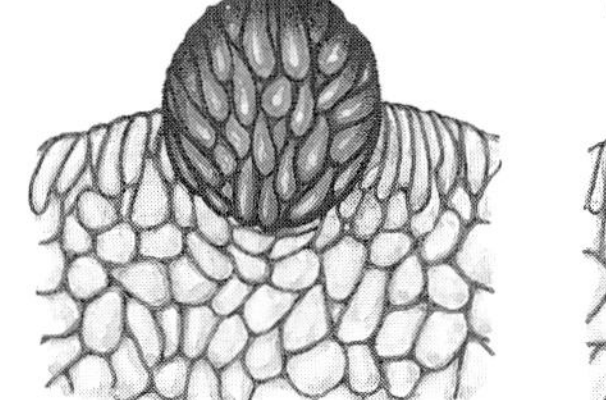

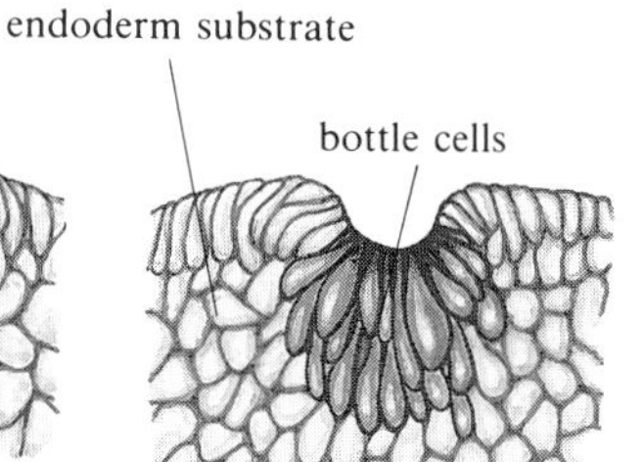

Figure 3.13 (a) Idealized side view into an early amphibian gastrula, showing the elongated bottle cells in section. (b) A graft of presumptive blastopore cells sinking into an endoderm substrate to form a groove, as revealed in cross-section.

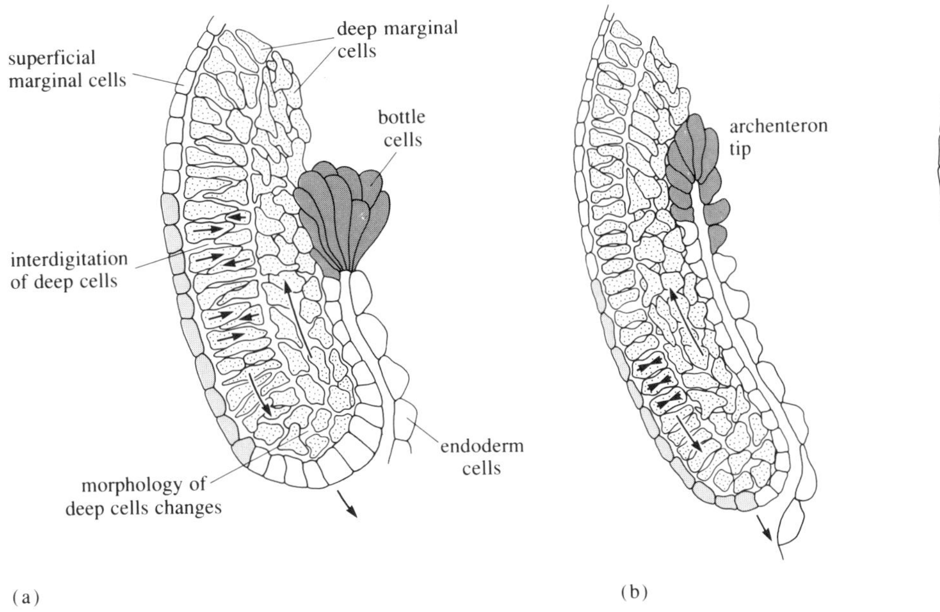

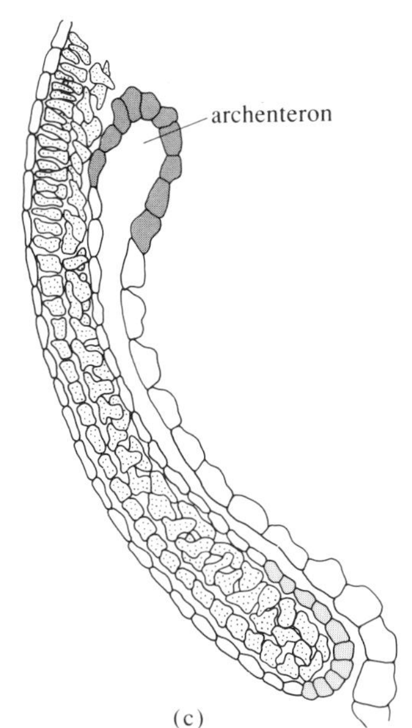

Figure 3.14 Integrative model of cell movements during gastrulation. Ten of the marginal cells have been coloured to show how they move. (a) Early gastrulation is characterized by the interdigitation of the marginal deep layers and by involution. (b) and (c) In later gastrulae, the deep marginal cells flatten and the formerly superficial cells form the wall of the archenteron.

cells as the implant. Although the cells grow into the host tissue, no elongated cells are formed, and no groove on the surface of the endodermal host tissue is seen.

◇ From these experiments, what factors would you suppose initiate blastopore development?

◆ It seems that inwardly migrating blastopore cells do not become greatly stretched or form a blastopore groove unless they remain attached to the surface layer. So the characteristics of the surface layer and the formation of bottle cells are responsible for blastopore formation.

Unfortunately, the picture is not complete. We have little idea about the mechanism responsible for the inward movement of the bottle cells. Possibly it is related to an increase in their adhesiveness, which might result in an inward movement. (We will return to this later.)

◇ What structure might integrate the pulling activity of the bottle cells?

◆ The surface layer. Note that the same structure might also communicate the pull to other more distant cells.

This is a nice story. But more recent work by Raymond Keller, working at the University of Illinois in 1981 indicates that the bottle cells may have a more passive role. Removal of the bottle cells, once formed, does not prevent adjacent cells moving into the blastopore. The cell movements during gastrulation seem to be associated with movement of the cells underlying the surface cell — the margin cells. Removal of these cells stops gastrulation.

Figure 3.14 illustrates the current model for cell movement during gastrulation.

From Figure 3.14, notice that following the formation of the blastopore by bottle cells, many of the surface cells begin to move towards the expanding blastopore, then roll over its lip and disappear inside. What cellular forces are responsible for this type of movement?

First, notice that the darkly pigmented ectodermal cells spread during gastrulation.

◇ If the entire ectoderm is removed from an embryo, cells still move towards the blastopore and invaginate more or less as normal. What does this result suggest about the mechanism of invagination?

◆ It suggests that invagination via the blastopore does not result from a push exerted by the spreading ectoderm.

We know of at least two factors that seem to be involved in invagination.

1 If groups of cells from the surface of the gastrula are removed before their invagination and cultivated in saline media, the individual cells stretch. This inherent ability of cells to stretch in the appropriate direction may contribute to invagination *in vivo*. (Compare this with invagination in the sea urchin.)

2 During gastrulation, the mesodermal cells of the archenteron roof spread on the inner surface of the ectoderm (immediately beneath the presumptive neural plate). This spreading appears to be due to a difference in the adhesive properties of the mesodermal cells and the inner ectoderm surface. (Again, compare this with sea urchin invagination.)

◇ The adhesive properties of the two cell types enables a spreading of mesoderm on the inner face of the ectoderm. Which is the less adhesive?

◆ Mesoderm cells are less adhesive than the cells which make up the inner ectoderm surface. So the mesoderm will tend to spread.

The migration of the presumptive mesodermal cells is associated with a lattice of **fibronectin** secreted by the blastocoel roof. This lattice provides a framework for the mesodermal cells to move on. The cells bind to specific regions of the fibronectin protein that contain a sequence of ten amino acids. When large amounts of this decapeptide were injected into the blastocoel just before gastrulation the mesodermal cells were found to bind to this rather than to the fibronectin, and gastrulation stopped. The mesodermal cells remained as a mass outside the embryo.

Finally we should briefly mention the spreading of the surface ectoderm at gastrulation. Although the ectodermal sheet spreads as a unit during gastrulation, the individual cells also have a tendency to spread. This is apparent if surface ectodermal cells, after their dissociation, are cultured on glass. At first, the individual cells adhere to the glass but soon afterwards begin to spread. It seems that the expansion of the ectoderm depends on the capacity of the individual cells to spread.

Summary of Sections 3.1–3.3

Rather than being static 'jigsaw puzzle pieces', an animal's cells are capable of movement during development. This movement may be amoeboid or 'gliding'. Under *in vitro* cell culture conditions cells can align themselves perhaps because of characteristics of the cells themselves or by properties of the matrix they are growing on. This alignment may be the trigger for future differentiation, as in muscle tube formation.

Cleavage divisions of the egg are influenced by the amount and location of yolk within the egg. In eggs with little yolk, cleavage results in a ball of equally sized cells. In eggs with more yolk, a ball of different-sized cells is formed. This ball of cells becomes hollow and is known as the blastula. In cases where the egg is virtually all yolk, such as some birds' eggs, cleavage divisions produce a disk-like region of cells, the blastodisc.

Gastrulation in both vertebrates (amphibians for example) and invertebrates (sea urchins for example) seems to involve simple cell properties like adhesion and change in shape.

The main features of sea urchin gastrulation are:

(a) Migration of primary mesenchyme cells to the blastocoel from the vegetal pole.

(b) Primary invagination may involve a change in adhesiveness of the adjacent cell membranes at the vegetal pole.

(c) Pseudopods of cells at the archenteron tip attach to the upper part of the blastocoel wall and then contract.

In amphibians invagination was thought to occur by inward migration of the bottle cells but it may be that movement of the cells underlying the surface coat is more important. The properties of the surface coat are important — as shown by the fact that explanted bottle cells sink into an ectodermal substrate but do not seem able to anchor to the outside of this coatless cell layer. Fibronectin is thought to provide a framework which organizes mesodermal cell movement.

Question 1 (*Objectives 3.2 and 3.3*) Which of the following statements are true and which are false?

(a) Amoeboid movement does not involve a change in cell shape.

(b) Fibroblast movement involves an organelle called the leading lamella.

(c) Contact inhibition is the inhibition of cell movement between two neighbouring fibroblasts in culture.

(d) The inherently precise machine is used as a metaphor to explain amoeboid movement.

Question 2 (*Objectives 3.5–3.7*) Which of the following experimental observations (a)–(e) fit in with Gustafson and Wolpert's theory about the mechanism for primary invagination in the sea urchin?

(a) Pseudopods formed by cells at the archenteron tip are attached to the animal pole and shorten as the archenteron elongates.

(b) Some of the vegetal cells detach, migrate into the blastocoel, and accumulate as a mass of primary mesenchyme.

(c) Changes of adhesiveness occur between cells at the vegetal pole.

(d) The sheets of cells at the vegetal pole bend inwards to form a more or less hemispherical structure, the cavity of which is the archenteron.

(e) Suppression of pseudopodial activity prevents invagination.

Question 3 (*Objective 3.4*) What accounts for the differences in the results of cleavage between sea urchins and amphibian eggs?

Question 4 (*Objective 3.5*) What factors initiate blastopore development in amphibians?

Question 5 (*Objective 3.5*) Does the volume of the blastocoel increase or decrease as gastrulation proceeds?

Question 6 (*Objectives 3.5 – 3.7*) What experiment shows the special invasive properties of the bottle cells?

Question 7 (*Objectives 3.6 and 3.7*) What observations indicate that the bottle cells may not be crucial for gastrulation?

Question 8 (*Objective 3.5*) What are the main factors that make vertebrate gastrulation different from sea urchin gastrulation?

3.4 MORPHOGENESIS IN CELL POPULATIONS

In the previous section, we looked at the behaviour of individual cells. How do cells cooperate and interact to form specific shapes? In this section we will deal with the phenomenon of cellular reaggregation and discuss types of recognition signals that may exist between cells.

3.4.1 Cellular reaggregation

The following experimental method was first used by H.V. Wilson at the beginning of this century in the USA. By squeezing pieces of sponge tissue through fine silk, he prepared a suspension of disaggregated cells from the sponge *Microciana prolifera* and allowed them to settle at the bottom of a dish. Soon, active movement of the individual cells led to the formation of numerous small clumps or aggregates of sponge cells. After about a day, some of the aggregates appeared as small but fully differentiated sponges, with an internal organization identical to that of the normal adult. The mechanism underlying this reconstitution soon became the subject of some controversy. Did the cells sort out according to their original type, or did their differentiated state change, so that those on the outside of a reaggregated clump, for example, would become cells normally associated with that position even if they were originally inner cells? It was difficult to resolve this problem because the histological identification of the different types of sponge cells is not easy and they change their state and function in normal, undisturbed life.

Fifty years later, Johannes Holtfreter and Philip Townes, at the University of Rochester, New York, discovered that the cells of an amphibian embryo will dissociate when exposed to saline solution at pH 10 and will reaggregate at pH 8. They tested the reaggregation of different combinations of ectoderm, mesoderm and endoderm cell suspensions. Because these types of cell are readily identifiable, the movement of each type could be followed within the aggregates. Initially, the cells were distributed haphazardly, but by their subsequent movement, each cell type eventually sorted out according to type, and each cell type took up a characteristic position within the aggregate (Figure 3.15). The separation of the various cell types seemed to mimic the situation seen in the real embryo. For example, if ectodermal and mesodermal cells are mixed, after reaggregation, the mesodermal cells form an inner core around which is a boundary of ectodermal cells. How does this sorting out occur?

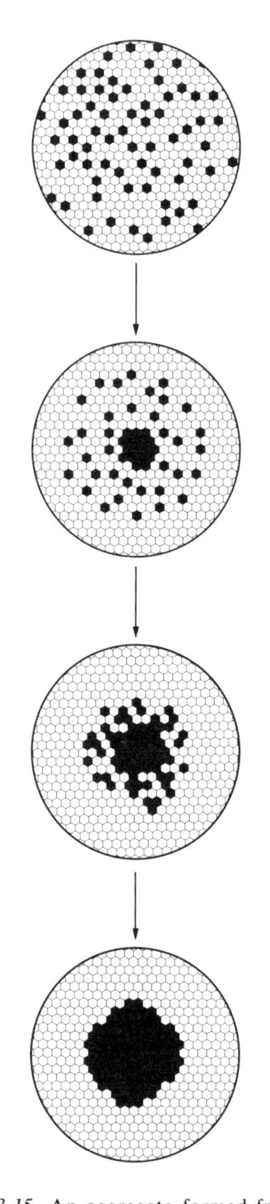

Figure 3.15 An aggregate formed from dissociated cells of two types arranged randomly; the cells sort out according to type giving the configuration of a sphere within a sphere. The same configuration can also be obtained when the two cell types are combined as tissue fragments.

3.4.2 The cellular basis of sorting out

Although sorting out is an *in vitro* phenomenon, it is important to establish the mechanisms at work. The underlying mechanism may well be significant in normal morphogenesis. The ability of cells to move into particular positions during development seems to be of importance, as we see later when we discuss morphogenesis in the embryo. There have been several explanations of sorting out.

First, Holtfreter and Townes proposed a chemotactic theory, suggesting that the migration of cells or groups of cells occurs along a gradient of some metabolite. For example, if two cell types A and B are present and A cells produce some substance that helps A cells to aggregate, then it is possible that these cells will form a central mass and that B cells will be left to aggregate as best they can around the central A cell core. Firm evidence against chemotactic theories is lacking, but there is no positive evidence either, so they are rather mistrusted.

One of the most elegant explanations for the sorting out phenomenon was proposed by Malcolm Steinberg in 1964. Consider a mixed aggregate containing two types of cell arranged randomly. What happens if there is a quantitative difference in the adhesiveness of the two cell types with respect to cells of their own type? Imagine a cell of the more adhesive type A surrounded by less adhesive B cells. Now if the cells can move around at random, then the probability of two A cells remaining in contact after a random encounter is higher, because of their greater adhesion, than is the probability of either two B cells or an A cell and a B cell. Steinberg described the various possibilities for selective adhesion — for example, another case would be where the greatest adhesion is between cells of unlike type, A + B — and predicted the sorts of cell groupings that would occur for each possibility. Several examples are shown in Figure 3.16.

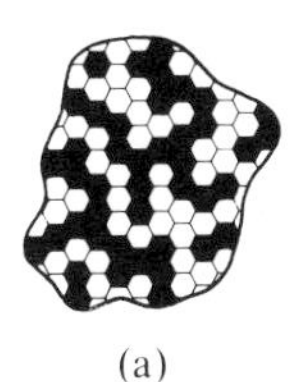

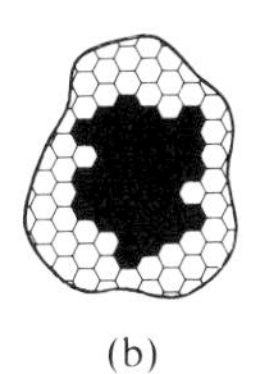

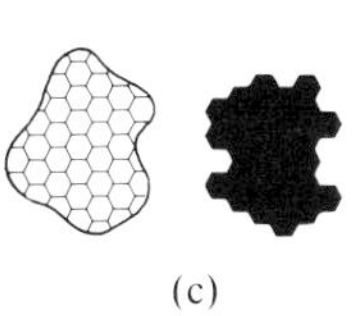

Figure 3.16 Relative adhesion between cells of two types. (a) The situation where maximum adhesion is between unlike cells. (b) The pattern formed if the black cells are the more attracted to one another. (c) No adhesion between unlike cells.

Much work has been done to try to find experimental evidence for Steinberg's hypothesis, ranging from sophisticated histological studies (to look for possible candidates for the role of 'stickiness' receptors on the cell surface) to complex computer simulations (designed to test the behaviour of abstract 'computer cells' programmed with Steinberg's adhesion rules).

Recent work on regeneration of amputated limbs in the salamander suggests that differential cell adhesion is involved in morphogenetic processes *in vivo*. If the limb of the salamander is amputated the stump produces a mass of undifferentiated cells (called the **regeneration blastema**) which divide and differentiate into the missing tissues. If the amputation is carried out at the wrist, the blastema regenerates just the tissues and bones of the missing digits. If the amputation is carried out at another position on the limb only the part of the limb that is removed is regenerated. This suggests that the blastema at each point of amputation 'knows' its location relative to the whole limb and thus regenerates only the appropriate missing parts. A more detailed examination of how this may be understood is presented in Chapter 4.

Work with isolated blastemas shows that if blastemas from the same region are placed together in a cell culture, they form a single mass but the cells do not move relative to one another and remain spatially segregated. If, however, blastemas from different regions are placed together, the more proximal one (e.g. upper arm) will surround the more distal one (e.g. wrist). This is shown in Figure 3.17.

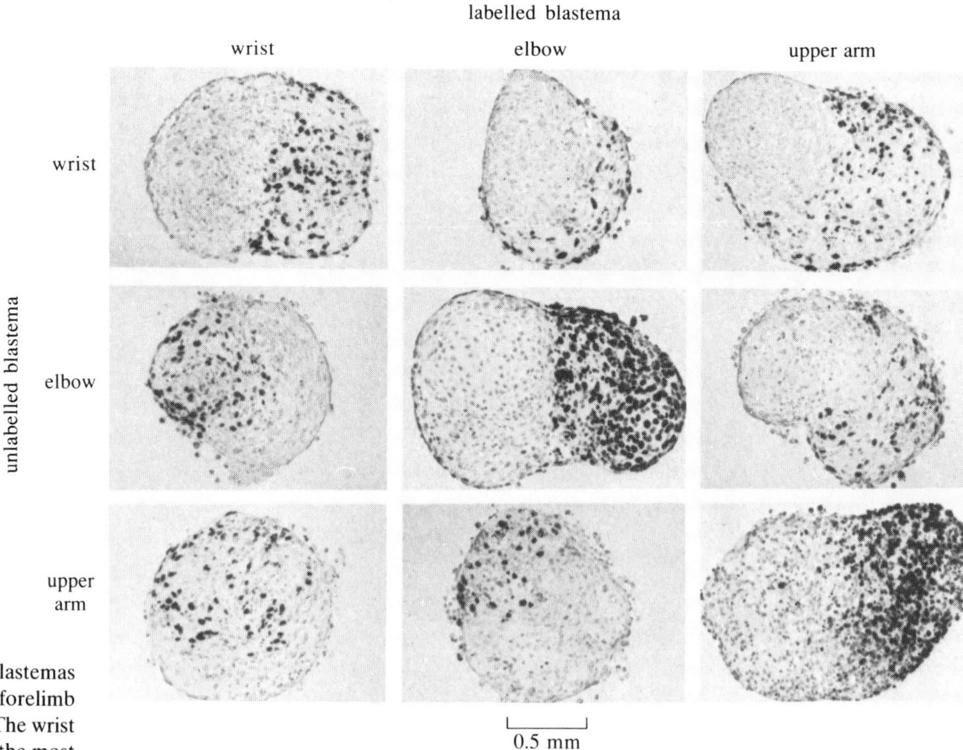

Figure 3.17 Sorting out when blastemas from different levels of the same forelimb are brought together in culture. The wrist is the most distal, the upper arm the most proximal structure tested. One member of each pair was marked with tritium to distinguish it from the other. After 3 days in culture, the aggregates were fixed and sectioned. Blastemas from the same level fused in a straight line. When blastemas were from different levels, the proximal blastema attempted to surround the more distal cells.

This rearrangement of cells fits in with the idea that there is a gradient of cell adhesiveness, decreasing from wrist to the upper arm. This can be related to the actual events in the regenerating limb. In the USA in 1988, David Stocum found that when they grafted blastemas from the wrist, elbow or upper limb to the junction between the stump and blastema of a regenerating mid thigh, the blastemas move during the regeneration process so as to take up a position in the host limb corresponding to their position of origin in the fore limb. Thus, for example the wrist blastema moved to the end of the regenerating hindlimb where it formed a wrist adjacent to the foot and the forelimb blastema migrated distally to form a forelimb on the thigh. It seems that the blastemas are migrating to regions in the hindlimb which have similar adhesiveness to their own cells. This is shown in Figure 3.18.

This fascinating experimental work does strongly suggest that the mechanisms of cell sorting seen *in vitro* are also important *in vivo*. We will now consider the nature of the molecules that can influence cell adhesion.

3.4.3 Cell adhesion molecules (CAMs)

One of the first cell adhesion molecules to be isolated and identified was from embryonic chick neural retina. This neural cell adhesion molecule (NCAM), a component of the cell surface, is a protein of relative molecular mass (M_r) about 140 000.

The technique of **fluorescence immunohistochemistry** has been very useful in locating the presence and position of CAMs in embryonic tissue. This involves making antibodies to the proteins in which you are interested and binding to the antibody an agent which will fluoresce.

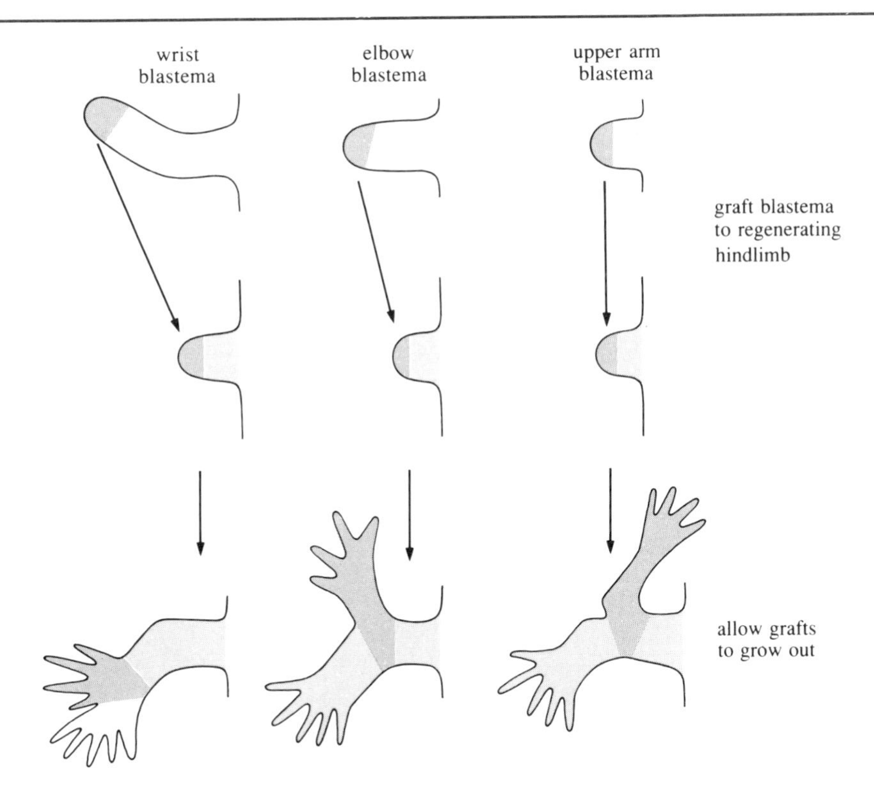

Figure 3.18 Sorting out *in vivo* (affinophoresis) whereby regenerating forelimb blastema (black) grafted into the mid-thigh blastema (grey) are displaced to the corresponding region of the regenerating hindlimb (upper arm to midthigh; elbow to knee; wrist to tarsus) and initiate forelimb formation from that point distally.

Thin sections of tissue are treated with the antibody and the fluorescing agent. When examined under a special fluorescence microscope the location of proteins binding to the CAM antibodies can be seen as areas of fluorescence.

Research shows that these CAMs are of three types: (a) general CAMs, which are found on a number of tissues at different stages of development, (b) restricted CAMs, which are only found on specific cells and (c) saccharide CAMs, proteins to which sugar molecules are covalently linked (glycoproteins), which can link sugar residues between adjacent cells.

The general CAMs are known to be of two types. The distinction here is whether or not calcium is needed for them to work in terms of cell adhesion. Those which require calcium you will find referred to in the literature as 'cadherins'.

Initially at gastrulation in the chick embryo CAMs for both neural cells (NCAMs) and liver cells (LCAMs) are found. (CAMs are named according to the circumstances in which they were first found, so the name does not necessarily describe their only function, or even their major function.) However, as the gastrula develops and neural tissue is formed NCAMs are found only on the developing neural tissue with LCAMs being restricted to surrounding ectodermal cells. In later stages of development other CAMs are expressed on specific tissues: for example adhesion molecules associated specifically with retinal cells. Since different cells are known to have different adhesion molecules this does provide evidence to support Steinberg's cell adhesion theory (see Section 3.4.2).

How do these molecules work in making similar cell types stick together? Most of the evidence on this comes from studies on NCAMs but the basic principles are likely to apply to other cell adhesion molecules. Current research suggests that the CAMs bind by the association of amino acid groups at the end of molecules on adjacent cells. The CAM protein is located in the cell cytoplasm but also spans the membrane and the remainder of the molecule resides in the extracellular space. Thus binding can be modified. Although CAMs are proteins, they are always found in association with carbohydrates, and specifically with large amounts of sialic acid (a complex 10-carbon saccharide). These sialic acid molecules are covalently bound to the regions of CAM protein outside the cell.

The sialic acid molecules are negatively charged and if the sialic acid content of the CAM is relatively large, the CAM binding strength is reduced. It has been found that CAMs from older cells have relatively lower amounts of sialic acid and, as a result, will stick more closely together. A reduction in the sialic acid content of the CAMs as the embryo gets older could have a major influence on maintaining the position of cells relative to each other, a factor crucial for the development of embryonic organs.

Some recent work (1989) in Japan has shown that the differential expression of two cadherins is essential for the development of lung tissue in the mouse. The scientists there developed antibodies against these two cadherins and then looked at how lung tissue developed in a tissue culture in the absence or presence of the antibodies.

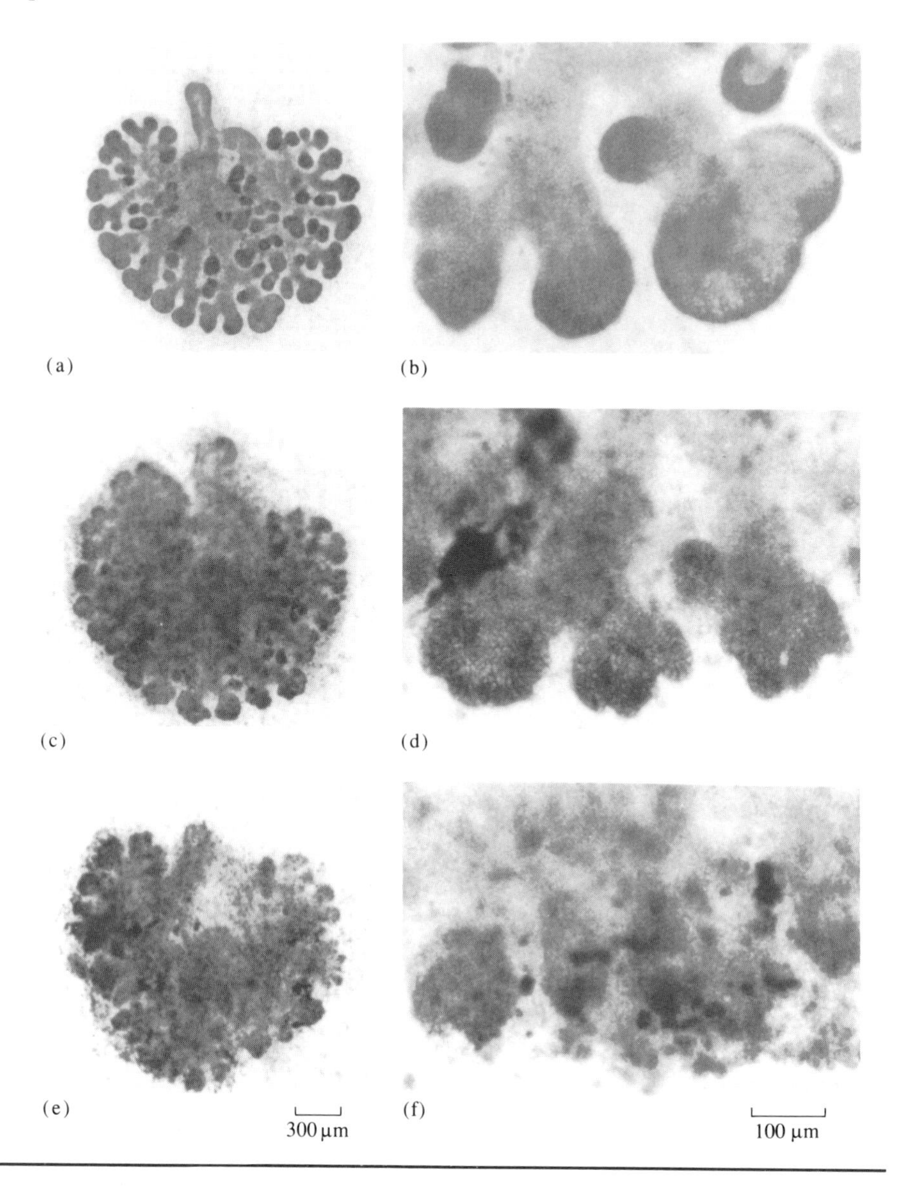

Figure 3.19 Mouse lung tissue development. (a) and (b) Sections of tissue cultured in the absence of antibodies against CAMs. (c) and (d) Sections of tissue cultured in the presence of antibodies against one CAM. (e) and (f) Sections of tissue cultured in the presence of antibodies to two CAMs.

◇ If you incubate tissue with antibodies to CAMs would you expect the cells in the tissue to become more adhesive or less adhesive — and why?

◆ Less adhesive, with respect to each other, because the antibodies would bind to the CAMs reducing the cell's adhesiveness towards each other.

In Figure 3.19, (a) and (b) are micrographs of how the lung epithelia developed under normal conditions, whereas (c)–(f) are micrographs from the cultures treated with antibodies.

◇ What do you notice about the lung structure when the CAMs have been blocked by antibodies?

◆ The branching of the epithelial cells is not seen so clearly when CAMs are blocked.

There is evidence from other species and tissues that different CAMs are expressed at different stages of development and this could explain relative movements of different cell types in developing tissues.

We mentioned also **restricted cell adhesion molecules**, really just for completeness. We will not consider them in any detail, if only because little is known about them. This lack of knowledge has a lot to do with the fact that they are expressed only in specific tissues and only for short developmental periods. So they are very difficult to isolate. Of particular interest, however, is a group of molecules known as 'fasciclins' which are involved in nerve growth. Experimental work on the development of the grasshopper nervous system indicates that restricted CAMs are involved in directing axon growth so that bundles of nerve fibres grow in the same direction.

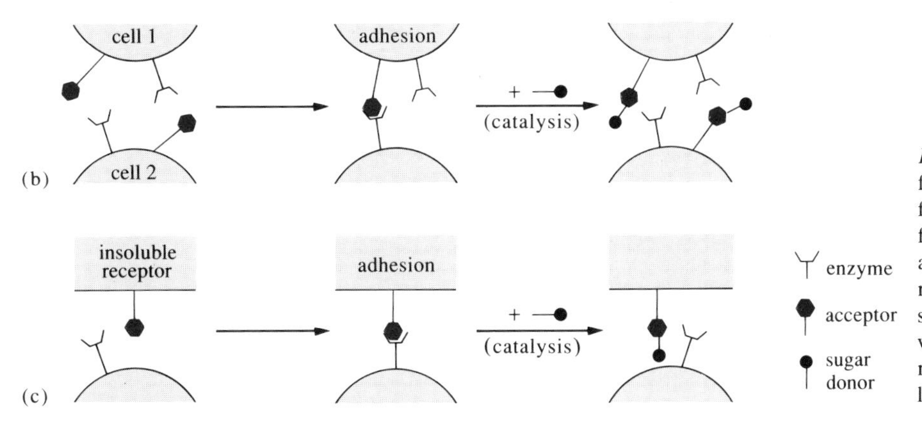

Figure 3.20 Reactions of glycosyltransferase. (a) The standard glycosyltransferase reaction, in which a sugar is transferred from a nucleotide carrier (NDP) to an acceptor. (b) and (c) Intercellular reactions that may be mediated by cell surface glycosyltransferases. If the activated sugar were absent, adhesion would result, as is thought to occur during fertilization.

Saccharide CAMs are thought to be involved in more general aspects of cell adhesion. In simple terms this involves linking the glycoprotein molecules on the surface of adjacent cells. Membrane-bound enzymes, **glycosyltransferases**, are responsible for joining sugar molecules to proteins to make glycoproteins. Figure 3.20 illustrates how the sugar residues on one cell can be linked to the enzyme on another cell leading to cell adhesion.

If free sugar molecules are present the enzyme will bind to these molecules, thus unsticking the cells.

◇ Why would breaking of cell adhesion be important?

◆ It would allow cells to migrate during developmental processes.

Evidence that saccharide CAMs are involved in the cell movements associated with gastrulation comes from experiments and observations on the cells derived from chick tissues.

Autoradiography shows that sugar molecules are found on the surface of migrating cells and there is also considerable glycosyltransferase activity, indicating that the glycoproteins are being repeatedly linked and unlinked.

We have concentrated here on adhesion between cells but saccharide mediated cell adhesion is also involved in linking cells to extracellular substrates such as collagen and other glycoproteins.

We will look at this again when we have considered one of the main processes whereby embryonic cells become different from one another — the topic of embryonic induction.

3.5 EMBRYONIC INDUCTION

Blastula and gastrula formation are the first steps in the chain of morphogenetic events that finally produce the adult organism. We have already looked at the sorts of mechanical properties of cells that bring about these changes. Do we have any information about later stages of embryogenesis? Can we recognize any of the control signals that initiate developmental events.

This section deals with a set of experiments that set out to explain developmental control mechanisms in terms of biochemical signals. The experiments are important for several reasons, particularly because they show the general or interspecific nature of development phenomena: tissue from one species of organism can often provide signals for a tissue from another species to develop normally. They also demonstrate that signals of some kind pass from tissue to tissue during development.

Here we discuss the phenomenon of embryonic induction, already introduced in Chapter 1. In particular we will look at development up to the stage of the neurula, the embryological term given to the developmental stage of the vertebrate embryo following the gastrula. At this stage, nerve tissue first appears as a flattened, dorsally located plate: subsequently the edges of the plate fuse to form a hollow neural tube.

If a small piece of presumptive epidermis (tissue which, if left undisturbed, forms epidermis) from an early newt gastrula is transplanted into the area of another embryo that is destined to form **neural plate**, the transplanted piece develops in conformity with its new surroundings, first as neural plate tissue and then as part of the neural tube. In the same way, presumptive neural plate from an early gastrula differentiates as skin epidermis following its transplantation into the presumptive skin region of a similar gastrula (Figure 3.21a). Evidently the neural plate tissue could not have been determined at the time of transplantation: its fate was eventually set by some factor related to its new position.

If similar experiments are carried out on late gastrulae, the results are strikingly different. A transplanted piece of presumptive neural plate develops as neural tissue irrespective of its new position (Figure 3.21b).

By the end of gastrulation, the neural plate has lost its former ability to develop into epidermis. How is the neural tissue determined? The trans-

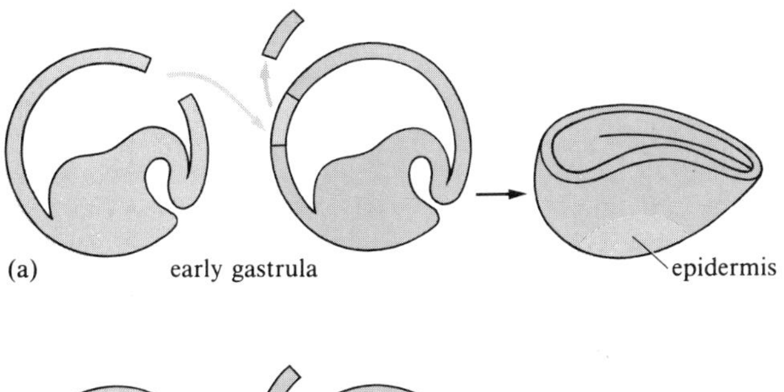

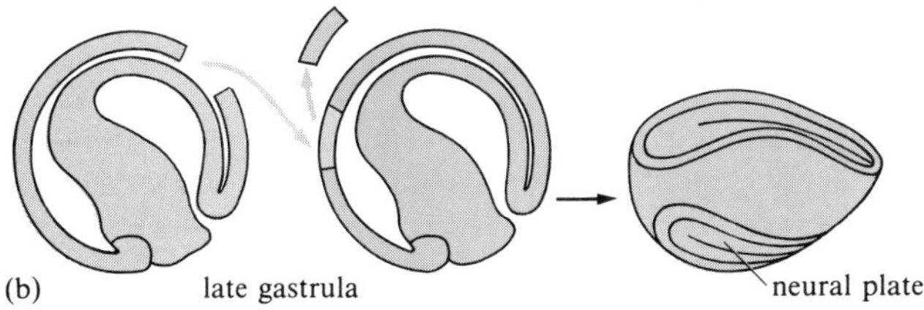

Figure 3.21 (a) Presumptive neural plate removed from an early gastrula and transplanted into a host gastrula at the same stage develops as belly epidermis. (b) A corresponding piece removed from a late gastrula differentiates as neural plate, according to its original fate.

plantation studies described above show that each cell is not endowed with a set role from the moment of its creation. Some evocative signal from the surrounding cells is required before differentiation can proceed normally. The determination of the neural plate thus provides an example of embryonic induction.

We can see from this experiment that embryonic induction is an interaction between one (inducing) tissue and another (responding) tissue, as a result of which the responding tissue undergoes a change in its pathway of differentiation. To quote from John Gurdon, a leading research worker in this field:

> 'Embryonic induction ... is probably the single most important mechanism in vertebrate development leading to differences between cells and to the organization of cells into tissues and organs.'

It is now nearly 70 years since the first experiments that attempted to characterize the nature of embryonic induction. In 1924, Hilde Mangold and Hans Spemann, working at the University of Freiburg in Germany, showed that the roof of the archenteron influences the determination of the neural plate. Mangold did a fascinating transplant experiment with two different species of newt, one colourless and the other heavily pigmented. When the **dorsal lip of the blastopore** (Figure 3.22) was transplanted from the early gastrula of one species into the ectoderm of the gastrula of the differently coloured species, a second embryo started to develop on the implant site, containing a neural tube as well as an embryonic **notochord** and blocks of tissue destined to become muscle.

Because the second embryo was formed from tissue pigmented like the host and not the implant, the grafted cells must therefore have influenced the host cells to change their fate. How does this happen? The problem was made even more interesting when it was realized that the transplant could not only influence presumptive neural plate, but also endoderm and mesoderm.

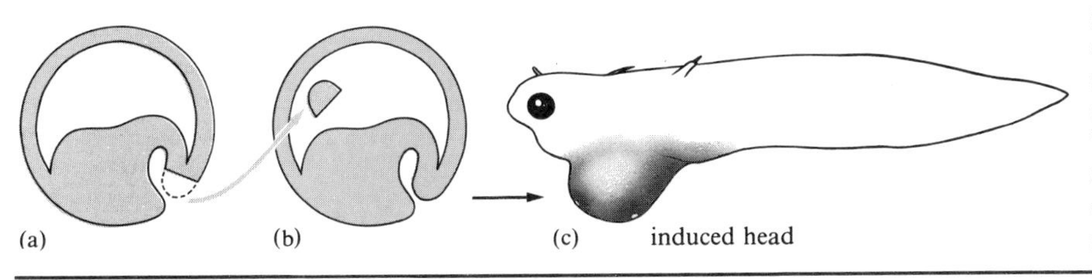

Figure 3.22 Transplantation of blastopore material. (a) Blastopore from colourless gastrula. (b) Transplant into ectoderm of darkly pigmented gastrula. (c) Resulting twin embryo development. The pigmentation pattern of the induced second embryo is that of the host and not of the graft, suggesting that the implant's role is organization.

Spemann subsequently called the dorsal lip of the blastopore the **organizer**, this term being meant to imply an inherent ability to organize the whole process of development.

The problem of the organizer, and of embryonic induction in general, dominated experimental embryology for almost 30 years following the Mangold and Spemann experiments. In 1935, Spemann was awarded a Nobel prize for his work on embryonic physiology. Most embryologists believed that if the substances responsible for the inductive powers of the organizer could be isolated, the key process of development could be identified. The fallacy in this argument is that even if an 'organizer substance' could be isolated (which has not been done to date), it would do very little to explain how the complex processes of development produce shapes and patterns.

3.5.1 The chemical identity of the inducer

The discovery of the organizer sparked off a dramatic search for the inducer. Many embryologists believed that it must be a chemical of great complexity — a 'master chemical' capable of inducing such elaborate neural structures. However, the search turned out to be far from straightforward. Firstly, the dorsal blastopore lip was able to induce neural plate formation even after quite drastic treatment such as freezing, crushing, heating, or after treatment with various solvents such as alcohol, ether and chloroform. Secondly, it was possible to induce neural plate development, or just neural plate using substances derived from other tissues, including vertebrate liver, kidney and muscle derivatives, besides some invertebrate tissues.

Some experiments seemed to indicate that a chemical is involved. In the USA in 1953, M. Niu and Victor Twitty incubated archenteron roof or dorsal lip tissue in special saline solution. After ten days incubation, the inducer tissue was removed and a very small piece of (undetermined) presumptive neural plate from an early gastrula was placed in the remaining conditioned medium (i.e. medium containing some kind of factor that came from the incubated tissue). After 24 hours exposure, most of the tissue fragments subsequently differentiated into neural plate. As a control, presumptive neural plate tissue was incubated in unconditioned medium — and remained as a sheet of undifferentiated cells. This shows that in neural plate induction at least, the inductive stimulus could be associated with a chemical.

The situation was shown by Niu to be even more complicated than a simple search for a single chemical that could do one specific job. If the archenteron roof of embryos that were 7–10 days old was used as the inducer, the receptive neural plate cells developed mainly into nerve tissue. If the archenteron roof was derived from embryos that were 12–15 days old, muscle cells were a major product. These experiments raise the possibility that there is a range of inducer molecules produced at different times in development and responsible for influencing tissues to develop along different developmental pathways.

Many embryologists thought that finding out the chemical nature of the substances that bring about induction would give clues to the process itself. Spemann discovered that inductive ability is not limited to the dorsal lip of the blastopore and the archenteron roof. Neural tissue, following its own induction, can itself induce neural plate differentiation if transplanted beneath presumptive neural plate. This ability is retained during development: for example, spinal cord of amphibian larvae and adults can also induce. Non-neural embryonic tissue (e.g. notochord) has the same effect. Vertebrate liver, kidney and muscle cells are also effective. With the newt as a host, various adult tissues from a variety of animal species act as inducers — *Hydra*, insects, fishes, reptiles, birds and mammals.

Niu tried to identify the active agent in the conditioned medium. Elaborate chemical analysis proved difficult because of the minute quantities of inducer released. Nevertheless, he succeeded in extracting a ribonucleoprotein (a complex of ribonucleic acid and protein) with strong inductive activity, and he concluded that the inductive properties of the substance lay in the RNA portion of the nucleoprotein. But Niu's claim has been disputed by other workers.

3.5.2 A whole range of inducers

Work with abnormal inducers (i.e. from tissues not present in the gastrula) has shown that different tissue preparations have varying inductive abilities. For example, liver and kidney tissues exclusively induce neural structures: in other words, they are neuralizing agents. On the other hand, alcohol-treated bone marrow from guinea-pigs induces mesodermal parts of the trunk and tail. Such tissues are generally referred to as vegetalizing agents, because they induce vegetal cells which in turn induce animal cells to form mesodermal cells. The situation is further complicated by the fact that certain tissue extracts induce specific neural structures. For example, alcohol-treated liver promotes differentiation of anterior neural structures (forebrain, eye and nose rudiments), while the kidney of the adder (!) specifically induces hindbrain and ear rudiments. The alcohol-treated kidney of the guinea-pig induces spinal cord, notochord and muscles.

In the 1950s in Helsinki, Lauri Saxén and Sulo Toivonen studied this problem of the multiplicity of inductive effects. They implanted two tissue pellets simultaneously into the blastocoel of a young newt gastrula. One pellet, prepared from guinea-pig liver, on its own could only induce forebrain, eye and nose rudiments. The other pellet, of bone marrow, on its own could only induce notochord and muscles. Implanted together, they induced the complete range of neural structures — forebrain, midbrain and hindbrain together with the ear rudiments and spinal cord. These results were explained by there being three different areas of competent ectoderm. Two respond to neuralizing and vegetalizing influences, while the third requires an interaction of the two signals.

There seems to be a chemical difference between the vegetalizing and neuralizing agents. Neuralizing factor is soluble in organic solvents and is relatively stable during temperature changes. Vegetalizing factor is insoluble in organic solvents and readily breaks down on heating. Exhaustive chemical analysis has provided some clues about their identities; the vegetalizing agent appears to be a protein, while there is some evidence (not supported by its thermostability) that the neuralizing agent isolated from guinea-pig liver might be a ribonucleoprotein.

It is possible that chemical gradients might set up patterns and hence direct morphogenesis. Do the results with the abnormal inducers described here give evidence for a gradient mechanism in the determination of neural structures? Saxén and Toivonen concluded that the full range of neural structures could only be produced if the type of structure formed depended on the interaction of each cell with both inducer substances. They proposed that, in the normal embryo, the neuralizing and vegetalizing agents are distributed unequally within the archenteron roof in the form of opposing

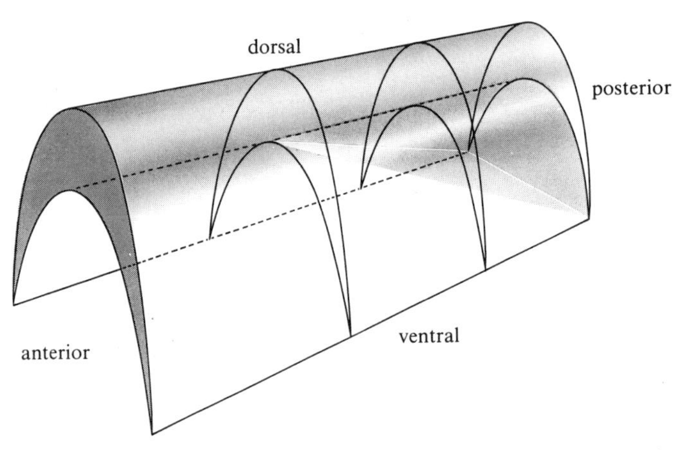

Figure 3.23 A schematic representation of Saxén and Toivonen's two-gradient hypothesis indicating the form of the gradients in the archenteron roof. Neuralizing factor is shown as a grey shade, vegetalizing factor as coloured.

gradients (Figure 3.23). The neuralizing agent was postulated to decrease in concentration laterally and ventrally, while the vegetalizing agent is absent in the most anterior part but its concentration gradually increases posteriorly.

Note that the central portion of the archenteron roof contains both neuralizing factor and small amounts of vegetalizing factor. It therefore induces both neural (hindbrain) and mesodermal (notochord) tissue. In addition, there is a supposed dorsal–ventral variation in neuralizing agent. The highest gradient level appears to induce neural plate, a lower level gives rise to skeletal elements, sensory nerves, the adrenal medulla and pigment cells. These tissues are collectively called the **neural crest** and are ectodermal in origin. Parts of the ectoderm not stimulated by the neuralizing factor differentiate as skin epidermis.

We now consider briefly the philosophy of the approach used by these embryologists. Development in higher organisms is a complex process: just consider the vast range of cell types and organizational complexity of something even so humble as a newt neurula. Identifying the substances involved will certainly advance our understanding, especially when their mode of action is clarified. We saw earlier how changes in adhesiveness, mediated by cell adhesion molecules helps in the understanding of how changes of shape may occur in cells and tissues. However, it is necessary always to remember that changes in shape of the type that occur during neurulation involve a hierarchy of processes, from the molecular to the cellular to the tissues of the whole organism. Only when all of these are understood can we consider that we have an explanation of embryonic induction.

3.5.3 An *in vitro* model for induction

In recent years, attention has been focused on an interesting *in vitro* experimental system, developed by Clifford Grobstein in the late 1950s, which promises to throw light on the mechanism of induction. This is the study of ectoderm–mesoderm interactions (the interrelation between ectoderm and mesoderm cells in development) and how new patterns of morphogenesis occur by their cooperation. To distinguish them from the primary inductions of early embryogenesis, interactions of this kind are called secondary inductions. They occur in a variety of situations. A tooth is formed by the interaction of the enamel ectoderm and the tooth mesoderm, the type of tooth formed being determined by the latter. Likewise, organs consisting of glandular epithelia, for example, pancreas, salivary tissue, mammary tissue and kidney, depend for their formation on the interaction between epithelium and mesoderm. As an example take the pancreas. Even at a very early stage of development, pancreas mesenchyme may be replaced with mesenchyme from many places in the body and a normal pancreas still results.

What is the basis of this interaction? Morphogenesis in the epithelium seems to be triggered by some inductive influence from the mesoderm. The experimental approach mentioned above for the pancreas shows a lack of mesodermal specificity. How can we find out more about the induction?

It has been known for many years that kidney tubules are formed as a result of the interaction of kidney mesoderm and epithelium. Grobstein separated the kidney mesoderm from the epithelial bud attached to it and placed it on a Millipore filter, a special filter with pore size of 0.45 μm. A piece of spinal cord epithelium was sealed to the opposite side with agar. After incubation for a number of days in a suitable medium, kidney tubules were formed in the mesoderm (see Figure 3.24).

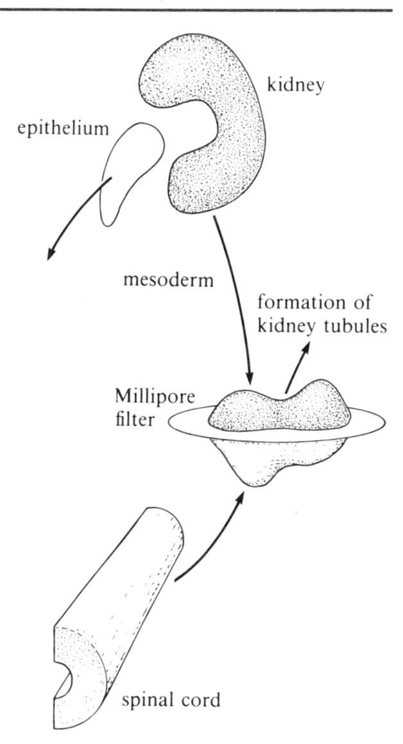

Figure 3.24 Epithelium–mesoderm interactions across a Millipore filter. Kidney mesoderm is separated from its adjoining epithelium and placed on a Millipore filter. Kidney tubules are induced by spinal cord tissue attached to the other side of the filter.

◇ As there was a narrow-pored filter between mesoderm and epithelium, is the possibility of induction mediated by cell–cell contact ruled out?

◆ No. Although the initial results of Grobstein's work were taken to imply that the inductive influence must occur by diffusion, electron microscopy showed that cell processes could enter the pores.

In fact, an inverse relation was found to exist between the pore size and the thickness of the filters across which the induction took place. Later, a minimum pore size of 0.15 μm was established for the inductive process to occur.

The transfilter method could also be used to measure the time needed for the inductive stimulus to be transmitted, because contact between the two sides could be broken at any time. From experiments of this kind, it was shown that the minimum induction period needed with one filter placed between the two tissue types was 12 hours. The diffusion rates of a number of substances were then tested with the filters. As all chemicals were found to diffuse across in less than 12 hours, this was taken as extra evidence that cell–cell contacts must be formed within the pores. We must therefore seriously consider direct cell–cell contact as a possible candidate for the inductive stimulus, at least for ectoderm–mesoderm interactions.

There is some information on which processes might be going on during ectoderm–mesoderm induction. It was noticed in electron micrographs of filters used for transfilter induction that collagen molecules are laid down in the pores during induction. Collagen is a protein, and a structural component of many biological tissues such as connective tissue and skin. It is an integral component of cell basement membrane, and treatment with collagenase, an enzyme that degrades collagen, has been shown to affect the development of some ectoderm–mesoderm systems.

More evidence exists on the role of **glycosaminoglycans** which are also present in the basement membrane. In the mouse embryonic salivary gland, freshly synthesized glycosaminoglycans accumulate on the surface of the epithelium, especially where there is epithelial branching in the mesoderm. If the epithelium of mouse embryonic salivary gland is treated with hyaluronidase (an enzyme that degrades glycosaminoglycans) and collagenase (to separate epithelial and mesodermal tissue), it loses its lobed appearance and becomes rounded. If it is cultured further with salivary mesoderm, it again becomes branched. Now study Figure 3.25, in which the changes in epithelial morphology can easily be seen. Biochemical analysis showed that the glycosaminoglycans were evenly distributed over the surface of the rounded explant, but on

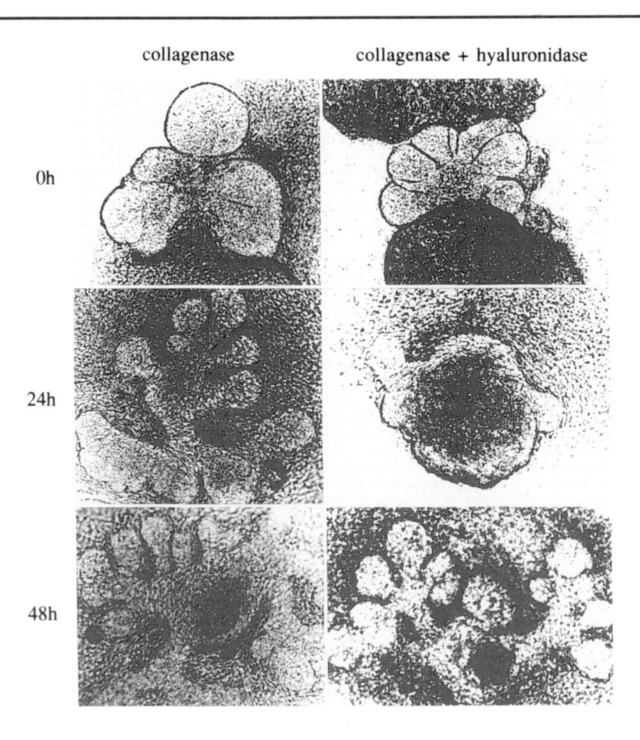

Figure 3.25 Mouse salivary gland epithelium shows normal development after collagenase treatment and subsequent growth in contact with salivary mesoderm. If treatment includes hyaluronidase, it initially rounds up, but later develops normally.

reacquisition of the branched pattern, these substances were again concentrated at the branching points. If this experiment is done with only collagenase in the separation procedure, the branching pattern is not lost.

It seems that normal salivary gland morphogenesis depends on the glycosaminoglycans. The mode of action is not known. One speculation is that the contractility of microfilaments in the epithelial cells might require the presence of glycosaminoglycans. These microfilaments, it is suggested, may control cell shape and promote the formation of branches in the growing salivary gland.

A problem with many investigations into the molecular basis of induction is that the evidence for induction may only be seen after a number of inducers have been involved. For example, it is known that development of lens tissue in the newt involves three sequential inductions by endoderm, heart mesoderm and retina. So it is important to find direct evidence for induction at the earliest stage possible. This means looking for molecular changes rather than for tissue changes.

◇ Can you think of a range of molecules which could serve as initial indicators of induction?

◆ Cell adhesion molecules. For example, evidence of neural induction could be shown by presence of NCAM.

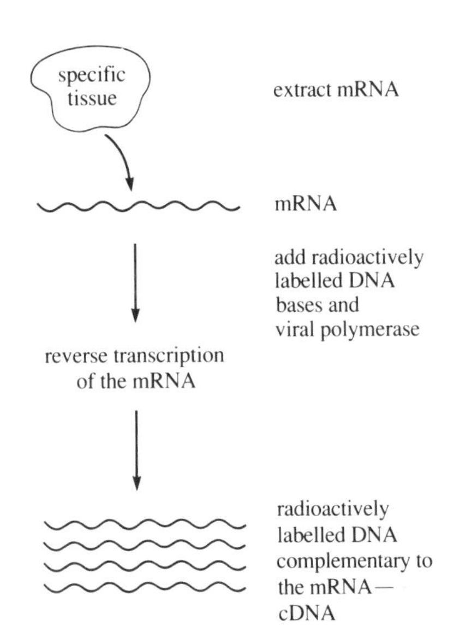

Figure 3.26 The production of cDNA from mRNA.

Inducing proteins can be detected by use of fluorescent antibodies while the RNAs that code for them can be identified by the technique of ***in situ* hybridization** using **nucleic acid probes**. This involves isolating the mRNA for the protein under investigation and making a complementary DNA (**cDNA**) copy; using an enzyme which is isolated from viruses and which copies the DNA in the reverse direction. This cDNA is then radioactively labelled. Since the cDNA is complementary to the mRNA that codes for the protein being investigated, it will bind specifically to the mRNA (Figure 3.26). If a thin section of tissue is cut and washed with a solution of cDNA, the cDNA will bind to the mRNA complementary to it. By autoradiographic techniques the presence of specific coding mRNA for specific proteins can be located in particular cells.

Figure 3.27 shows a cross-section of a tadpole head, treated with cDNA for mRNAs for neural tissues. You can clearly see in the autoradiograph the probe as white areas around the retina and neural tube. This particular probe is not highly specific for neural cells as it has also labelled the epidermis.

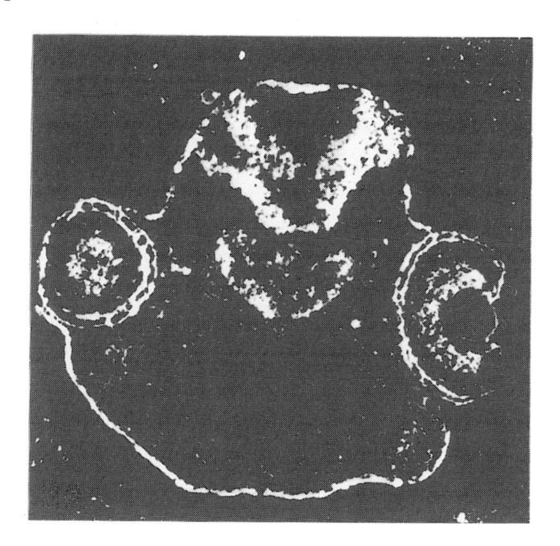

Figure 3.27 Cross-section of tadpole head. The white areas are the nervous system, the brain and the neural retina. The epidermis also shows as white.

Such techniques enable us to see that induction is occurring but do not pinpoint the chemical nature of the inducer itself.

Before continuing with this aspect of the story we should look at some general features of the inductive process. Some workers have thought it useful to distinguish between **permissive induction** and **instructive induction**. Permissive induction is said to occur when the tissue is so committed to its pathway of differentiation that any signal will initiate the process. In effect the differentiation is due to the state of the tissue rather than to the nature of the inducer. On the other hand instructive induction occurs in tissue that is 'uncommitted' in terms of differentiation and requires a specific signal from the inducing tissue to allow it to differentiate in a specific direction.

A good example of a permissive induction is differentiation of mouse pancreas. The differentiation is normally induced by pancreatic mesenchyme but at certain stages of development pancreatic mesenchyme can be replaced by mesenchyme from a variety of tissues which still allow differentiation of the pancreas.

Instructive induction is well illustrated when chick corneal epithelium is placed next to mouse mesoderm. Mouse mesoderm normally induces ectoderm to form hair. The chick corneal epithelium normally forms the transparent cornea of the eye. However, the interaction of those tissues has the consequence that chick endoderm forms feathers. Thus the mesoderm is directly responsible — it instructs the chick epithelium to alter course. But this induction is also dependent on the ability of the tissue to respond. It is these instructive inductions which have stimulated most work in identifying the inducer.

Identifying the chemical nature of inducers has proved very difficult. Even then, this work only provides information on unnatural (heterogenous, meaning of different origin) inducers. There is no reason to assume that the natural inducer has an identical chemical composition. However the use of heterogeneous inducers does allow more material for analysis and is at least a starting point.

Although some workers in the past claimed that inducers could be RNA molecules, most current work suggests that inducing molecules are small protein molecules.

A fibroblast growth factor (FGF), a small protein molecule, has been identified in both bovine and *Xenopus* tissues. FGF is thought to be important as a morphogen responsible for inducing mesoderm in early *Xenopus* development.

In 1988 a heat-stable, soluble protein with $M_r = 23\,500$ was isolated from *Xenopus* cells grown in tissue culture. This protein acts as a mesodermal inducing factor (MIF). The cells release the protein into the tissue culture medium which can then be treated with a sequence of chromatographic and electrophoresis techniques to purify the MIF. Since the cell line used for the tissue culture is known as XTC, the inducing factor is known as XTC–MIF. Work suggests that XTC–MIF is composed of two protein molecules of approximately equal M_r. XTC–MIF may well be one of the first natural inducers to be characterized.

The importance of this is that the induction of the mesoderm is really the first induction in amphibian development. So this work is a recent example of the continuing search for the identities of embryonic inducers.

3.5.4 Conclusion

To end this section we can do no better than to give you the conclusion to a recent review on embryonic induction by John Gurdon.

> 'The overall impression conveyed by this survey of embryonic inductions is that these processes are very complex, probably consisting of several reciprocal interactions between inducing and responding tissues, each step enhancing, little by little, progression towards the eventual differentiation. Most inductions seem to take place as follows. Both competence and inducing ability are acquired by restricted populations of cells for a limited time. The proximity of the first, apparently instructive, inducer further restricts the number of cells within the competent population that will respond. Especially with inductions that take place late in development, several further, usually permissive, inductions from other inducing tissues enhance the initial response. Tissue structure may be finalized by interactions among responding cells with each other.
>
> 'In retrospect it is not surprising that so many steps should be required for the reliable formation of something as complicated as an embryo. Any manufacturing process in which a complicated machine is assembled requires controls to ensure the coordinate operation of different steps. If the formation of an embryo were to take place without frequent interactions between different regions, errors in the final product would often arise.
>
> 'I believe that the greatest obstacle to the molecular analysis of induction over many decades may have been the imprecision and late appearance of the assays used, which often depend on morphological assessment many days after the inductive response has started. It may also have been necessary, *faute de mieux*, to concentrate on identifying inducer molecules; but compared to ligands for other cell interactions,

embryonic inducers seem less specific, since they can be substituted for by other substances, and therefore harder to purify. For embryonic induction to be accessible to the powerful methods of molecular analysis now available, it seems essential to use, as an assay, a single early response, such as the expression of one gene. Nucleic acid technology has probably now reached a sufficient level of precision and efficiency of operation to be usefully applied to the analysis of inductive responses, working from the response backwards, rather than from the inducer forwards.'

Summary of Sections 3.4 and 3.5

Both adult and embryonic cells can be dissociated and persuaded to clump together into a mixed aggregate where different cell types are randomly distributed. Subsequently, cells sort out as they move within the aggregate to become surrounded by cells of their own type. Each reconstituted tissue eventually adopts a characteristic position within the aggregate, even when the two cell types are combined as tissue fragments.

The behaviour of ectodermal, mesodermal and endodermal cells during aggregation bears some resemblance to their normal arrangement in the embryo. Sorting out may, therefore, have considerable morphogenetic significance.

Sorting out appears not to occur by chemotaxis, although this cannot be ruled out altogether. Steinberg's hypothesis proposes that sorting out in an aggregate can be accounted for by qualitative differences in cell adhesiveness. Adhesions between unlike cells are exchanged for adhesions between like cells, and the more adhesive cells move towards the centre. If these ideas are correct (and there is some evidence for this from work on the regenerating limb of the salamander), it would mean that ectoderm, mesoderm and endoderm cells have different degrees of adhesiveness and these differences could be of great importance in morphogenesis.

A wide range of cell adhesion molecules have been discovered on different tissues at different stages of development. General CAMs are associated with sialic acid which can influence the binding of the amino acid groups at the end of the CAMs. Changes in sialic acid concentration on cells during development would make the cells more adhesive or less adhesive. Saccharide CAMs (membrane-bound enzymes) link glycoprotein molecules between adjacent cells.

Embryonic induction is the process whereby cells of one type have their fate determined by cells of a second type. Some signal passes between the two types of cell. Induction is seen in early development, where blastopore tissue induces neural tissue from ectoderm. The dorsal lip of the blastopore has a particular ability to induce or 'organize' other tissues.

The search for the chemical basis of the inducer has occupied many embryologists for many years. Many types of embryo extract have been shown to have inductive ability, among them dorsal blastopore lip that has been frozen and substances derived from vertebrate liver or kidney. The fact that the inductive effect may be transferred from embryonic tissue to culture medium after some days suggests a chemical nature for the inductive process.

Different tissue preparations have different inductive abilities. These have been separated into neuralizing agents and vegetalizing agents. Gradients of these agents could account for the induction of tissues in the developing embryo.

Much work on inductive systems involves looking at *in vitro* tissue interactions to build models for *in vivo* systems.

Glycosaminoglycans have been proposed as being important in salivary gland morphogenesis. The presence of specific cell adhesion molecules provides evidence to confirm that induction has occurred. Inducing agents are generally thought to be small protein molecules.

Question 9 (*Objective 3.8*) If a mixture of ectoderm and mesoderm cells are allowed to undergo reaggregation, which would you expect to take up the core position?

Question 10 (*Objective 3.8*) Three types of tissue, A, B and C, are isolated and allowed to reaggregate in pairs. If B cells form a ball inside an outer covering of C cells, and C cells take up the core position when mixed with A cells, what would you expect to be the result of mixing A and B cells?

Question 11 (*Objective 3.9*) Summarize the conclusions about the role of cell adhesion which are obtained from the experimental work on the regenerating salamander limb.

Question 12 (*Objective 3.10*) What do experiments on lung tissue development reveal about the role of CAMs?

Question 13 (*Objective 3.11*) The following statements (a)–(e) are possible results on embryonic induction. Which are true and which false?

(a) When a graft of blastopore cells is made from an amphibian gastrula to a second gastrula of different pigmentation, the twin embryo formed has the pigmentation pattern of the grafted cells.

(b) Late gastrula neural plate develops according to its original fate when transplanted into a second late gastrula.

(c) Early gastrula presumptive neural plate develops according to its original fate when transplanted into a second early gastrula.

(d) Presumptive epidermis from an early amphibian gastrula can form neural tissue if transplanted into presumptive neural plate region of a second gastrula.

(e) The blastopore has a strong inductive effect on other regions of the embryo.

Question 14 (*Objective 3.11*) Identify the following material (a)–(e) as either abnormal inducer, normal inducer, or non-inducer.

(a) The dorsal lip of the blastopore of an early newt gastrula.

(b) A tissue extract of notochord from newt larva.

(c) The archenteron roof of a late gastrula.

(d) Alcohol-treated liver.

(e) Niu's unconditioned medium.

Question 15 (*Objectives 3.11 and 3.12*) Slices of bone marrow were heated for varying times: 25, 40, 60 and 150 seconds. Following each heat treatment, the inductive ability of the preparation was tested, and the resulting primary inductions were classified as A mesodermal, B hindbrain characteristics, and C forebrain characteristics. Table 3.1 gives the induction obtained from each heat treatment.

In what way are these results related to Saxén and Toivonen's interpretation of primary induction?

Table 3.1 Induction caused by heated bone marrow

Time of heat treatment/s	Induction/%		
	A	B	C
0	97	0	0
25	0	17	4
40	0	13	44
60	0	0	46
150	0	0	8

Question 16 (*Objective 3.13*) Are the following statements on ectoderm–mesoderm interactions true or false?

(a) Kidney tubules are induced by interaction between kidney mesoderm and spinal cord tissue.

(b) Transfilter induction needs to take place throughout the whole process of differentiation of the induced tissue to be effective.

(c) Cell surface molecules are thought to be important in transfilter induction.

(d) In ectoderm–mesoderm interactions, mesodermal tissue is influenced by epithelial tissue.

(e) Hyaluronidase inhibits the induction of salivary gland morphogenesis.

3.6 GRADIENTS

The spatial patterns that emerge during development depend upon long-range spatial ordering influences that coordinate local interactions between cells (e.g. induction and differential adhesion) so as to produce coherent structures such as limbs and nervous systems. In Section 3.5.2 it was suggested that gradients of inducers might be involved in the latter. Evidence was also presented in Chapter 1 that gradients of different substances are implicated in generating spatial order in organisms such as *Acetabularia* and *Hydra*. In this and the next section we will examine a body of evidence that suggests how gradients can give rise to different aspects of spatial order in developing systems.

The organisms shown in Figure 3.28 were all, at some stage in their development, simple systems without any of the discrete parts that make up the adult form. In the case of *Acetabularia*, the newt, and the fruitfly, this was a zygote, a single cell. Asexual reproduction in *Hydra* started from a bud, a mass of cells all of which are initially very similar.

As development proceeds, different regions become progressively more distinct from one another, while the whole retains an overall coordination. For example, tentacles do not normally arise at random over the surface of a growing *Hydra*, even though the potential is there and can be stimulated by a substance such as diacylglycerol (DAG), as described in Chapter 1. We have already encountered evidence that this generation of parts within a whole involves the formation of gradients of substances within the developing organism. For example, we saw that there is a gradient of calcium in the growing tip of *Acetabularia* that changes into a pattern with maxima within

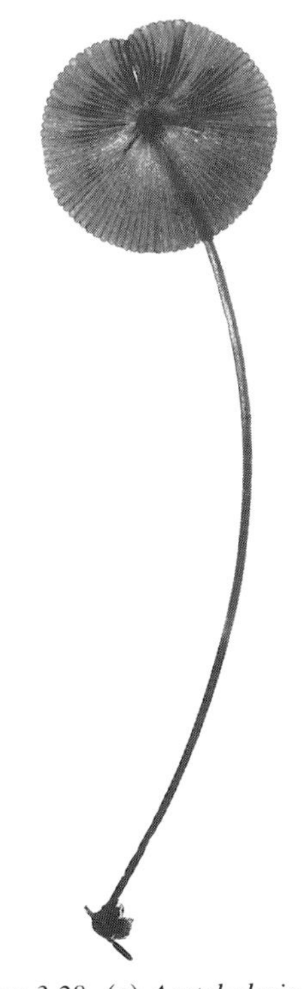

Figure 3.28 (a) *Acetabularia*.

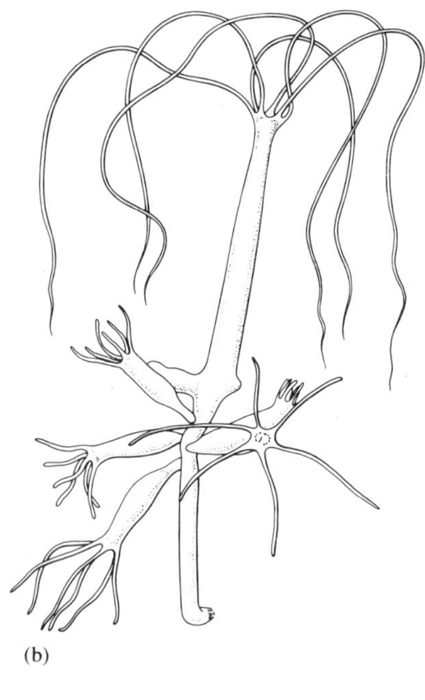
(b)

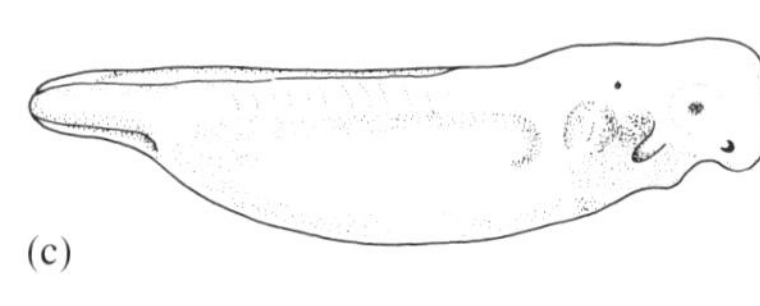
(c)

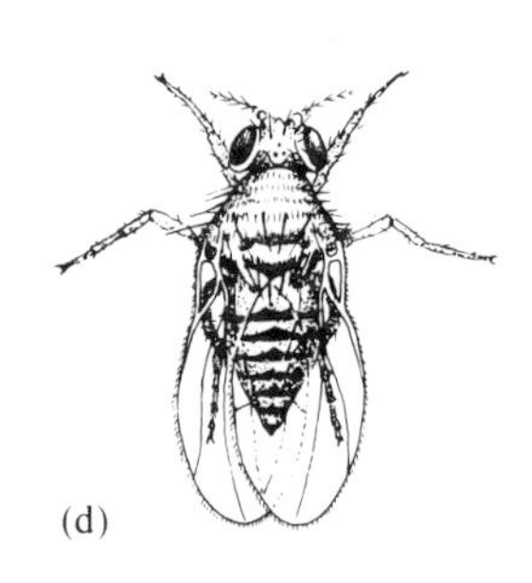
(d)

Figure 3.28 (b) *Hydra littoralis*. (c) Newt embryo. (d) Fruit-fly.

each hair of a whorl (Chapter 1, Section 1); that head activator and inhibitor increase towards the head in *Hydra* (Chapter 1, Section 2); and that in developing limbs there is a proximal–distal gradient of adhesiveness (Section 3.4.3). It has long been believed that such gradients play a primary role in the emergence of spatial patterns in developing and regenerating organisms, so they are regarded as basic components of morphogenetic fields. This idea goes back to the work on *Hydra* by Charles Manning Child in the USA in the early years of this century. He used dyes that are indicators of metabolic activity to identify metabolic gradients in *Hydra*, and found the maximum at the head and the minimum at the foot. On the basis of these and other studies he developed a theory of pattern formation which is based entirely upon such gradients, a theory that is now bearing fruit in terms of the identification of possible biochemical candidates for the crucial roles of primary pattern initiators, referred to as morphogens.

However, it is important to bear in mind that observing a gradient in a substance or activity that correlates with a morphogenetic pattern is not sufficient to identify it as an initiator of this pattern. It could simply be a secondary consequence of primary pattern formation by another substance. Even a substance such as the head activator in *Hydra* may not be the primary initiator of head formation, since its effect is to accelerate the rate of head formation in animals that are already engaged in the process of making a head. DAG, on the other hand, does have the property of initiating heads in places they would not normally form. But there is as yet no evidence that DAG has an increased concentration in developing head regions, so qualifying as an endogenous inducer of head formation.

◇ Bearing these qualifications in mind, how would you rank calcium as a primary initiator of tip and whorl formation in *Acetabularia*? (See Chapter 1.)

◆ Calcium increases towards the tip in *Acetabularia* and is distributed in a pattern similar to a whorl when it first forms. Also, by changing calcium concentration in the medium, tip and whorl formation can be controlled. These properties suggest, but do not establish, that calcium may be involved as a primary initiator of morphogenesis in *Acetabularia*.

3.6.1 Reaction, diffusion and cell communication

In investigating which type of control process is likely to be active in the formation of a head or a bud in *Hydra* (Chapter 1, Sections 1.2.3 and 1.2.4), it emerged that both negative and positive feedback appear to be involved. Positive feedback, whereby a substance stimulates its own production, is implicated in initiating the formation of a structure; while negative feedback, or inhibition, is required to prevent overproduction. Diffusion of these postulated substances or morphogens was also suggested as an important aspect of the production of a gradient in a tissue. These coupled processes of reaction, interaction and diffusion have been studied extensively in the context of pattern formation, and it is well established that interacting biochemical reactions of particular kinds, together with diffusion of the products, can spontaneously generate gradients of products. These are known as reaction–diffusion systems. In their simplest form they are characterized by a pair of substances, one of which stimulates its own production as well as the production of second substance, which in turn inhibits the first. The stimulator must also diffuse slowly relative to the second substance so that it has a localized activating influence, whereas the inhibitor has a longer-range effect. The result is the production of spatial patterns of the two substances, including the formation of gradients. The first person to analyse such processes mathematically and to demonstrate their capacity to generate spatial patterns was the British mathematical genius Alan Turing, in 1952, when he coined the term 'morphogen'. (He is much better known for his work on the theory of computation, the Turing machine bearing his name.)

In a single cell such as *Acetabularia*, a reaction–diffusion process can generate a gradient in the cytoplasm. However, in a multicellular organism such as *Hydra* there needs to be communication between cells for the production of a gradient. In animals, continuity between cells arises from the existence of gap junctions, regions where the opposed membranes of two cells are connected by tiny channels which allow the direct transfer of small molecules from one cell to the next. Direct evidence for this communication link is obtained by injecting a cell with a fluorescent dye and observing its passage to neighbouring cells. Substances such as *Hydra* head activator and inhibitor are small enough to pass through such junctions.

3.6.2 From continuity to discontinuity in patterns

Although it is necessary to be careful in interpeting the data that connect particular substances with morphogenetic processes in developing organisms, Child's proposal that gradients of some kind are essential to the initiation of spatial patterns in developing organisms is sound. This is virtually the only way in which an initially uniform system can develop spatial patterns. What it does not explain is how something that varies continuously, such as a gradient, can result in a spatial pattern with sharp differences of structure, as in the cap–stalk discontinuity in *Acetabularia* or the localization of tentacle formation to the head in *Hydra*.

The simplest way to account for such phenomena is to use the concept of a threshold — where the concentration of a substance is below a critical concentration, one structure is formed; where it is above this critical or threshold value, another structure is formed. Recall the results described in Chapter 1. As the external calcium concentration is varied in the seawater in which *Acetabularia* grows, different structures are formed in regenerating algae. At 1.5 mmol l^{-1} calcium, no tips form, while at 2.0 mmol l^{-1} they do; between 2.0 mmol l^{-1} and 3.0 mmol l^{-1}, whorls are initiated; while above 3.0 mmol l^{-1} cap formation occurs.

Such results show that cells are capable of making sharp discriminations between the concentration ranges of external signals. There are many other such examples that come from biochemical and physiological studies. For example, when the glucose level in the blood rises above a critical level (5 mmol l^{-1}), the β-cells in the pancreas start to secrete insulin. Such threshold-type responses of cells to external signals is one of their common properties, so it is reasonable to assume that they operate in developing organisms to discriminate between different concentrations of substances that are distributed in continuous gradients.

3.7 PATTERN FORMATION WITH ONE DIMENSION OF ORDER

We can now apply these ideas to the analysis of a variety of examples of pattern formation of increasing complexity, starting with the blue–green alga *Anabaena*. This consists of an unbranched filament of cylindrical cells, as shown in Figure 3.29. Most of the cells are rather cuboidal in shape when seen under the microscope, but at regular intervals, another cell type, larger and

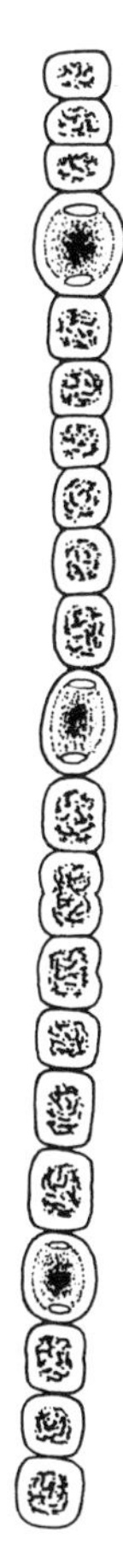

Figure 3.29 A single filament of *Anabaena* (×200).

more rounded, occurs. These are specialized cells called **heterocysts**, which have the capacity to fix atmospheric nitrogen. Mature heterocysts do not divide, while the other cells do.

Normally, the interval between heterocysts is made up of about eight cuboidal cells. As the cells divide, the interval widens until it reaches about 16 cells, when a new presumptive heterocyst (pro-heterocyst) develops in about the middle of the interval. This process is repeated in each interval along the filament, and in this way, a spaced heterocyst pattern is maintained as the filament grows. Obviously some mechanism ensures that new heterocysts do not develop near existing heterocysts. It appears that there is an inhibitory zone around each heterocyst; further heterocysts differentiate only when they lie outside these zones.

Could this simple pattern formation mechanism be explained by a gradient? One theory proposed that heterocysts might produce an inhibitory substance that diffuses outwards from each heterocyst along the filament. As it diffuses, it may be absorbed or destroyed by the surrounding vegetative cells. There would thus be a gradient of inhibitor concentration leading away from each heterocyst. The concentration of the proposed inhibitor substance would be lowest midway between each pair of heterocysts. There may be a threshold inhibitor concentration; any cell experiencing an inhibitor concentration below this will 'switch on' and differentiate into a heterocyst. Of course, at this stage, this is just a model that is used to interpret the observations.

Suppose we take a chain of eight cells between two heterocysts (see Figure 3.30a). The curve represents an inhibitor concentration in the cells in this filament. The heterocysts are very close together, so that all the cells in the interval between them have an inhibitor concentration above the threshold. As these cells divide, the interval widens and the concentration of inhibitor in the middle of the interval drops until eventually it falls below the threshold in some of these cells (see Figure 3.30b). This is the signal for those cells to begin differentiation into a heterocyst (Figure 3.30c) and to start producing inhibitor (Figure 3.30d).

An apparent problem arises with such a simple mechanism, for it is not easy to see how it could be made sufficiently accurate to ensure that single cells rather than groups of cells are picked to differentiate and thus to prevent a whole group of adjacent heterocysts from being formed. The gradient is, after all, quite shallow in the middle region between two mature heterocysts and Figure 3.30b suggests that more than one cell will experience an inhibitor concentration below threshold and will start to form heterocysts. However, the gradient–threshold model, together with a further property observed in cells, is sufficient to ensure that only one mature heterocyst is formed. It

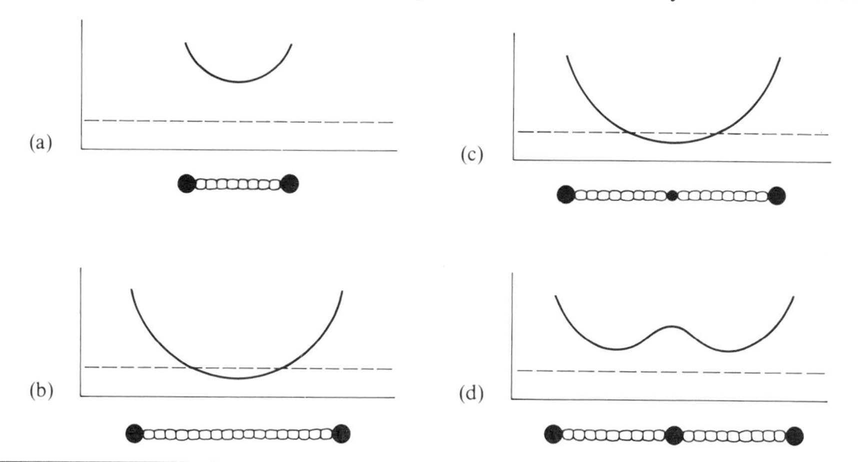

Figure 3.30 New heterocyst formation in *Anabaena*. (a) Inhibitor concentration above threshold. (b) Cells divide and inhibitor concentration in the middle falls. (c) New heterocyst begins to form. (d) Inhibitor concentration rises above threshold again.

seems that during the early stages of development, a pro-heterocyst can be inhibited (or 'switched off') if the inhibitor level in it becomes too high. If two pro-heterocysts are developing close together, each is liable to be inhibited by inhibitor produced by the other — a sort of competition occurs for the one available 'slot'. Experimental results demonstrate this rather dramatically. A pro-heterocyst will inhibit itself if it is isolated from the filament together with a small number of vegetative cells (see Figure 3.31).

The pro-heterocyst, which would normally have developed into a heterocyst within a few hours, divides and reverts to the vegetative state. This is because these few vegetative cells cannot absorb all the inhibitor produced by the pro-heterocyst; consequently its concentration builds up until it inhibits the pro-heterocyst itself. So each pro-heterocyst creates an inhibitory region near itself, with the result that only one cell differentiates into a mature heterocyst. This is the only stable solution to the spatial patterning process.

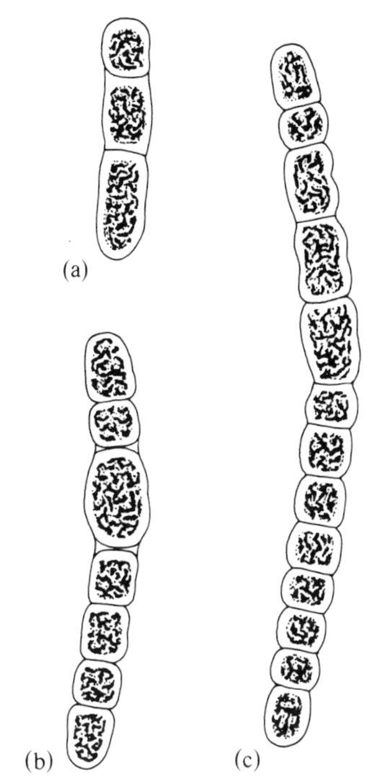

Figure 3.31 (a) A pro-heterocyst in a short filament. (b) Early stage of reversion. (c) Late stage of reversion.

3.7.1 Gradients and segmental patterns in insects

Elegant studies that strongly reinforced the belief in the existence of gradients as basic determinants of developmental patterns came from work on insect segments.

An important example that illustrates the type of observation made and the interpretation in terms of gradients is to be found in Peter Lawrence's work in the late 1970s on the milkweed bug *Oncopeltus*. The adult bug has a dense mat of anterior–posterior oriented hairs, each of which acts as a polarity indicator with respect to the anterior–posterior axis.

Oncopeltus has a segmented cuticle with a well defined segment margin between each segment. Occasionally one finds an individual with a gap in one of the segment margins. The hair pattern around the break is different from normal (Figure 3.32): it is as if something has pushed through the border and spread out around it to form a new pattern. How could a gradient acting across the *Oncopeltus* cuticle segment explain the peculiar behaviour of the hairs around the broken segment border?

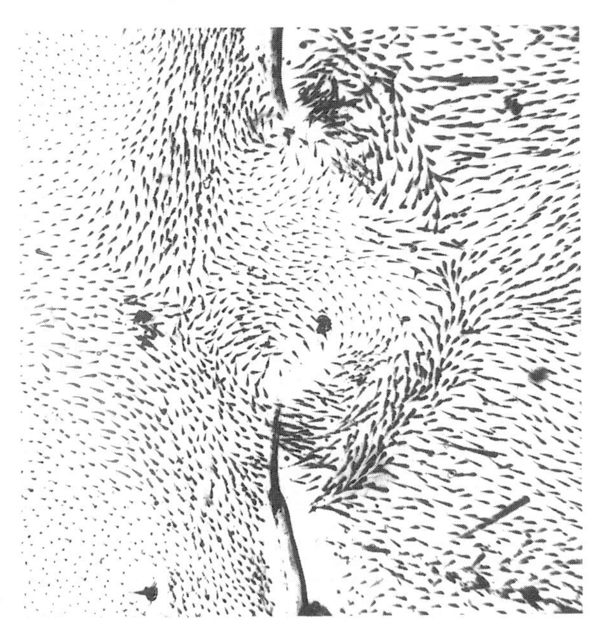

Figure 3.32 The hair pattern surrounding a natural discontinuity in a segment margin of *Oncopeltus*. Anterior is to the right.

(a)

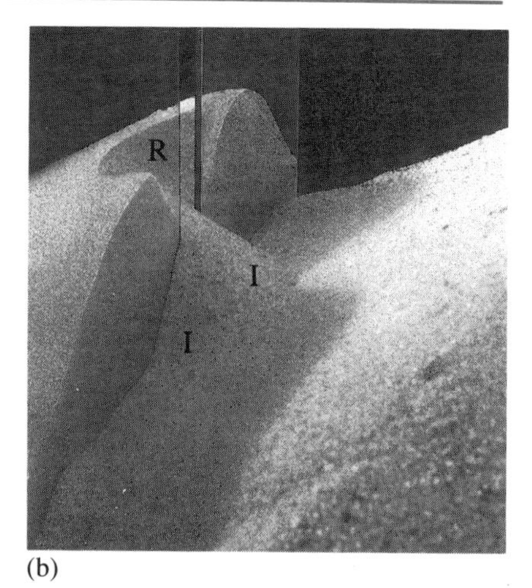

(b)

Figure 3.33 Sand model of serially repeating gradient of inhibitor concentration.

Lawrence constructed a sand model to represent the form of a serially repeating gradient. This is shown in Figure 3.33a where glass plates are used to represent the segment borders, and sand is used as an indicator of gradient level: a stable sand slope between each glass plate represents the gradient.

If a break in the glass plate develops (see Figure 3.33b) sand flows from one 'segment' to another and continues until new stable slopes are established. There is now a region of reversed slope in the region of the break. Imagine that lines are drawn all over the landscape of the model to indicate the direction of the steepest sand gradients. Suppose that these lines mark the orientation of *Oncopeltus* hairs, where the hairs point down the steepest sand slopes. Compare Figure 3.32 with Figure 3.33b, and note how such a pattern marked on the landscape of the sand model would conform to the hair pattern at the gap in the segment margin. In the centre of the glass separation the sand slope is reversed relative to the rest of the segment, so the hair direction would be reversed here, just as it is in Figure 3.32.

Because the model fits so neatly, we can postulate that the polarity of the epidermal cells depends on the orientation of the steepest gradient slope. When an unstable situation develops as a result of cells that were previously at different gradient levels finding themselves together, as occurs when the segment margin is broken, a flow of gradient material occurs until stable slopes are produced, as with the sand in the model. The polarity of the hairs then reflects the orientation of the steepest slopes in the new gradients.

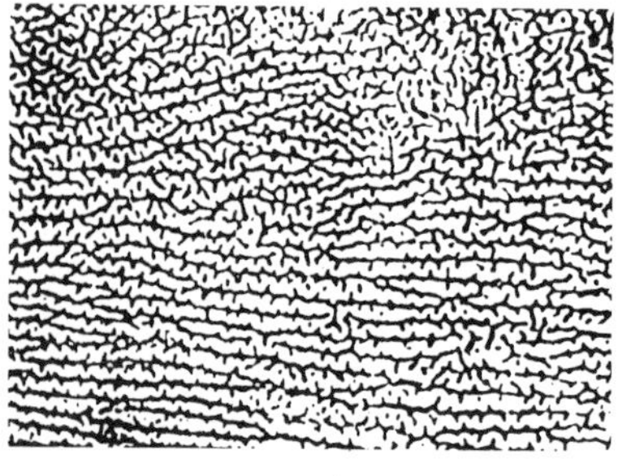

Figure 3.34 Ripples in the cuticle of adult *Rhodnius* (×150).

Another set of studies that strongly reinforced this type of interpretation of insect segment patterns was carried out by Michael Locke in the 1950s on the cuticle of the blood-sucking insect *Rhodnius*. The abdominal segments of this insect bear surface folds that run laterally across the dorsal surface of each segment parallel to the anterior and posterior segment margins (Figure 3.34). These folds occur in the cuticle which is produced by secretion of materials from the underlying single layer of epidermal cells.

Rhodnius has five larval **instars** (stages of larval growth), and a series of experiments may be performed on fifth instar larvae as follows. If a small square of the cuticle, together with the underlying epidermis, is cut out and replaced in the same site in the same orientation, the adult cuticle pattern developing from this individual is normal (Figure 3.35a, ripple pattern 1). If a square from one segment is swapped with a square from the same position in

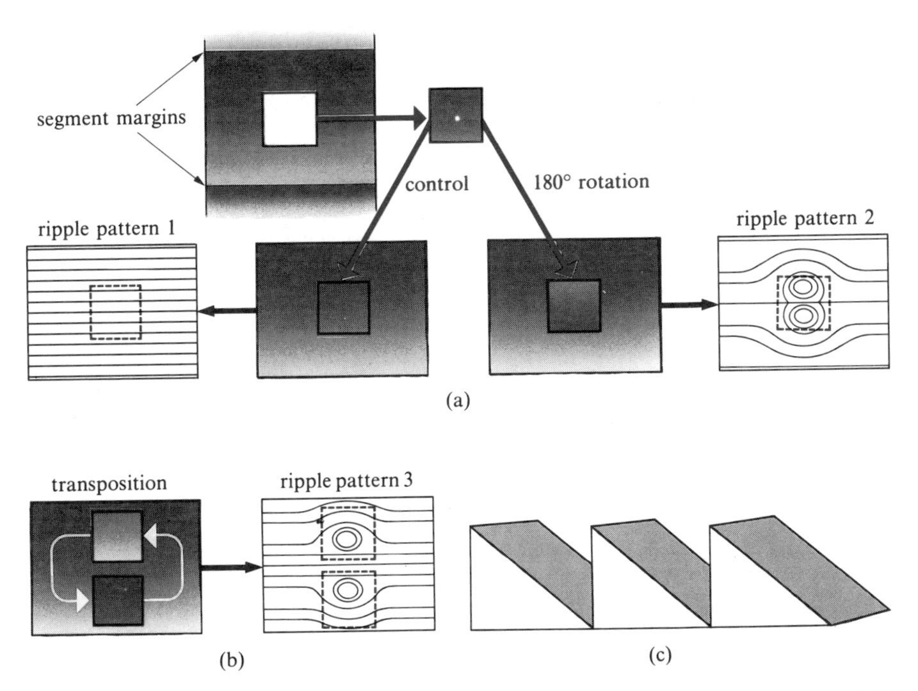

Figure 3.35 (a) The effect of transplantation of pieces of *Rhodnius* cuticle — control and 180° rotation. (b) The effect of exchange of anterior and posterior portions of the segment. The gradient in the larval epidermis is represented as a graded grey tone. Where cells at different gradient levels are adjacent, a disturbance of the adult ripple pattern emerges. (c) The serially repeated segmental gradients. The vertical lines represent segment margins.

an adjacent segment, then there is also no alteration in the adult pattern. However, if the cuticle square is replaced after rotating it through 180°, the pattern of ripples in the adult is disturbed by the formation of closed rings of ripples, or whorls (Figure 35a, ripple pattern 2). Also, if anterior and posterior portions of the same segment are exchanged, two whorls of ripples appear (Figure 3.35b, ripple pattern 3).

These experiments seemed to indicate some form of communication between the cells of the insect cuticle, as the placing of 'foreign' cells beside one another clearly induced the whorling. What was the underlying mechanism? Locke initially suggested that a biochemical gradient was present along the anterior–posterior axis of each segment (Figure 3.35c). Cells from different parts of the segment, being in different states, interacted in such a way as to result in change in ripple orientation.

Could the same principles as those used to explain hair polarity in *Oncopeltus* also be used to understand the oriented cuticular structures of *Rhodnius*? Analogues of the *Oncopeltus* hairs exist in *Rhodnius*. These are the tubercles (small scale-like cuticular outgrowths) of the ventral abdomen. (Remember that the ripple patterns described previously are on the dorsal side.)

Like *Oncopeltus* hairs, these tubercles are polarity indicators that might have their orientation determined by the direction of the steepest gradient slope. Suppose that in the larva, a piece of anterior ventral cuticle from one abdominal segment is exchanged for a piece of posterior ventral cuticle from an adjacent segment (see Figure 3.36a). Figure 3.36b shows a hypothetical cross-section of the gradient landscape immediately after the transplantation.

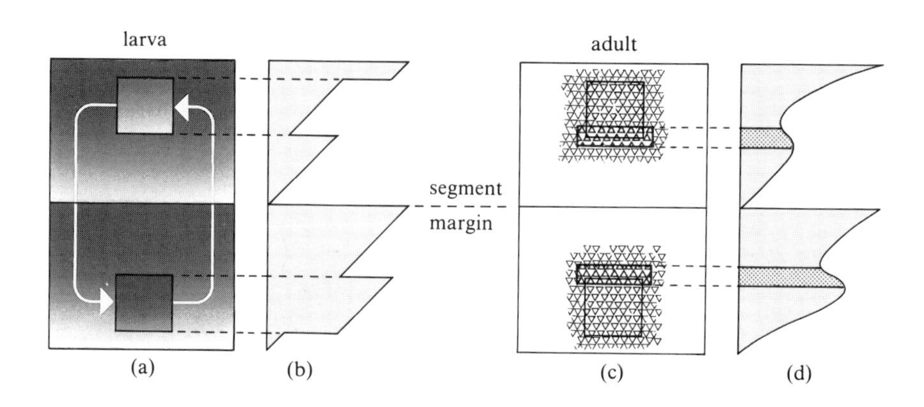

Figure 3.36 (a) An anterior square of larval cuticle from one ventral segment was exchanged with a posterior piece from an adjacent segment. The gradient in the epidermal cells is shown as a graded grey tone. (b) A cross-section of the gradient landscape immediately afterwards. (c) The resulting adult pattern showing in colour and tubercles pointing towards the anterior rather than the posterior margin. (d) The gradient landscape following diffusion, showing in colour the regions where the direction of the gradient is reversed.

Obviously, these gradients are unstable, and as in the sand model, flow of gradient material should eventually produce a stable landscape of maximum steepness (Figure 3.36d).

Note that the model predicts that in the two regions shown in colour, the orientation of the tubercles will be reversed, so that they point down the gradient slope towards the anterior margin. Just such a result was obtained experimentally (see Figure 3.36c).

Perhaps then the ripples on the dorsal surface of the abdomen of *Rhodnius* are also polarity indicators. If so, their orientation might be determined in much the same way by the direction of the gradient slope. But the ripples run transversely across the segment, so they cannot point down the gradient slope as do *Oncopeltus* hairs or tubercles on the ventral surface. Instead they must run perpendicular to the lines of the steepest gradient. So the ripples represent contour lines, joining epidermal cells at the same level in the gradient landscape; that is, joining cells that contain identical concentrations of gradient substance (see Figure 3.37a).

A circular whorl of ripples, then, implies that the epidermal cells immediately beneath are all at the same gradient level. If this is so, we should be able to predict the ripple pattern following transplantation by using a gradient model as shown in Figure 3.37. Consider, for example, Locke's experiment involving an interchange between anterior and posterior portions of a segment (see Figure 3.35b). Immediately after the change of the cuticle pieces, the gradient landscape resembles Figure 3.37b. After a flow of gradient material by diffusion, the profile that results is a smooth valley and a rounded hill (see Figure 3.37c). The contour lines drawn on this three-dimensional landscape, joining points at equal height, form a pattern with two whorls. Viewed from above (Figure 3.37d), it precisely resembles the ripple pattern that resulted from Locke's experiment (see Figure 3.35b ripple pattern 3).

◇ In Figure 3.37a, put numbers on the horizontal lines, starting with 0 at the bottom and 10 at the top. These represent hypothetical concentrations of the gradient substance, like contour lines indicating the height of the surface. In Figure 3.37b, what are the values of the middle contour lines of the lower and upper 'grafts'?

◆ The lower 'graft': middle line has value 8. The upper 'graft' middle line has value 4.

◇ Now put numbers on the contour lines in Figure 3.37d. What number does the lower circle (the one raised from the surface) have? What number is the upper circle — the one sunk into the surface?

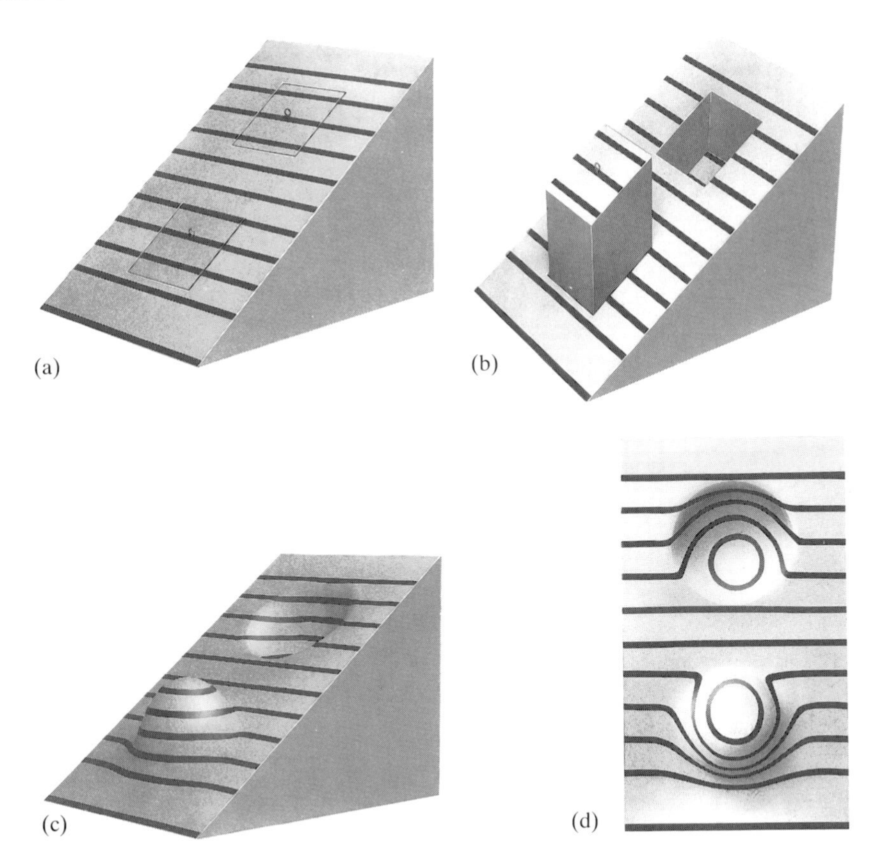

Figure 3.37 Three-dimensional gradient model. (a) Initial gradient. (b) Immediately after transposing anterior and posterior portions. (c) After diffusion. (d) Pattern of contour lines from above.

◆ The lower circle is numbered 5. The upper circle is numbered 6.

Suppose that the gradient is of some substance. How is it established and maintained? A simple source–sink model proposes that one segment margin, say the anterior, produces the gradient substance, and so represents the source, while the posterior margin breaks it down and so represents the sink. Thus, the segment margins maintain two different concentrations of a substance that diffuses passively through the epidermal cells. But consider the following experiment. Adult *Rhodnius* derived from larvae that have been operated on and therefore bear disturbed ripple patterns can be made to form an 'extra' cuticle by injecting moulting hormone (ecdysone). The disturbed pattern in the extra cuticle is very little different from the previous pattern, even though 3–6 weeks elapsed between the deposition of the two cuticles.

◇ How is this result incompatible with a gradient model relying solely on diffusion between anterior and posterior margins?

◆ If the gradient was maintained by a simple source–sink model, any local disturbances in the gradient would soon be overcome by continued diffusion, and the normal gradient landscape would soon be re-established.

Because the ripple pattern persists, it seems to represent a steady-state landscape, where the forces responsible for maintaining the landscape have come into some sort of equilibrium. It seems that there is some cellular force that can limit the extent of diffusion.

Alternative explanations therefore propose that the epidermal cells play an active part in maintaining the gradient. One such theory suggests that the epidermal cells actively pump the gradient substance up the gradient slope. A stable situation develops where the forces of diffusion down the gradient slope are in equilibrium with the active movement of the substance against the slope. If the direction of the gradient slope is changed, the epidermal cells react by actively transporting the substance against the new concentration gradient, which is therefore maintained in its new orientation. According to this model, the segment margins are considered to make no contribution to the maintenance of the gradient.

Another explanation was proposed by Peter Lawrence and his colleagues, who imagine that cells can 'remember' their original gradient concentration following their removal to a new site, and that cells attempt to maintain their internal concentration of that substance at the original level. Cells were assumed to change their states, resulting in smoothing of the patterns as shown in Figure 3.37, only during cell divisions that occur when the insect moults and lays down a new cuticle. There is now evidence that this is in fact what occurs. The ripple patterns produced by experiments and by computer simulation based on this model are remarkably similar.

Whatever the precise mechanism, note that there is likely to be some active contribution by the cells, which must therefore interact and communicate as part of a cell population to maintain the gradient.

3.7.2 The segmental gradient and threshold values

The experiments just outlined suggest that the polarity of the epidermal cells, as expressed in the orientation of cuticular features, appears to be locally determined by the steepest slope of the gradient. But a gradient carries two sorts of information. As we have seen, the direction of a gradient slope could indicate polarity. In addition, at different positions along a segment the gradient substance is present at different concentrations as you saw when you put numbers on the gradient in the first question of the last section. Cells could respond to this variable by differentiating in different ways depending upon threshold values of the gradient substance. This idea that cells in a tissue have their relative spatial positions specified by a gradient was called **positional information** by Lewis Wolpert, and the occurrence of thresholds specifying ranges of gradient values to which cells respond by differentiating in genetically prescribed ways was called the interpretation of positional information by the cells in different positions in the gradient.

An experimental example illustrating this idea is provided by the abdominal segments of *Galleria*, the wax-moth. Adult segments have a variety of differentiated cell types arranged in characteristic banded patterns (see Figure 3.38) while the larval cuticle is more or less homogeneous.

Small pieces of cuticle can be cut from various parts of a larval segment and cultured in isolation in the abdominal cavity of a mature caterpillar. If after

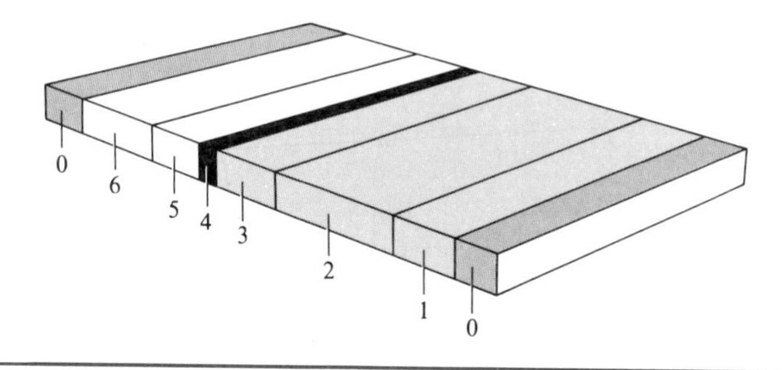

Figure 3.38 Diagrammatic representation of an abdominal segment of the adult *Galleria*. The main area of the segment surface bears three kinds of scale; regions 3, 2 and 1 (grey). Just anterior to this is a thin ridge of tanned cuticles; region 4, (black). The segment margin consists of three types of cuticle; regions 5, 6 and 0. So each segment margin has a posterior and an anterior aspect; region 0 (coloured) and regions 5 and 6 (white), respectively.

metamorphosis of the host the pieces of cuticle are removed from the adult moth, they are found to have developed into adult cuticle according to their presumptive fate, that is, larval cuticle from region 3 develops into adult type 3 cuticle. So the various parts of the segment appear to be determined relative to this operation of cutting and culturing before metamorphosis. However, would cells develop according to their presumptive fate if moved to a different location in a segment when they interact with cells of a different presumptive fate? In *Galleria*, cells from males and females can be distinguished by characteristics of their nuclei, so when presumptive tissue is transplanted between male and female larvae, the cells of the graft and of the host can be distinguished. A small piece of larval cuticle from presumptive region 0 (see Figure 3.38) was transplanted into larval presumptive region 2. In the adult, all the cells of the graft had produced cuticle according to their origin (type 0), while the host cells adjacent to the graft (which would normally differentiate into type 2 cuticle) produced a ring of type 1 scales around the graft. In addition, the orientation of these host type 1 scales had altered. Instead of pointing posteriorly, they pointed towards the graft.

◇ How would you account for these observations in terms of the segmental gradient?

◆ Graft region 0 seems to maintain a low gradient position and so causes inflow of gradient material. A local valley is therefore formed in the gradient landscape. As the polarity of the scale-secreting epidermal cells is determined by the steepest gradient slope, this results in the reorientation of the scales. But notice that the developmental fate of some of the host cells within this dip is also changed. The gradient level within presumptive type 2 cells seems to have been changed by the dip in the landscape to a level corresponding to that of presumptive type 1 cells. This causes presumptive type 2 cells to develop as type 1 cells. So, it appears that it is the position of a cell within the gradient, rather than its position in the segment or its developmental history, that determines its developmental fate.

These examples show how the concept of a gradient of some substance over a region of developing cells that are communicating and interacting with one another can, in principle, account very well for a range of normal and disturbed patterns in organisms. As we have seen, a gradient can account for a number of different aspects of these patterns:

(a) the polarity of pattern elements such as hairs or tubercles, determined by the slope of the gradient

(b) patterns that are organized along level contours of the gradient, without discriminating between different levels

(c) patterns in which different types of cell are produced in different ranges of the concentration gradient, with threshold values defining these ranges.

All of these patterns are based on one dimension of order, determined by the gradient. On a level contour of a gradient, there is no way of knowing where you are relative to some other point on the same contour. The only information available at any point is the height of the contour and the direction of steepest slope (polarity). So this gradient theory is restricted to organization of patterns with only one dimension of order, even though the tissue organized may have two spatial dimensions, such as the insect cuticle. In every example studied, the order is along the anterior–posterior axis of the organism. An extension to two dimensions of order will be considered in Chapter 4.

Summary of Sections 3.6 and 3.7

The gradient hypothesis for the initiation of spatial patterns in developing organisms was introduced in the early years of this century and has become well established, both theoretically and experimentally, as a primary pattern generator in developing organisms. Biochemical reactions involving activation and inhibition, together with diffusion of products, have been shown mathematically to generate spatial gradients. In multicellular organisms, continuity between cells exists in the form of gap junctions which allow small molecules to diffuse from one cell to the next. Different structures can be produced along a concentration gradient if it is assumed that there are threshold points between different concentration ranges of the morphogen.

The spacing pattern of heterocysts in *Anabaena* provides a model which can explain pattern formation with one dimension of order. Heterocyst spacing may be controlled by a gradient of inhibitor. The polarity of the epidermal cells of some insects, as expressed in the orientation of cuticular structures, is locally determined by the steepest slope of a segmental gradient. The maintenance of the gradient involves active participation of the epidermal cells. In *Galleria*, which has different types of scale on the abdominal segment, the type of cuticular structure formed depends on the position of the scale-secreting epidermal cell within the segmental gradient. The orientation of the scales and the sequence of cuticular types in the segment have a common gradient basis.

Question 17 (*Objectives 3.14 and 3.15*) Suppose that a mutant *Anabaena* is found in which heterocysts are separated by about 12 vegetative cells rather than the normal 8. Give possible explanations of this in terms of the gradient model and predict the expected pattern of inhibitor in comparison with the wild type.

Question 18 (*Objectives 3.16–3.18*) Which of the experimental procedures (a)–(g) illustrated in Figure 3.39 would result in a distortion of the ripple pattern in *Rhodnius*? In each case, A marks the site from which the graft was removed, and B its final location. If the graft was rotated, the extent of rotation is indicated.

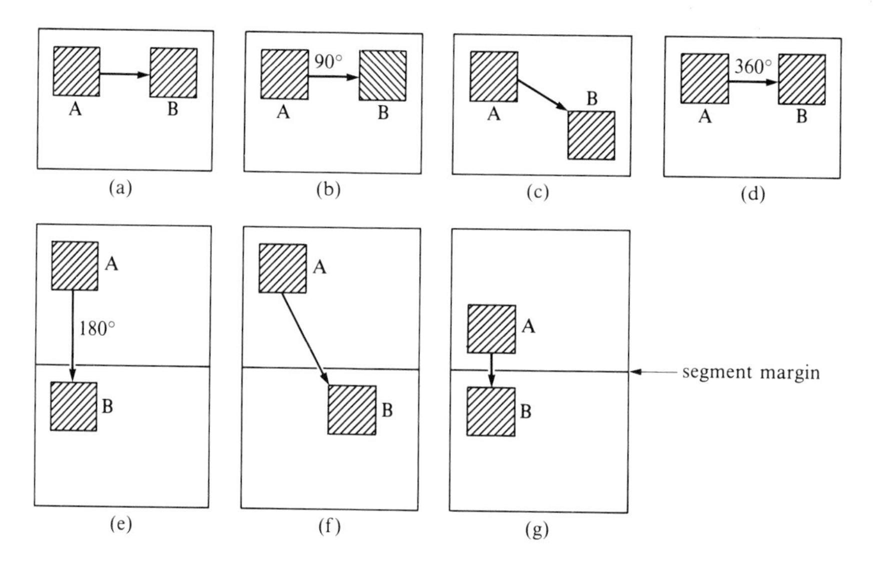

Figure 3.39

OBJECTIVES FOR CHAPTER 3

Now that you have completed Chapter 3 you should be able to:

3.1 Define and use, or recognize definitions and applications of each of the terms printed in **bold** in the text.

3.2 Summarize how fibroblasts and other specified cells move and are oriented *in vitro*. (*Question 1*)

3.3 List the properties of single cells that may be important in morphogenesis. (*Question 1*)

3.4 Write some notes which describe the different pattern of cleavage divisions of yolky and non-yolky eggs. (*Question 3*)

3.5 Briefly summarize the process of gastrulation in amphibian and sea urchin embryos. (*Questions 2, 4, 5, 6 and 8*)

3.6 Explain why cell properties such as adhesiveness and movement are thought to be important in the process of gastrulation. (*Questions 2, 6 and 7*)

3.7 List the differences between gastrulation in 'simple' and 'complex' animals by comparing a sea urchin and an amphibian. (*Questions 2, 6 and 7*)

3.8 Critically evaluate the various theories proposed to account for sorting out. (*Questions 9 and 10*)

3.9 Give examples of the experimental work which shows that differential cell adhesion is involved in morphogenetic processes. (*Question 11*)

3.10 Give examples of the role of cell adhesion molecules (CAMs) in morphogenesis. (*Question 12*)

3.11 Summarize the experimental approaches used in attempts to characterize inducers. (*Questions 13–15*)

3.12 Give examples of the regional specificity of inductive effects. (*Question 15*)

3.13 Summarize how ectoderm–mesoderm interactions provide a model for studying cell induction. (*Question 16*)

3.14 Describe how a continuous gradient can give rise to a pattern consisting of distinct parts. (*Question 17*)

3.15 Summarize the main features of pattern formation in *Anabaena*. (*Question 17*)

3.16 Provide a brief account of experimental evidence for the existence of an insect segmental gradient. (*Question 18*)

3.17 Explain the basic properties of the sand model and describe how the gradient might determine the polarity of cell markers. (*Question 18*)

3.18 Predict and interpret the result of experiments involving the transplantation of pieces of insect cuticle. (*Question 18*)

PATTERN FORMATION AND MORPHOGENESIS

◆ CHAPTER 4 ◆

In this chapter we continue an enquiry introduced in Chapter 3 — are there similar developmental principles underlying the emergence of complex form in different species of organism, despite the great diversity of their morphologies? In Section 3.6 of the last chapter the notion of gradients as primary spatial organizers of pattern formation was developed in relation to a range of organisms and their parts. The conclusion was that a gradient could indeed serve a variety of purposes in generating different types of order such as polarity of pattern elements (e.g. hairs or bristles), orientation of cuticle patterns, or spacing of differentiated cells and their organization along an axis. However, these observations failed to reveal any actual gradient of a particular substance that could act as a control signal. Here we pursue this search in the context of a study of *Drosophila* development. Within the last decade or two, there have been some remarkable achievements in molecular biology, which have made it possible to identify some of the primary gene products involved in morphogenesis. This adds a significant dimension to the study of gradients and pattern formation in developing organisms. It begins to make a link between general morphogenetic principles and the specific genetic activity in individual species.

We will develop this theme further by examining how the simple gradient idea of previous chapters can be extended to two dimensions and applied to the regenerative properties of the imaginal discs which give rise to wings and limbs in *Drosophila*. We will also examine how the gradient idea can be applied to the regeneration of limbs in other insects and in amphibians. This study again takes us to the molecular level of influence, with the identification of a morphogen which has the most striking effects on morphogenesis yet discovered.

4.1 DROSOPHILA AND DEVELOPMENT

The fruit-fly *Drosophila melanogaster* has become perhaps the most intensively studied of all organisms in the past 20 years. This is partly because its development is distinctive, partly because it is easy to rear and has a very short life cycle, and also because it is already the subject of a vast body of genetic studies. In *Drosophila*, there is a huge collection of mutations affecting various developmental events. So a good way of trying to understand genetic influences on development is to study developmental processes in *Drosophila* mutants.

The life cycle of *Drosophila* is shown in Figure 4.1. Let us consider first the developmental events that underlie the transformation of the fertilized egg into the first instar larva. The normal egg (Figure 4.2) has a characteristic shape which allows the axes to be identified: the dorsal surface is slightly

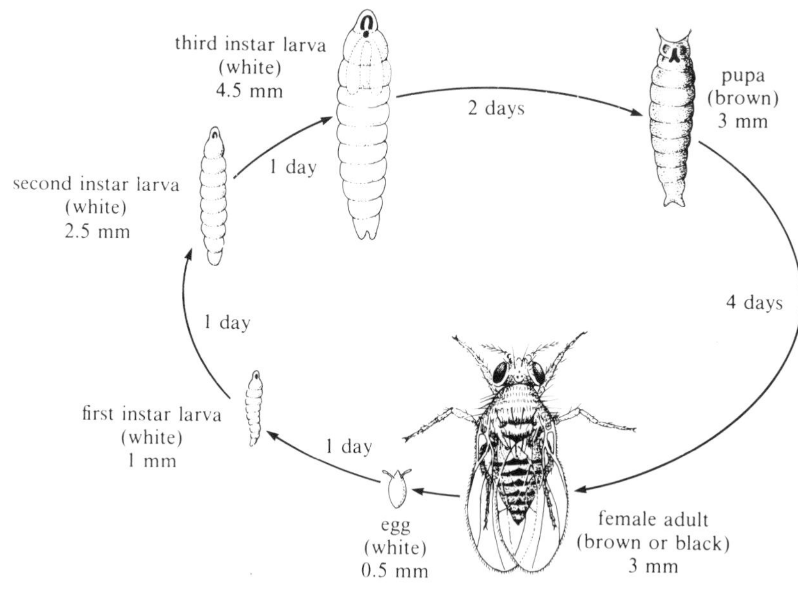

Figure 4.1 Life cycle of *Drosophila*. The durations of the stages are for a temperature of 25°C.

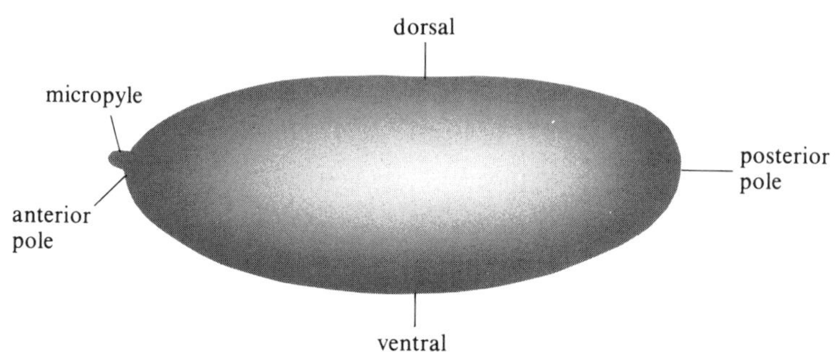

Figure 4.2 The *Drosophila* embryo.

flattened, while the ventral surface is more curved. The anterior pole (the head end) can be recognized by a little protruding structure called the micropyle. Fertilization takes place internally within the female oviduct, so the egg at laying has a diploid nucleus. It is located centrally within the yolky cytoplasm (Figure 4.3a). Towards the periphery, the cytoplasm forms a thin clear layer next to the plasma membrane.

The first stage of development is a rapid series of synchronous mitotic divisions which, at 25°C, occur at intervals of about 9 minutes (Figure 4.3b–j). Unlike the cleavage process in amphibian embryos, there is no cell division during these mitotic cycles. The synchronously dividing nuclei all share the same cellular space which is called a **syncytium**. The early *Drosophila* embryo is therefore a single multinucleate cell. Cytoplasm accumulates around each nucleus. These are fairly evenly distributed throughout the centre of the egg until after the 7th division, when a distinct redistribution occurs (Figure 4.3h). By the end of cycle 8 (Figure 4.3i), the majority of the nuclei have begun to migrate towards the egg surface, leaving a small population behind in the interior of the embryo.

During the 9th mitotic cycle, the nuclei migrate closer to the egg surface and lie within a continuous cytoplasmic layer known as **periplasm** (Figure 4.3j). The **syncytial blastoderm** is thus established, a continuous layer of periplasm containing nuclei around which the plasma membrane has begun to fold. At the posterior pole of the embryo, some nuclei become fully surrounded by a plasma membrane to form the first distinct cells, known as the pole cells.

These are the germ cells, later developing into oocytes (unfertilized eggs) or spermatozoa within the gonads (sex organs), and so are destined to give rise to the next generation.

Synchronous mitoses continue without cell division until the 14th cycle, during which membranes form around the blastoderm nuclei by progressive infolding of the egg membrane to create a continuous layer of individual cells around the periphery of the embryo. This layer, one cell deep, is called the

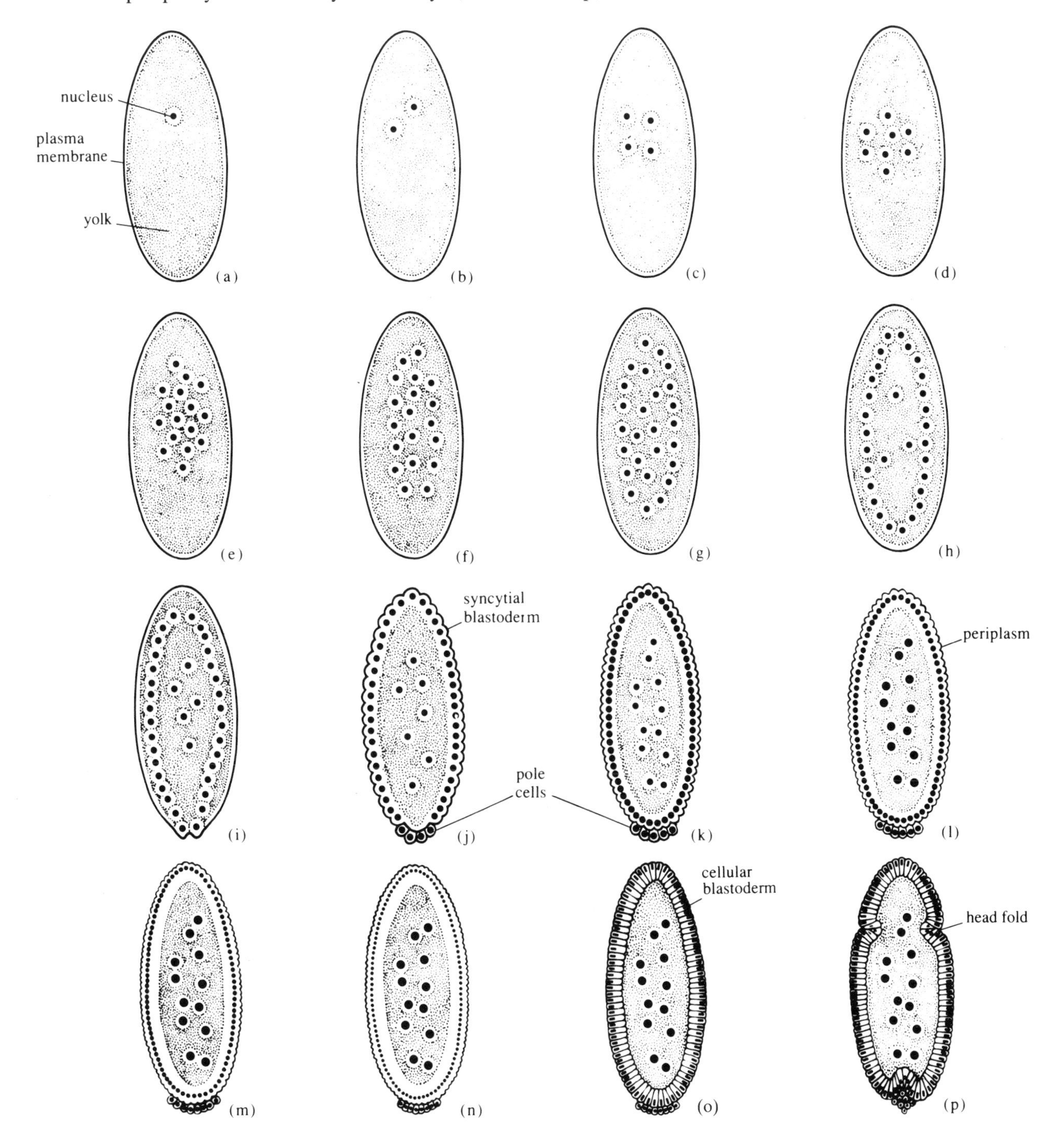

Figure 4.3 Development of the *Drosophila* egg.

cellular blastoderm (Figure 4.3o). The pole cells lie outside it at the posterior pole. When the embryo develops at the standard temperature of 25°C, it takes about $3\frac{1}{2}$ hours to reach this stage.

The cellular blastoderm is a layer of cells around a central space, which in the *Drosophila* embryo is filled with yolk and a few nuclei. The embryo then proceeds to turn itself into a multilayered structure, which is achieved by buckling, folding, expansion and contraction of different regions of the cell layer (see Figure 4.3p).

The form that emerges from these processes is shown in Figure 4.4b, known as the segmented germ band stage, by which time the embryo is 8 hours old. Above it, for comparison, is the cellular blastoderm stage (Figure 4.4a), the time interval between these being about $4\frac{1}{2}$ hours. As a result of the extensive morphogenetic movements that transform one form into the other, the segmented germ band embryo has not only all its surface elements delineated, but also all the internal structures such as gut, nervous system, muscle regions, and so on. The basic plan of the organism, with anterior–posterior (front–rear) axis, dorsal–ventral (top–bottom) axis and a segmented structure, is clearly visible. But, as we shall see, the transformations of shape that result in this form are all consequences of the spatial patterns of many substances, including particular gene products, which are already present at the cellular blastoderm stage though they are invisible to a scanning electron microscope, which shows a uniform layer of cells.

The 8-hour embryo of Figure 4.4b has a further 14 hours of development before the first instar larva hatches, whose ventral surface is shown in Figure 4.4c. Its most distinctive feature is the series of segments, each of which has

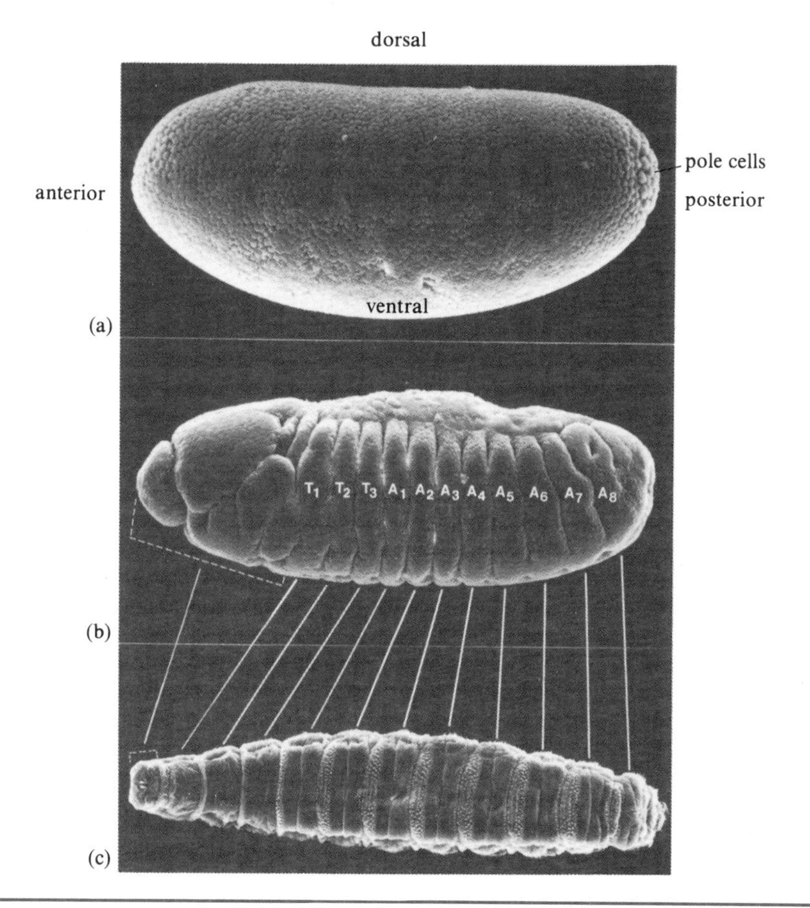

Figure 4.4 Scanning electron micrographs of *Drosophila*. (a) The cellular blastoderm stage. (b) The segmented germ band stage. (c) The first instar larva stage. The lines show the segmental correspondence and the broken line shows the head.

what is called a denticle band made up of little elements called setae. These bands vary in width, those on the thoracic segments (T_1, T_2 and T_3) being narrow, while they get progressively broader on the abdominal segments (A_1 to A_8). The larva feeds and grows, then pupates and goes through the remarkable process of **metamorphosis**, which results in the adult form (Figure 4.5).

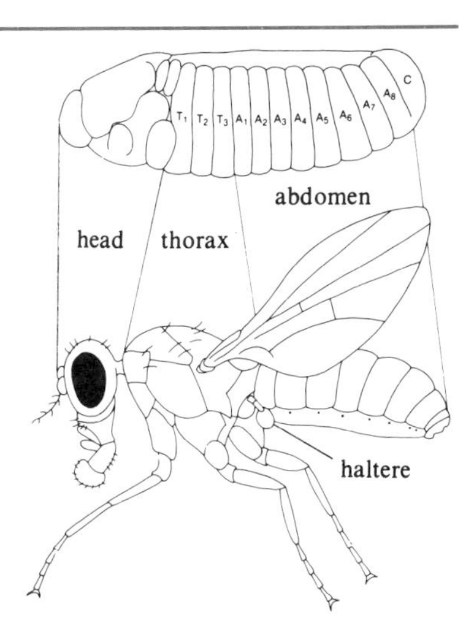

Figure 4.5 An 8 hour embryo and an adult *Drosophila*, showing the correspondence between them.

Drosophila segments have a similar spatial organization to those of other insects such as *Rhodnius*, *Oncopeltus* and *Galleria*, as described in Chapter 3. But because of the very small size *Drosophila* is much more difficult to study by direct surgical methods. However, it is a superb organism for pursuing problems of spatial organization at the level of genes and molecules, as we shall now see.

4.1.1 Formation of a gradient in a *Drosophila* embryo

Mutations in a gene called *bicoid* result in dramatic changes of form in the *Drosophila* larva: the entire anterior half (head and thorax) is missing, replaced by a posterior region (the telson) joined to abdominal segments with a normal telson at the posterior pole. Evidently the product of this gene is involved in the spatial organization of the anterior half of the developing embryo. *Bicoid* belongs to a category of what are called **maternal effect genes**, because it is the activity of the gene in the cells of the maternal ovary that is essential for normal development of progeny.

Fifteen genes of this type have been identified, all playing a role in the specification of the anterior–posterior axis. They have been classified into three groups based on their domains of influence along this axis. The first group, including *bicoid*, is active in organizing the anterior part of the axis. The second group influences the formation of the posterior axial parts. The third group is involved in specifying the extreme ends of the body, the anterior acron (head region) and the posterior telson (abdominal region).

All of these maternal effect genes are active in the maternal ovary. Figure 4.6 shows a normal follicle within an ovary with the growing egg surrounded by ovarian cells. The large nurse cells are known to transfer mRNA of particular kinds to the egg, whose anterior pole is adjacent to the nurse cells. The techniques of molecular genetics now make it possible to study directly the spatial distribution of both the mRNA produced by the *bicoid* gene, and of the protein translation product. As described in Chapter 3, these techniques depend upon procedures for extracting mRNA from embryos, then using this to prepare complementary DNA (cDNA). This is achieved by the enzyme 'reverse transcriptase' which uses an mRNA template to generate DNA whose base sequence is complementary to that of mRNA. The required cDNA is then isolated by means of a suitable probe, and radioactively labelled. This can then be used to identify the distribution of *bicoid* mRNA in eggs and developing embryos by autoradiography. To examine the distribution of *bicoid* translation product, an antibody to the *bicoid* protein is prepared, then covalently bonded to a staining or fluorescent substance that makes it visible.

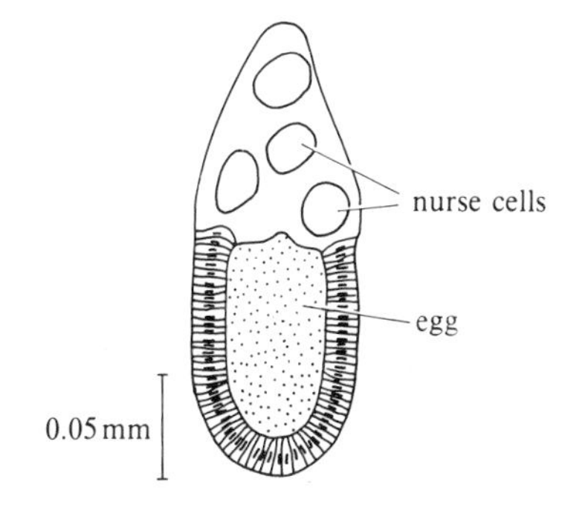

Figure 4.6 Normal follicle and nurse cells.

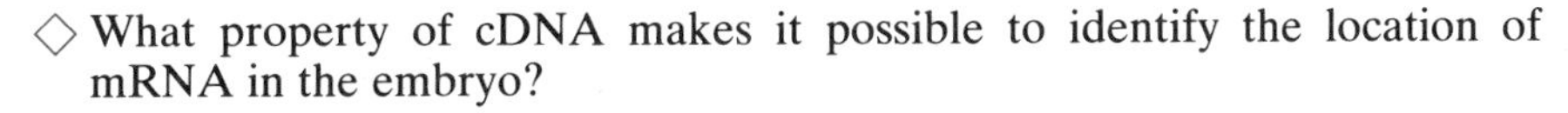

◇ What property of cDNA makes it possible to identify the location of mRNA in the embryo?

◆ Since the cDNA has a base sequence that is complementary to the mRNA, it will bind to the mRNA by complementary base pairing. Autoradiography then makes its location visible.

Using these methods, it was shown that *bicoid* mRNA is present in the anterior region of the oocyte well before fertilization. However, at this stage there is no *bicoid* protein so the mRNA is not being translated. It is only after fertilization and egg deposition that the *bicoid* mRNA is translated within the oocyte. A translation control process is evidently in operation here. The protein then begins to accumulate, with a maximum in the anterior part of the egg where the mRNA remains localized. The *bicoid* protein diffuses from this source and results in a smooth curve that decreases exponentially (Figure 4.7), reaching background levels about two-thirds of the distance from anterior to posterior poles, shown by the dotted horizontal line. These results were obtained by Wolfgang Driever and Christiane Nüsslein-Volhard, working in Germany.

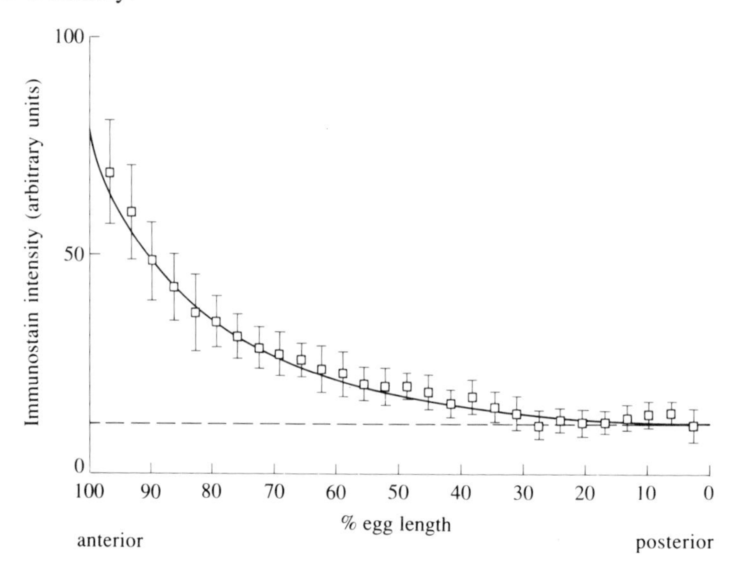

Figure 4.7 *Bicoid* protein distribution.

Here then is an actual gradient of a gene product of the type that was anticipated from previous studies of developing embryos. This gradient of protein persists throughout the stages of early development up to the formation of the cellular blastoderm, after which it gradually decays. The protein is localized primarily within the nuclei, where it plays a regulatory role by activating the expression of another gene, called *hunchback*, which is involved in segmentation. This gene is a member of a group known as gap genes, whose effects we will consider shortly.

The details of the *bicoid* protein gradient can be altered in a number of ways, the simplest being to change the number of copies of the gene within the developing embryo. There is a mutant allele of the gene which produces no detectable protein. Embryos that are homozygous for this allele (i.e. have two identical copies, one on each of the two chromosomes) give the dotted line in Figure 4.7. Such embryos fail to develop head and thorax. By means of genetic crosses, embryos with different numbers of this and the normal *bicoid* allele can be produced.

When the distribution of *bicoid* protein was studied in a heterozygote carrying one copy of the normal allele (bcd^{+}) and one of the mutant (*bcd*), the result was as shown in Figure 4.8a. The gradient is significantly reduced anteriorly, to about half of the normal level measured in the control.

◇ Assume that the *bicoid* protein gradient determines anterior structures (head and thorax) in a direct manner by means of fixed threshold values, so that different concentration ranges of the protein in the normal embryo correspond to different head and thoracic structures. What is your prediction for the expected phenotype of the bcd^+/bcd larva with the *bicoid* protein levels shown in the lower curve of Figure 4.8a?

◆ All the head structures corresponding to concentrations of *bicoid* protein in the normal embryo that are above those reached by the mutant should be missing from the latter. This corresponds to the region of the normal embryo from about 85–100% egg length. (Draw a horizontal line from the top of the mutant bicoid protein gradient to the point of intersection with the normal, and read off the % egg length it corresponds to, which is about 85%.)

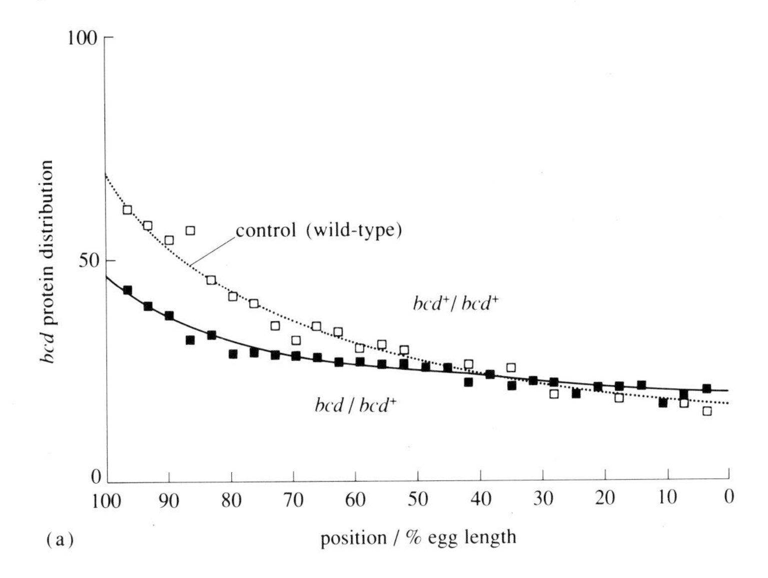

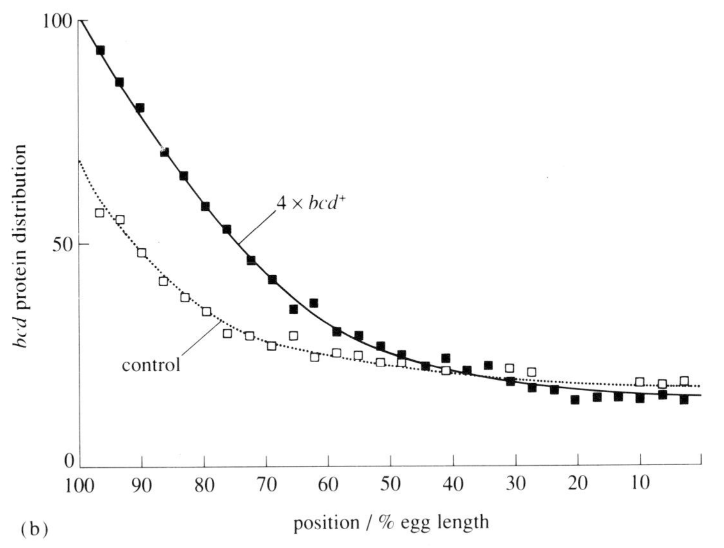

Figure 4.8 Bicoid protein distributions in *Drosophila* mutants. (a) Heterozygote with one mutant (*bcd*) gene and one normal (bcd^+) gene. (b) Embryo with four copies of the bcd^+ gene.

In fact, this is not what happens. The heterozygous embryos with a significantly reduced anterior gradient of *bcd* protein result in perfectly normal hatching larvae. The position of the head fold in these embryos (see Figure 4.3p) is shifted forward by about 8%. So there is some phenotypic effect, but it is significantly less than the change in concentration of the protein. Evidently the embryo compensates for the modifications during subsequent development, becoming fully normal; that is, it regulates.

What about the opposite effect — increasing *bcd* protein level by increasing the number of copies of bcd^+ above the normal two? Figure 4.8b shows the protein gradient in embryos with four copies, which arise if the chromosome region carrying the bcd^+ gene is duplicated during meiosis. The whole anterior region now has significantly increased levels. Again, if there were fixed threshold values determining the range of *bicoid* protein concentrations corresponding to particular structures, then we would expect that the most anterior head parts, corresponding to the highest region of the normal gradient, would occur at about 88% of the mutant egg length. The very high values towards the anterior pole, never reached in the normal embryo, might be expected to produce either nothing, or something strange.

With the increased *bcd* protein gradient the head fold in the embryo shifts posteriorly by about 9%. But once again the hatched embryo is perfectly normal. Evidently there are compensatory processes that result in regulation, despite the substantially elevated level of gene product.

These results warn us not to impose too simple an interpretation on the relationship between gradients and pattern formation. On the one hand it is clear that gradients of gene products, and gradients of ions such as calcium, play a significant role in the generation of spatially organized form in developing organisms. However it is not clear precisely how these substances act in the determination of the transitions from one structure to another — the transition from head to thorax in *Drosophila* or from cap to stalk in *Acetabularia*. In influencing these morphogenetic events, the gradients of gene products must interact with many other variables. And the end result can be relatively insensitive to significant changes in any one variable. Rather than simple concentrations of gene products, it is perhaps the ratios (or some more complex function) of gene products, key metabolites and ions that determine threshold points for morphogenetic events.

4.1.2 The dorsal–ventral pattern

An examination of gene influences on the dorsal–ventral axis of the *Drosophila* embryo provides further insight into the complexity and the subtlety of the processes involved. Eleven genes have been identified by the effects that their mutant alleles have in disturbing this axis, nine of which act maternally. The most dramatic of these results from the failure of a gene called *dorsal*. Embryos from eggs produced by homozygous *dorsal* mothers produce none of the characteristic ventral structures seen in Figure 4.4. They develop into a tube that is entirely surrounded by dorsal elements, completely lacking any sign of dorsal–ventral organization.

◇ From the example of the spatial pattern of *bicoid* and its influence on the establishment of the anterior–posterior axis, what would you expect to be the spatial distribution of *dorsal* transcripts in the early embryo, and what pattern of *dorsal* protein would you expect to find?

◆ Since ventral elements fail to form in a *dorsal* mutant, it is to be expected that *dorsal* transcripts would be found in the ventral region of the embryo, giving rise to a gradient of *dorsal* protein with a maximum ventrally and a minimum dorsally.

The surprising result is that *dorsal* mRNA was found to be uniformly distributed throughout the early embryo. Nevertheless, *dorsal* protein forms a gradient, as expected, with a maximum ventrally and a minimum dorsally. Of the other genes involved in establishing the dorsal–ventral axis, none has transcripts that are non-uniformly distributed in the egg or the embryo, so none of them is the initiator of the gradient of *dorsal*. Therefore we are left with two possibilities. Either the crucial gene has not yet been identified or there is another variable, other than a gene product, that is involved in triggering the asymmetric translation of the uniformly distributed *dorsal* mRNA to generate the dorsal protein gradient. There is some evidence that calcium, electrical currents and the state of the cytoskeleton might be involved in this. These observations remind us not to lose sight of the general physiological context within which genes act, and the capacity of the morphogenetic field to initiate patterns of change leading to spatial organization in which gene activity is involved as an amplifier and a stabilizer of spontaneous pattern-forming processes.

4.1.3 Making segments

As seen in Figure 4.4b, by 8 hours of development the *Drosophila* embryo has become spatially organized into a segmented structure which bears a direct correspondence with the first instar larval form (Figure 4.4c). How this comes about is the focus of intense research activity in laboratories around the world, for the molecular data that is emerging on the patterns of gene activity is regarded by many as the Rosetta stone whose deciphering will give the key to the whole of embryonic development. A considerable number of genes which are involved in this process have been identified. The spatial distributions of their products have been measured, both in normal and in many mutant embryos, by the types of molecular technique used in the study of *bicoid*.

What became apparent from the first systematic study of the genes involved, by Christiane Nüsslein-Volhard and Eric Wieschaus in 1980, is that this process, like all aspects of embryonic development, is hierarchically organized. The first stage of this process involves the establishment of the anterior–posterior axis, which is formed as the gradients of maternal effect gene products emerge. Some genes, such as *bicoid*, participate in the organization of the anterior half of the embryo (head and thoracic segments). Other genes participate in the organization of the posterior half. A primary example is *oskar*, whose mutant phenotypes lack posterior structures (abdominal segments). You can see in Figure 4.4b that the boundary between thoracic segment 3 (T3) and abdominal segment 1 (A_1) is just about the middle of the embryo. This divides the domains of influence of the anterior and posterior genes like *bicoid* and *oskar* which each act over approximately half of the embryo, the equivalent of about eight segments. However the segments anterior to T_1 are not so clearly defined as those posterior to it.

Another group of genes involved in segmentation goes under the name of **gap genes** because their mutant alleles result in extensive gaps in the larval phenotype. An example is shown in Figure 4.9, which is the mutant called *Krüppel*. Unlike the maternal genes such as *bicoid* and *oskar*, these genes actively transcribe mRNA in the embryo. When the gap gene transcripts, at the stages described in Figure 4.3 m and n, were studied by the use of molecular probes, they were observed to extend in primary bands over the presumptive domains of about four segments. Figure 4.10 shows the results for *Krüppel*. However, the phenotypic effect is more extensive than four segments. Each gene in this group is active over a distinct region of the segmentation domain. The translation products (proteins) have more extensive distributions than the mRNAs but cover smaller domains than a maternal gene product such as *bicoid* protein.

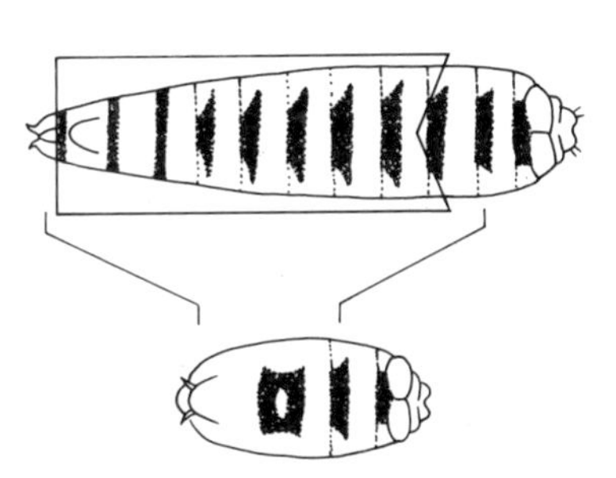

Figure 4.9 Schematic drawing of *Krüppel* mutant.

The next category of segmentation is the **pair-rule genes**. Mutants in this group result in pattern deletions that are regularly spaced over every pair of

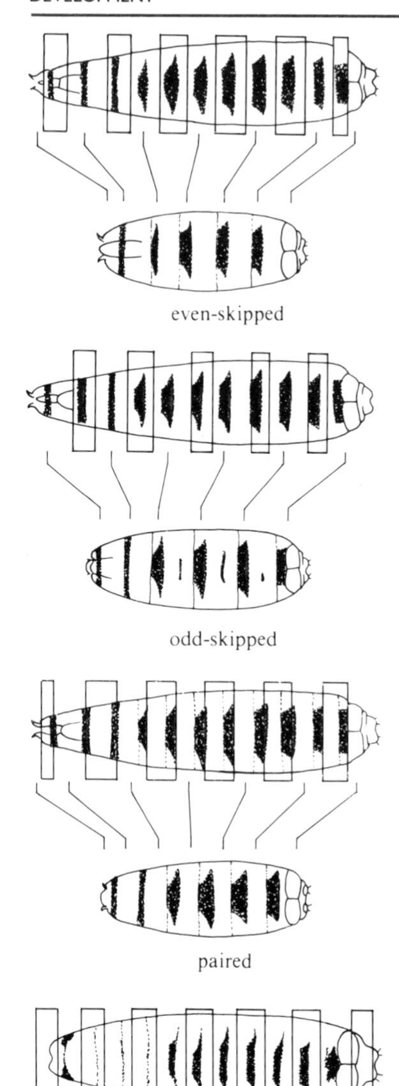

Figure 4.11 Pair-rule genetic mutations.

Figure 4.10 *In situ* localization of *Krüppel* transcripts.

segments; hence their name. Four different examples are shown in Figure 4.11. In each instance the regularly repeated deletions involve about one segment's worth of cuticle and associated structure from every pair. The different members of the group are distinguished by the precise positions of the deletions. It is to be expected then, that the products of those genes are distributed periodically in the positions of the deleted regions, failure to produce the product in a mutant resulting in a failure of the corresponding cuticle region to be differentiated. This turned out to be the case. Figure 4.12 shows the bands of mRNA produced by the gene *fushi tarazu*. Other pair-rule gene products showed similar striped patterns, but differently positioned in accordance with the deletion patterns.

◇ Can you predict what the pattern of protein distribution is from genes in the next group involved in segmentation?

◆ The protein distributions appear to be decreasing systematically in extent and increasing in number. So the next group of genes involved in segmentation is likely to have proteins distributed in bands within every segment.

The fourth group of genes involved in the specification of segments goes under the name of **segment polarity genes**. As the name suggests, these are involved in determining detailed features of segmental patterns, such as the polarity of the setae that make up the denticle bands. Mutant phenotypes include larvae with denticle band patterns which have segmental regions of reversed polarity, resulting in domains of mirror-symmetry, as shown in

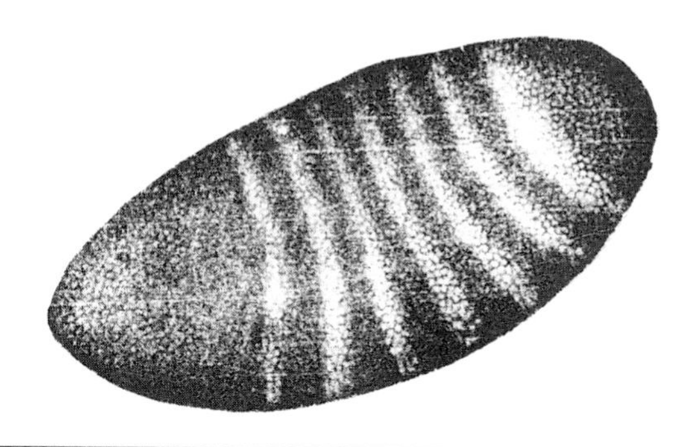
Figure 4.12 The mRNA bands produced by the gene *fushi tarazu*.

Figure 4.13 for the mutant *gooseberry*. As already anticipated, the transcripts and the protein products of these genes are distributed in bands in characteristic positions within each segment. They arise shortly after the pair-rule gene products have assumed their characteristic two-segment distributions.

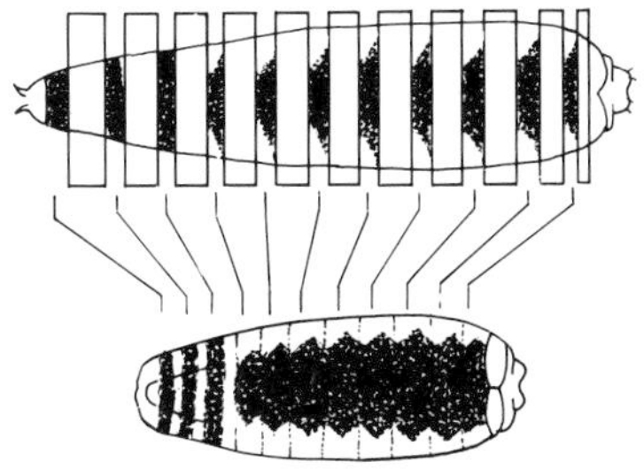

Figure 4.13 The segment polarity genetic mutation *gooseberry*.

What these studies have revealed about gene activity is a detailed confirmation of the general deduction that was drawn from classical embryological studies and described in Chapter 1. Development is a process that is organized hierarchically. Large-scale spatial patterns are established first and more local detail emerges progressively within the global order. The sequence of transcripts in the four major categories of segmentation genes shows this very clearly. Large-scale pattern over the anterior–posterior axis is established by the protein products of the maternal effect genes, which are the first to show a stable distribution in the early embryo. The gap gene products then exert their effects over more restricted domains, followed by the pair-rule and segment polarity transcript and protein patterns acting over successively narrower regions of the anterior–posterior axis.

4.1.4 Organizing the hierarchy

How do these spatial patterns of gene activity come about? We have seen that the gradient of *bcd* protein can be understood as a result of a restricted localization of *bcd* transcripts and diffusion of translated protein from this restricted source. The *bcd* transcripts are transferred from the ovarian nurse cells into the end of the developing egg that becomes the anterior pole (see Figure 4.6). There they remain localized by binding to components of the cytoskeleton. There are actually mutations that result in a loss of this binding of *bcd* transcripts, which then diffuse through the egg with the consequence that no gradient of *bcd* protein is produced and the anterior–posterior axis fails to develop. However, the pattern of *Krüppel* transcripts arises from the activity of zygotic nuclei (those arising in the embryo by mitotic division of the diploid nucleus produced in the fertilized egg, the zygote). How is their location regulated? It has been shown in a series of detailed studies of mutant embryos, particularly by Herbert Jäckle and his colleagues working in Germany, that *bcd* protein is directly involved in this by acting as a repressor of *Krüppel* transcription. By comparing Figures 4.7 and 4.10, you can see that *Krüppel* transcripts appear where the *bcd* protein gradient has fallen to a low value.

◇ What is your prediction about the distribution of *Krüppel* transcripts in embryos homozygous for *bcd*$^-$?

◆ *Krüppel* transcripts will be present throughout the anterior part of the embryo.

Repression of *Krüppel* by *bcd* accounts for the absence of the former gene transcripts from the anterior part of a normal embryo. But what restricts its expression posteriorly? The maternal gene *oskar* was mentioned in Section 4.1.3 as a member of the set complementary to *bicoid*, acting in the organization of the posterior part of the embryo. It is the protein product of a gene *manos* with a product distribution similar to that of *oskar*, that represses *Krüppel* posteriorly, restricting its transcripts to a central band. Other members of the gap gene set have transcripts that are expressed in different regions of the embryo. One of these, with a region of transcription anterior to *Krüppel*, is called *hunchback*. It turns out that *bcd* protein acts as an inducer of this gene. So the role of the maternal genes is crucial to the spatial patterning of gap gene activities.

The gap genes in turn affect the spatial patterns of expression of the pair-rules, but in this case the interactions are more complex and not yet fully understood. The problem is how three separate domains of primary gap gene products (the third being call *knirps*), with two slightly overlapping regions of protein expression, can generate a pattern of seven stripes in pair-rule gene activity. It appears that there are either additional influences or subtleties of interaction that have not yet been identified, which are involved in this patterning process.

The pair-rule genes interact with one another by both repressive and inductive effects, resulting in stable patterns in which different gene products are distributed in spatially periodic patterns over the two-segment intervals in which they are expressed. And finally, the pair-rule pattern influences the domains of expression of the segment polarity genes, which are systematically ordered in their activity over domains corresponding to single segments. So it emerges that activities of the whole hierarchy of segmentation genes is organized by interactions within and between members of the different major groups, together with the action of spatial organizing influences that are yet to be understood. These are likely to include basic variables, such as ions, metabolites and the cytoskeleton, that define the context within which gene action occurs.

This description provides only the roughest outline of the segmentation process in *Drosophila* and the genes involved in it. It is not yet clear exactly how to interpret in molecular terms the properties represented by the segmental gradients used in Figure 3.35c of Chapter 3 to describe polarized patterns in insect segments. We have seen that the segment polarity genes, active in every segment, are involved in determining the orientation of the setae within the denticle bands. However, they are not distributed in gradients of the saw-tooth type. This, it must be remembered, is a formal description of segmental patterns that provides a way of interpreting their response to perturbation. In the segments themselves there may be no single substance distributed in this pattern. The ordered sequence of segmental polarity gene products within each segment define polarity and pattern by some means that is not yet fully understood. However, it remains a possibility that some other substance, not necessarily a direct product of a gene, does have a spatial distribution of saw-tooth character in each segment at some stage of the segmentation process. The options remain open.

4.2 METAMORPHOSIS AND IMAGINAL DISCS

How is the dramatic transformation of shape from pupa to adult fly (Figure 4.1) achieved? The general correspondence between the segmented structure of the 8-hour germ band embryo and the adult is shown in Figure 4.5, but it is clear that there are very extensive changes of form that occur during the process of metamorphosis. The adult structures come not from the whole segment to which they correspond, but from special groups of cells that are present in the larva and which form structures called **imaginal discs**. All the adult epidermal derivatives such as wings, legs, head, eyes, genitalia, and so on, derive from specific imaginal discs (Figure 4.14). There are, for example, three pairs of leg discs, one pair in each of the three thoracic segments, which give rise to the six legs of the adult fly. The discs are relatively simple in form but give rise to complex structures, and so they have been extensively studied for insight into pattern-forming processes.

The cells that will produce the imaginal discs become committed to this process soon after cellular blastoderm formation (Figure 4.3o). By the beginning of larval development, each leg disc consists of a small nest of 20 or

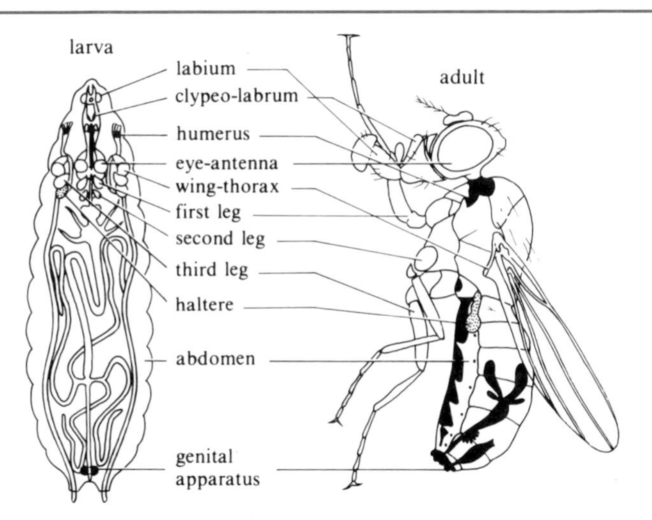

Figure 4.14 Larval organization and the location of the different discs, with the corresponding adult derivatives.

so cells which will increase in number to about 30 000 by the onset of metamorphosis. Just as in the segmentation process, the development of a leg or any other disc proceeds hierarchically. First there is a general commitment to leg and then the progressive development of the leg parts and their detailed structure (Figure 4.15). The detail identifies the whole as one of the three thoracic appendages which, though similar to one another, have distinctive features.

As the 4-day period of larval growth proceeds, the disc grows in size and takes on its characteristic shape. It is only during metamorphosis, however, that the actual leg structure begins to become apparent. How are the cells determined to become specific parts of the adult leg? Why do they differentiate to give certain structures?

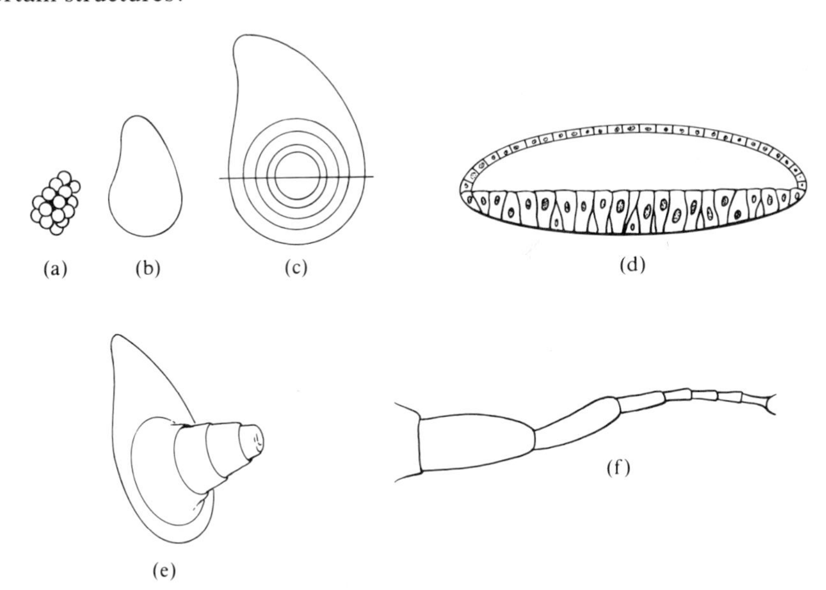

Figure 4.15 States in leg disc development. (a) Nest of cells at the blastoderm stage. (b) Disc primordium. (c) Mature disc. (d) The end of larval development. (e) Evagination of leg structures. (f) Adult leg.

4.2.1 Imaginal disc transplantation

Like other insects, *Drosophila* has an open circulatory system with blood filling the primary body cavity. In the 1930s, it was discovered that it was possible to remove imaginal discs from the larva and to place them either in the body cavities of other larvae or else in the abdomens of adult flies.

Because of the nutritious blood acting as a kind of tissue culture medium, such transplanted discs will grow quite successfully.

Before the discs differentiate into adult structures, they must be stimulated by the hormones normally present in the blood at metamorphosis. Discs transplanted from mature larvae into adult abdomens undergo cell proliferation; that is, the cells grow and divide but they will not differentiate. If the discs are transplanted back to mature larvae which are then allowed to metamorphose, the transplanted discs also metamorphose, and the adult structures produced may be dissected out of the body cavities of the host flies.

4.2.2 Commitment of discs

Transplants can be used to investigate both the time and the spatial pattern of disc determination. If discs from immature larvae of various ages are transplanted into larvae that are about to metamorphose, then the ability of discs at particular ages to differentiate, or their commitment to a particular pathway of development, can be studied. The mature larva immediately metamorphoses, making the implanted immature disc metamorphose along with it. Experiments showed that young discs transplanted at the mid-larval stage into metamorphosing larvae form adult tissues that have limited features of the normal adult tissues formed from the discs. Progressively older discs can form more and more of the normal adult structures on transplantation. So it seems that the determination of discs proceeds in stages.

The spatial pattern of determination can be worked out by transplanting fragments of discs at a stage when these parts have become committed to forming adult structures. After metamorphosis, only the adult structures derived from these parts of the disc will be formed. **Fate maps** (also called presumptive maps) have been constructed in this way for all the various discs, one example of which is shown in Figure 4.16.

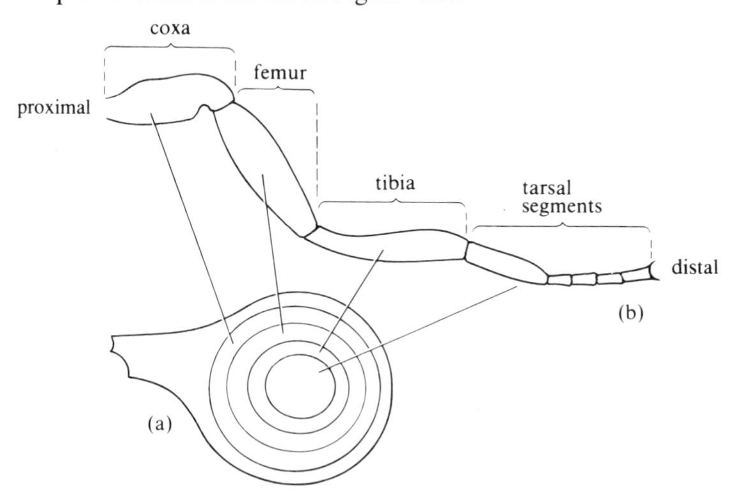

Figure 4.16 (a) A fate map of the first leg disc in *Drosophila*. (b) Corresponding segments of the adult leg.

4.2.3 Regeneration and duplication of disc fragments

We have seen what happens if disc fragments are made to metamorphose earlier than normal. What happens if they are allowed to go through extra cell divisions before metamorphosis? To find out, some experiments were done involving the transplantation of disc fragments into adult abdomens, where extra cell divisions occur. Each disc is cut into two halves, and after further culture transferred back to larvae and allowed to metamorphose. From fate

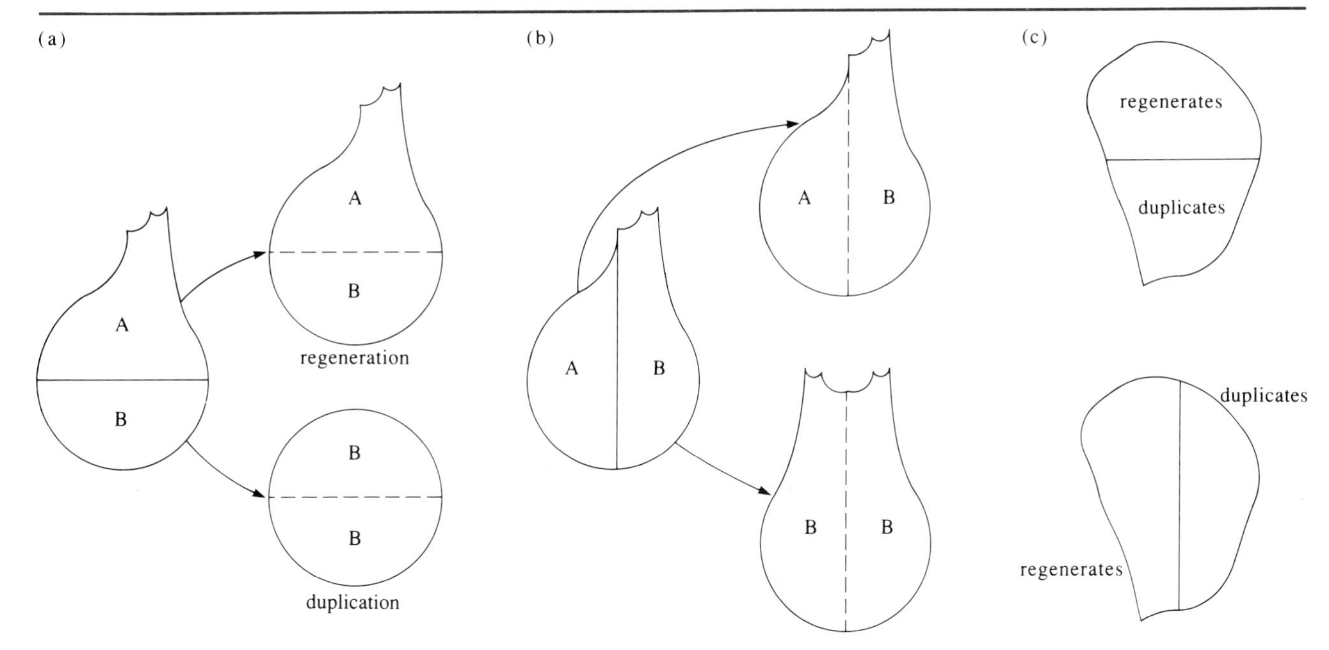

Figure 4.17 Disc regeneration experiments. (a) Leg disc. (b) Leg disc. (c) Wing disc.

maps, we know what structures should be formed by a particular disc fragment. What happens when extra time is spent in the adult abdomen, resulting in extra growth of the fragment?

After metamorphosis of the larva there are two possible results, depending on which of the two disc fragments is studied. One fragment regenerates so that the removed part is restored, giving a complete disc that produces all of the adult structures. The complementary fragment produces a mirror image duplicate of itself. These experiments were carried out extensively by Edgar Schubiger in 1971 working in the United States, and he noticed the following points. The upper part of a bisected leg disc (which normally forms the dorsal side of the adult leg) can regenerate the lower part (ventral in the adult). However, the complementary lower part of the disc undergoes duplication (Figure 4.17a). Bisection of the leg disc in the other direction again gives one part that regenerates and the complementary part that duplicates (Figure 4.17b).

Similar results were obtained in experiments with the wing disc by Peter Bryant in 1971, also in the United States. When the discs were cut in half by straight lines, one half regenerated while the other half duplicated, as shown in Figure 4.17c. We shall return to a discussion of these and related observations in Section 4.2.5.

4.2.4 Transdetermination

Serial culture is a technique that allows further interesting aspects of commitment and determination of imaginal discs to be revealed. It involves the transfer of discs from an older adult host (see Figure 4.18), making it possible to extend the period of culture beyond the lifetime of an individual fly (normally 2 weeks). If this transfer is carried out for several generations, a new phenomenon emerges when the discs are finally transferred back to larvae to allow metamorphosis to take place. It is found that in such instances a few discs have undergone **transdetermination**, a switching from one determined state to another. These changes may be very striking. For example after, say, 20 transfers from one adult abdomen to another, what was a wing disc before transfer may develop into eye or leg tissue.

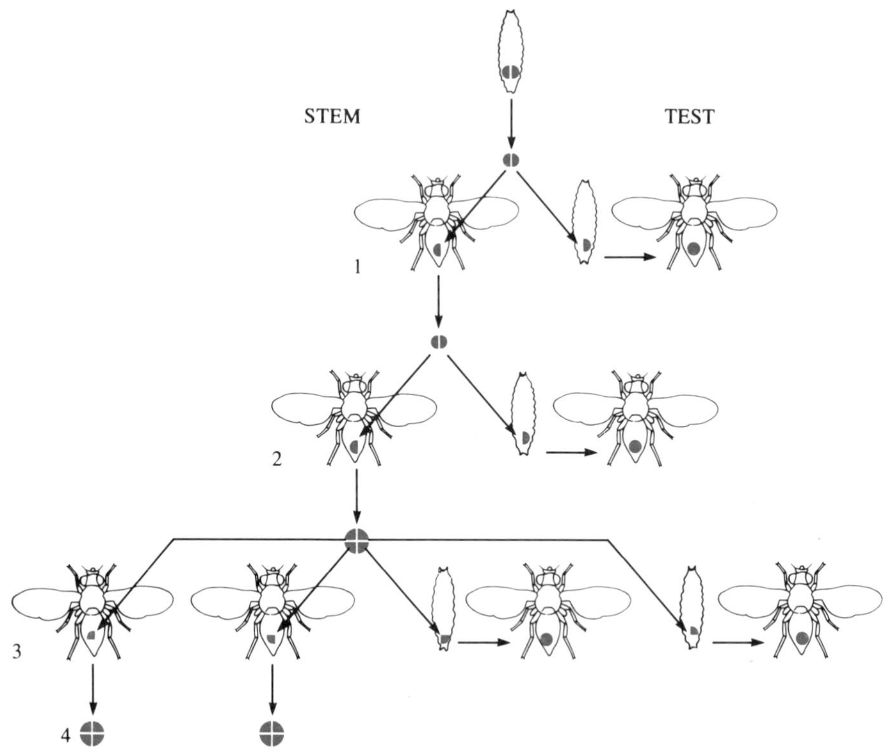

Figure 4.18 Serial transfer of cells from a single imaginal disc. A bit of genital disc from a larva is divided between an adult host (left) and a larva which is allowed to mature so that the condition of the transferred cells can be checked (right). The process is repeated (2) by retransplanting of the disc cells. By the next transplant (3) the disc cells have increased sufficiently to provide two transplants for testing and two for control. In the next generation, (4) the number quadruples. Disc cells have survived and multiplied in this way for more than 150 transfer generations.

Not all transformations, however, are possible. Figure 4.19 shows those that have been observed. The frequencies of these transdeterminations differ from each other. Thus the states of commitment are related to one another rather like metabolites in a complex metabolic pathway — each one has a particular 'energy' level and there are thresholds that separate the states, like hills between valleys in a landscape.

Those results illustrate again the relative nature of the property of determination. Under the perturbation of bisection and growth of the fragments in an adult abdomen, discs remain committed to the formation of structures of the type they would normally produce, whether eye or leg or wing, despite spatial

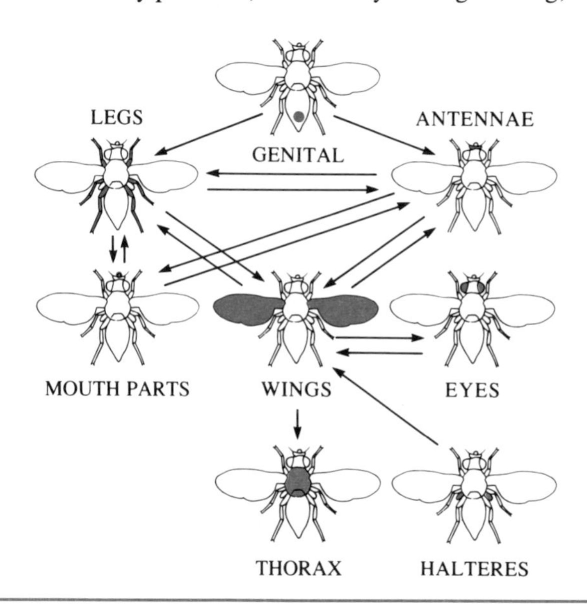

Figure 4.19 Transdetermination sequences for seven types of imaginal disc cell. Genital cells, for example, may only change into leg or antenna cells, whereas leg and antenna cells may become mouth part or wing cells. In most instances, the final transdetermination is from wing to thorax cell. The change to thorax appears to be irreversible.

reorganization as in duplicated fragments. They are therefore 'determined' in relation to this treatment. However, the stimulus of serial transfer reveals that one disc type such as leg has the potential to transform into mouth parts, wing, or antenna, and so to alter what is by other criteria a determined state: hence the term transdetermination. The causes of these transformations are not understood, but they tell us something interesting about the relative stability of the different disc states. Thorax is evidently the most stable, while wing is more stable than haltere.

◇ Are transdeterminations restricted to structures that are adjacent to one another in the normal fly?

◆ No. Genitalia and antennae are at opposite ends of the organism, while genitalia and legs are also spatially separate (see Figure 4.19). Yet the discs can transform between these states.

4.2.5 Gradients in imaginal discs

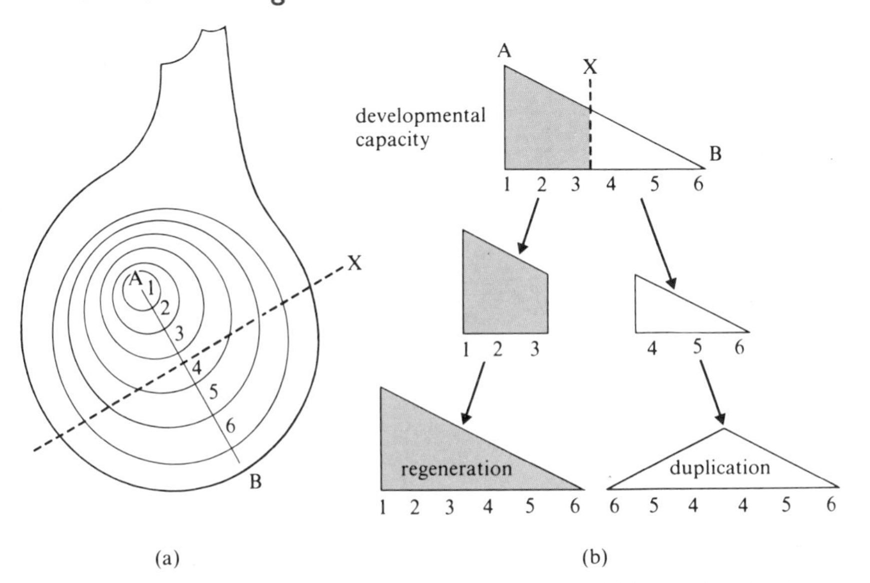

Figure 4.20 Gradients of developmental capacity in *Drosophila* imaginal discs. The high point of the gradient is A. (a) A cut is made through the disc along X. (b) Subsequent behaviour of the disc fragments after adult culture. Note that one half regenerates while the other half can only duplicate itself. (The contour lines in (a) are of developmental capacity and do not represent fate map areas.)

How can the regeneration and duplication results of bisected and cultured discs be explained? The disc fragments heal in a characteristic manner, the cut surfaces closing by a kind of zipper action, followed by the production of new cells by division in the region of the closed wound. These new cells then undergo differentiation and produce the structures characteristic of the disc in either a regenerated or a duplicated pattern. A possible explanation for this, following similar reasoning to that used in explaining the segmental patterns in *Rhodnius*, *Oncopeltus* and *Galleria*, was suggested by Peter Bryant. He proposed that there is a gradient of developmental capacity in the disc with a maximum at position A (Figure 4.20a), falling off radially in all directions from this point, as shown in Figure 4.20b. Disc regions with higher values of developmental capacity can generate ones with lower values, but not vice versa.

◇ Suppose that a leg disc is cut into two fragments by the dotted lines shown in Figure 4.21. Which part will regenerate and which duplicate?

◆ The smaller part will regenerate the whole disc, since it contains the high point of developmental capacity, while the larger fragment will duplicate. This result was obtained by Schubiger.

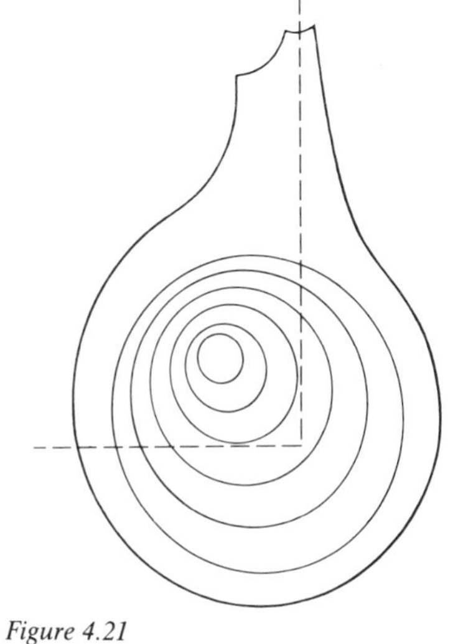

Figure 4.21

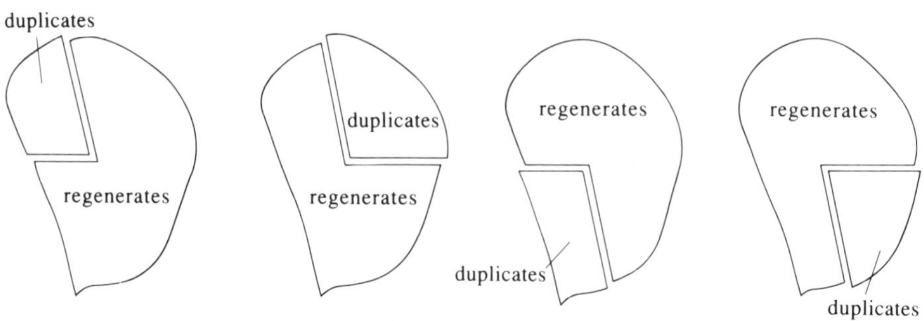

Figure 4.22 Regeneration experiments with imaginal wing discs.

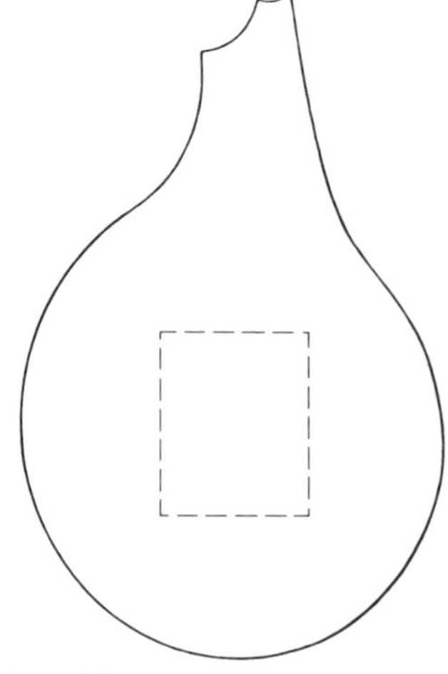

Figure 4.23

This gradient model accounts for many of the observations on regeneration and duplication of disc fragments. However, a further series of results obtained by Bryant with wing discs suggested that there was something fundamentally wrong with his own model. He cut the wing disc into two pieces in the manner shown in Figure 4.22 with the results indicated. No matter where he located the cuts, the larger fragment always regenerated. There seemed to be no point in the disc where a high point of developmental capacity could be located that would be entirely contained within the larger regenerated fragment. Then Bryant performed the coup de grace: he cut the disc into two pieces by isolating the central region from the periphery, as shown in Figure 4.23.

◇ What is your prediction about the outcome of such an experiment?

◆ The central fragment should regenerate since it contains the high point of the gradient, and the peripheral fragment should duplicate.

But the central fragment duplicated while the periphery fragment regenerated! So the idea of a single gradient of developmental capacity in the disc had to be abandoned.

It is always a good strategy to push an idea to its limits in order to see just how much it can explain, and precisely where it breaks down. We have done this with the concept of a gradient as the spatial pattern that gives order to morphogenetic fields, and it is remarkable how much it can provide by way of explanations of simple patterns. In an organism such as *Anabaena*, which consists of a filament of cells, a gradient can explain the uniform spacing of the heterocysts. Applied to patterns in insect segments, which are two-dimensional sheets but have only one dimension of order (anterior–posterior), the gradient concept can be used to explain polarity (the orientation of hairs) in terms of slope, ripple patterns in terms of contour lines, and regions of cell differentiation in terms of concentration ranges of some hypothetical gradient substance, separated by threshold values. The reality of an actual gradient was then revealed in this chapter by the distribution of a gene product in *Drosophila* (*bicoid* protein), necessary for the normal organization of the anterior–posterior axis. Many other genes are also involved in organizing this axis, and their patterns of distribution reveal at the level of gene activity the hierarchical nature of the spatial patterning process in developing organisms. Extending the analysis to imaginal discs, generators of adult morphology, it looked as if the same idea of a gradient could be used to explain the patterns of regeneration and duplication observed in discs that had been cut in two and allowed to grow in adult abdomens. However, this is where the idea of a single morphogenetic gradient fails. So in the next section we shall see how to go further with an extension of this productive idea.

Summary of Sections 4.1 and 4.2

The *Drosophila* embryo develops by a distinctive process into a segmented body pattern which is characteristic of insects. First, a multinucleate syncytium is formed, within which synchronous mitoses take place, but initially no cell division. Nuclei migrate to the egg surface to form the syncytial blastoderm. Then the egg membrane folds round the nuclei, forming a layer of single cells, the cellular blastoderm. This is followed by the morphogenetic movements that lead to the segmented larva.

Many genes are involved in segmentation but they fall into four hierarchically related categories called maternal effect, gap, pair-rule, and segment polarity genes. The spatial patterns of the products of these genes can be studied using molecular probes. The products of these genes are distributed over progressively smaller domains — roughly half the embryo for some maternal genes, a quarter to a third for gap genes, double segments for pair-rule genes and single segments for segment polarity genes. Interactions between the different categories of gene product are involved in generating those spatial patterns.

The protein product of the maternal effect gene *bicoid* is distributed as a gradient over the anterior half of the embryo, resulting from anteriorly localized mRNA. The number of normal copies of the gene can be altered, resulting in changes in the protein product concentration gradient. But even with a four-fold change in concentration gradient, the resulting larvae are all normal. Therefore embryonic patterns are not determined simply by gene product concentration. They seem to be determined by some more complex function of gene products and other variables.

Nine maternal genes influence the formation of the dorsal–ventral axis. The protein product of *dorsal*, a major gene in this group, is distributed in a gradient with a maximum ventrally but the mRNA is uniformly distributed. The initiator of the gradient is unknown.

The spatial patterns of segmentation gene activity are highly organized, each having distinctive distributions over characteristic domain sizes. These patterns are dependent upon inductive and repressive actions of gene products of one segmentation group acting on the next level of the hierarchy. There are also interactions between the genes within each category, via their products, which stabilize the relative positioning of the domains of different gene activity.

The major body parts of an adult *Drosophila* arise from larval tissues called imaginal discs, from which the legs, wings, eyes, genitalia, etc. develop and emerge during metamorphosis. These discs can be dissected from larvae and grown in adult abdomens, then transplanted back to larvae where they undergo metamorphosis with the hosts. Studies of discs and disc fragments by this means allows the construction of fate maps and gives information about the capacity of discs to regenerate missing parts. Results show that, in general, if a disc is cut into two parts, one part regenerates the whole structure while the other duplicates, resulting in a mirror-symmetrical pattern.

If whole discs are cultured for prolonged periods (several weeks) by serial transfer from one host to another, they may change their fate so that, for example, leg becomes wing or genital becomes antenna, a process called transdetermination. This gives useful information about the stability of committed tissue.

Some of the regeneration and duplication behaviour of bisected discs can be explained by the existence of a gradient of developmental capacity in the discs, running from a central point to the periphery. However, a crucial set of experiments showed that this one-dimensional model of spatial patterning is unable to account for the phenomena of regeneration and duplication.

Question 1 (*Objectives 4.2–4.6*) Are the following statements true or false?

(a) The early stage of *Drosophila* development is essentially the same as that of the newt (Figures 1.26–1.30).

(b) The nuclei undergo synchronous mitotic divisions in the syncytial blastoderm stage of *Drosophila* development.

(c) Changes of shape in the embryo (i.e. morphogenetic movements) occur only after the formation of the cellular blastoderm.

(d) There is a precise relationship between the level of *bcd* protein and the anterior structures of the embryo that are subsequently produced, as in a gradient model with fixed threshold levels for different structures.

(e) The gene products involved in early development are produced after fertilization of the egg.

(f) The products of the major categories of genes involved in segmentation are distributed in spatial gradients.

(g) The morphological effects of segmentation gene mutants reveal a hierarchical pattern of action, proceeding from large-scale defects in the maternal effect genes to disturbances in each segment in the segment polarity genes.

(h) During metamorphosis, insect imaginal discs transform directly into adult appendages under the action of hormones.

(i) When an imaginal disc is cut in half, one fragment regenerates the whole while the other duplicates to produce a mirror-symmetrical structure. So on both fragments, the same part of the disc is regenerated.

(j) Once an imaginal disc in a larva has become committed to forming a particular appendage, its fate is determined.

4.3 TWO-DIMENSIONAL ORDER

Drosophila is a bilaterally symmetrical organism, its appendages developing in right and left handed pairs. A glance at the structure of a leg shows that it has a complex three-dimensional pattern which arises from specific positions in the imaginal disc as identified by the fate map (Figure 4.16). There is a well defined pattern of bristles that distinguishes different regions of each component of the leg, such as anterior and posterior, dorsal and ventral. There are also differences between the components: coxa, femur, tibia, tarsus. The different positions on the disc that correspond to the surface of the limb can be described by two variables since it is a two-dimensional surface. Therefore the morphogenetic field that gives rise to the detailed surface pattern of the limb must have two dimensions of order. So it is not surprising that a one-dimensional model fails to explain the observations on imaginal disc regenerative properties. Now we have to consider how to introduce a second variable to get a two-dimensional morphogenetic field.

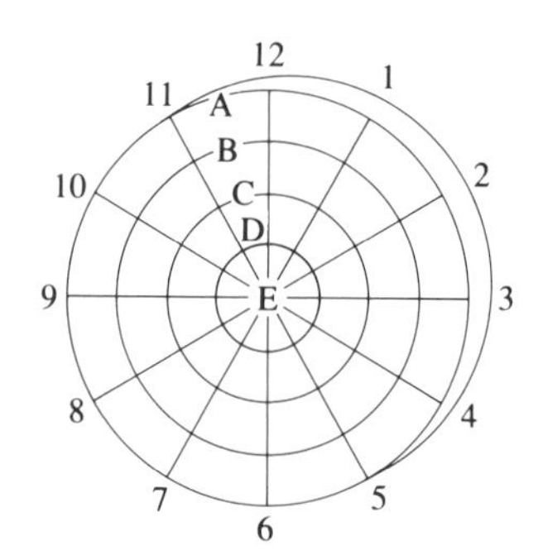

Figure 4.24 The polar coordinate model.

Let us start from Figure 4.16a and idealize the shape of the disc to a circle. The simplest way to add another dimension of order to a pattern of concentric circles is to draw radial lines at different angles and number them, as shown in Figure 4.24. These are numbered 1–12, like the angular positions on the face of a clock. (Note that the discontinuity between 12 and 1 does not reflect any discontinuity in the developmental field. The numbers are just arbitrary labels for angular positions.) The concentric circles are labelled from A to E, with A on the outside and E at the centre. It will become evident in the next section why the labelling starts at the outside and not as you might expect at the centre. The two variables are called **polar coordinates**, since the reference point is at the centre (the pole). The model that uses this description of position is, not surprisingly, called the **polar coordinate model**.

4.3.1 The rules of the polar coordinate model

The rules describing how the variables defining state in the disc vary in response to cuts come from the same principles that have been used in the one-dimensional gradient model. It is assumed that sharp discontinuities of state are smoothed out by a process like that occurring in the diffusion of substances. This means effectively that missing values are inserted in proper order whenever there are gaps in the coordinate system produced by cuts. In the actual disc, cuts heal by union of the exposed surfaces, then growth and cell division occur, followed by cell differentiation that restores missing parts, a process called **intercalation**. In the model, the process of filling in missing coordinate values is accordingly called the rule of intercalation. Let us see how this works in a simple example which corresponds roughly with the experiment shown in Figure 4.22.

Suppose that the lower right hand quadrant is cut from the disc (Figure 4.25a). The fragments heal together along their cut surfaces (b and b′) This is followed by intercalation in which missing values are replaced by filling in the numbers smoothly wherever there is a discontinuity. Both fragments intercalate the same tissue. The result (c and c′) is that the larger fragment regenerates while the smaller duplicates, as observed experimentally.

Consider next an example that is similar to the experiment described in Figure 4.17a. The disc is cut horizontally into upper and lower fragments (Figure 4.26). Again both parts intercalate the same missing tissue after wound healing, with the result that the upper part regenerates and the lower duplicates, conforming to observation.

We come now to the experiment that invalidated the one-dimensional version of the model (Figure 4.23). So far we have been considering intercalation only around the circumference of the disc. For the next experiment it is necessary

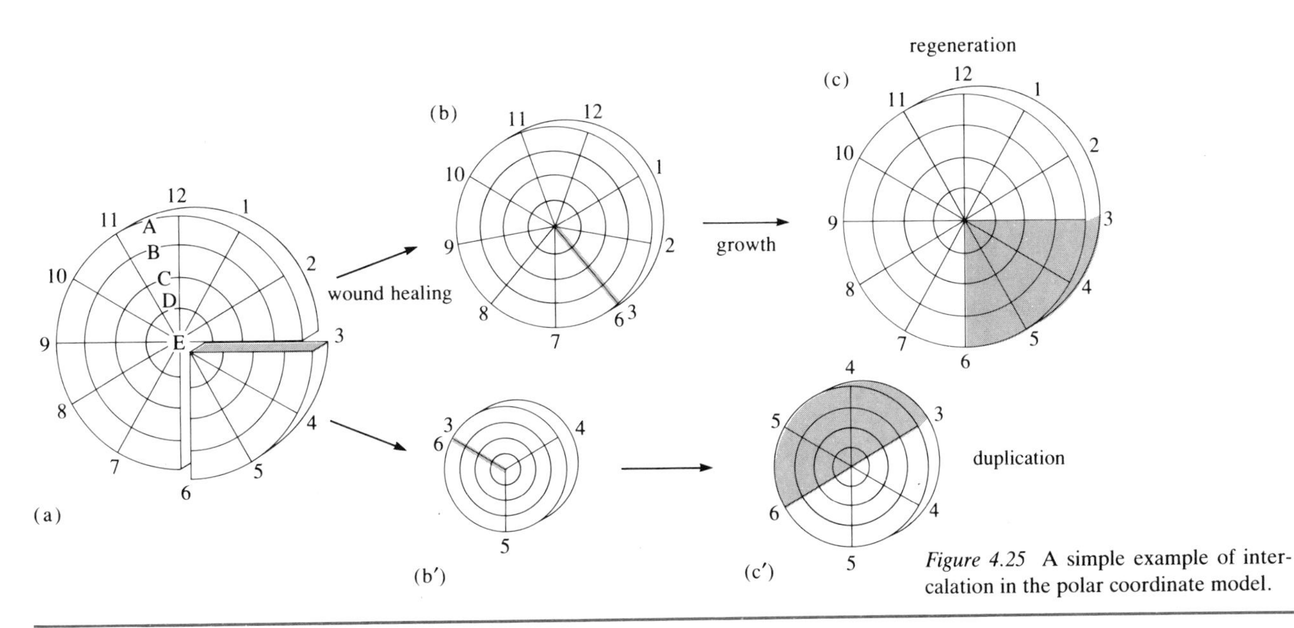

Figure 4.25 A simple example of intercalation in the polar coordinate model.

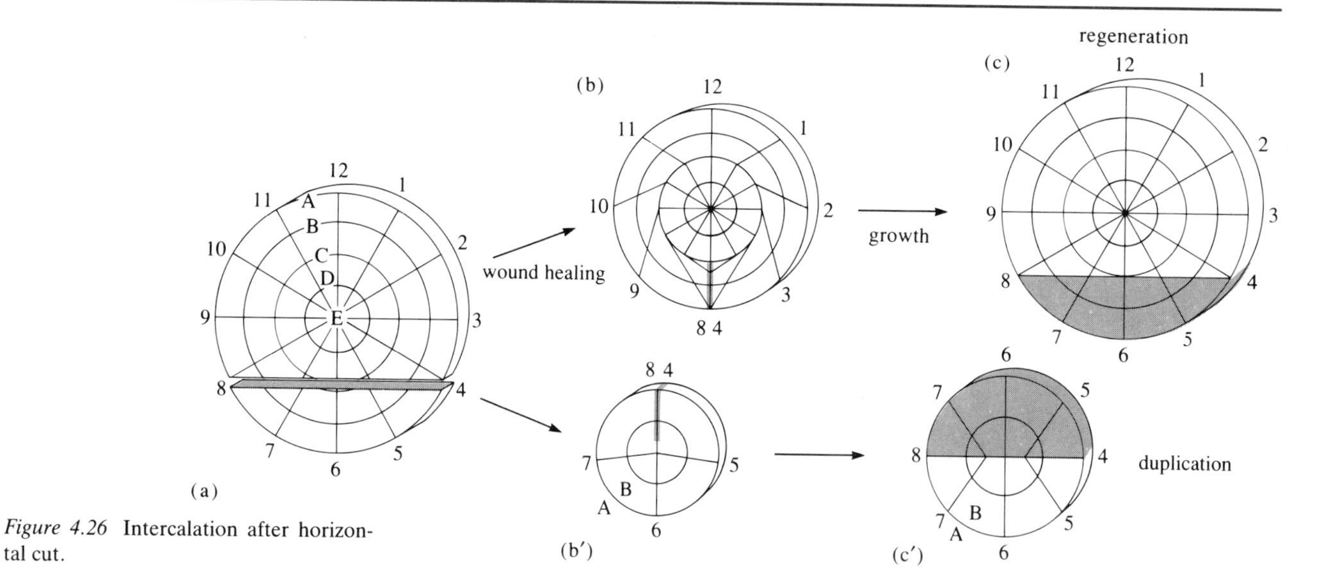

Figure 4.26 Intercalation after horizontal cut.

to consider the other dimension of order, from the periphery to the centre, and how this interacts with the circumferential values. The experiment is described in Figure 4.27. After wound healing the two tissues have rather different geometries, but they both have complete circles of numbers because the cut simply separated inner and outer regions of the disc, each part retaining a complete circumference. So the behaviour of the two fragments depends upon the specification of regeneration from the periphery to the centre. In the previous one-dimensional gradient model, this is all that was specified in terms of a gradient of developmental capacity, with a maximum at the centre of the disc. Clearly such a rule would result in a duplication of the upper fragment and regeneration of the lower, which is the opposite of what is observed. So the gradient of developmental capacity in the polar coordinate model is reversed: in terms of the letters, level A can give rise to all the others; B to C, D and E; C to D and E; and D to E; but the opposite is forbidden. For reasons that will be apparent shortly, the rule is that regeneration from periphery to centre in the disc can occur only where there is a complete circle of numbers present. Since the periphery of the disc

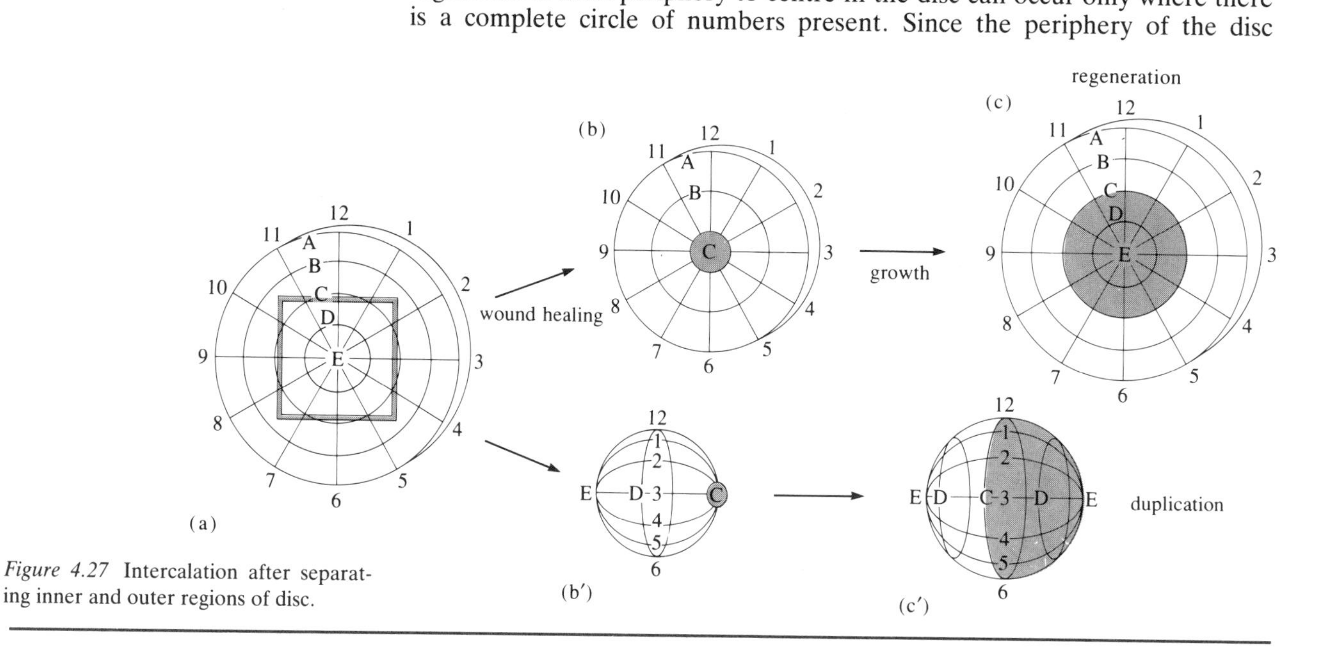

Figure 4.27 Intercalation after separating inner and outer regions of disc.

corresponds to proximal in the limb or other appendage (Figure 4.17) and centre to distal, this is called the **complete circle rule of distal transformation**. When applied to actual limb regeneration, it describes what is observed as we shall see in Section 4.4.

With this rule, which couples together the circumferential and the proximal–distal dimensions of the morphogenetic field of the disc, we can complete the analysis of the experiment described in Figure 4.27. The upper fragment undergoes distal regeneration from C to E, resulting in a complete regenerate. The lower fragmentation does the same: i.e. from C, which is the union of the cut edges coming together as in a purse-string action, with a complete circle of numbers, arises the tissue marked D and E, resulting in a duplication.

◇ What are the proximal–distal levels on the circles in the lower disc fragments of Figure 4.26 b′ and c′?

◆ A and B

◇ Why do these not generate the lower proximal–distal levels C, D and E?

◆ Because the fragment does not have a complete circle of circumferential values.

These results justify the model in its application to imaginal disc behaviour, since it allows us to describe a result that was previously inexplicable in terms of a single gradient. Clearly this does not mean that the model is correct, since there are other ways of drawing two-dimensional maps on surfaces and designing rules for their behaviour to give correct results. However there is a natural elegance to the model that is attractive, and it gives some very significant insights into another major area of developmental research. This is the study of limb regeneration itself, to which we now turn.

4.4 LIMB REGENERATION

Certain species of organism can regenerate their limbs, while others are unable to do so. Cockroaches and newts, for example, have this capacity, while *Drosophila*, frogs and humans do not. We noted in Chapter 1 that humans have the rudimentary capacity to regenerate fingertips. Frog tadpoles are capable of full limb regeneration, but this capacity is lost after metamorphosis to the adult form. The reasons for these differences between species are not understood, but they belong within the regulative–mosaic spectrum of developmental behaviour. With **regulative development**, when part of a limb is removed, the remaining cells will form new cells to replace the missing portion. With **mosaic development**, when one group of cells is removed, the remaining cells cannot replace the missing cells and a gap is left in the pattern.

We have seen that, in general, the hierarchical character of development is accompanied by progressive determination as development proceeds, so that the organism tends towards a mosaic of determined parts as it approaches adulthood. An example of this is the loss of regenerative capacity in limbs following metamorphosis from tadpole to adult form in the frog. However, different species proceed along the pathway towards the mosaic condition to differing degrees. Urodeles such as newts and salamanders retain more regulative capacity in the adult form than do anurans such as frogs and toads.

Cockroaches, newts and salamanders are very convenient organisms to use for studies of limb regeneration because they are relatively large, are easily kept in the laboratory, and have excellent regenerative capacities. The structure of their legs is shown in Figure 4.28. When a leg is amputated at some level along the proximal–distal (shoulder to digit) axis, the wound first heals over. The cells around the periphery of the stump migrate to the region beneath the wound epidermis and accumulate to form a mass of tissue known as a **regeneration blastema**. The cells in the blastema grow and divide, and gradually there develops from this tissue all the parts of the limb that were removed by the cut, as described in Chapter 3.

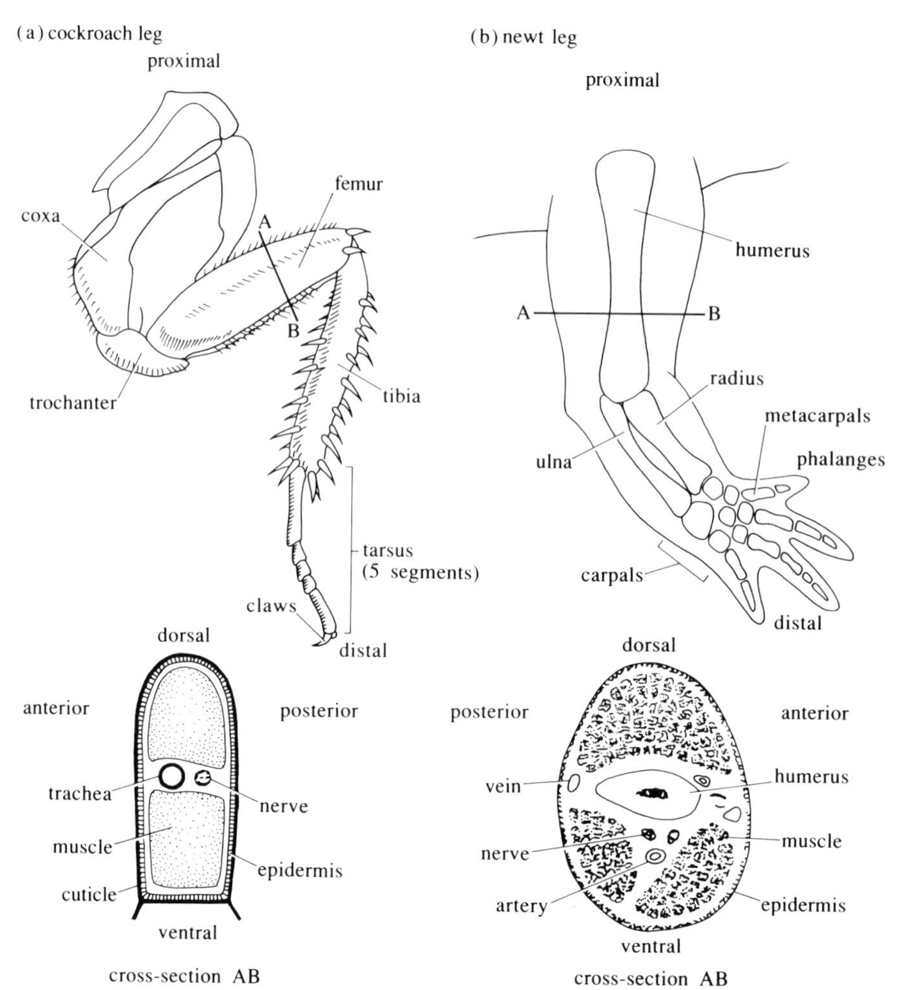

Figure 4.28 Comparative anatomy of a cockroach leg and a newt leg.

4.4.1 Intercalary regeneration

A variety of interesting experiments were carried out in the 1970s on cockroaches by Vernon French in the UK and on newts and salamanders by Susan Bryant in the USA. Together with a very extensive body of previous experimental studies, and Peter Bryant's imaginal disc observations, these actually laid the foundations of the polar coordinate model. An example is shown in Figure 4.29. Using the letters A to E as previously to describe proximal–distal levels of the tibia of the cockroach limb, a transplant is carried out in which a distal part (E) is grafted to a proximal stump (A) in an early instar cockroach. During subsequent moults there is a regenerative process involving intercalary growth in which the missing tissue is restored,

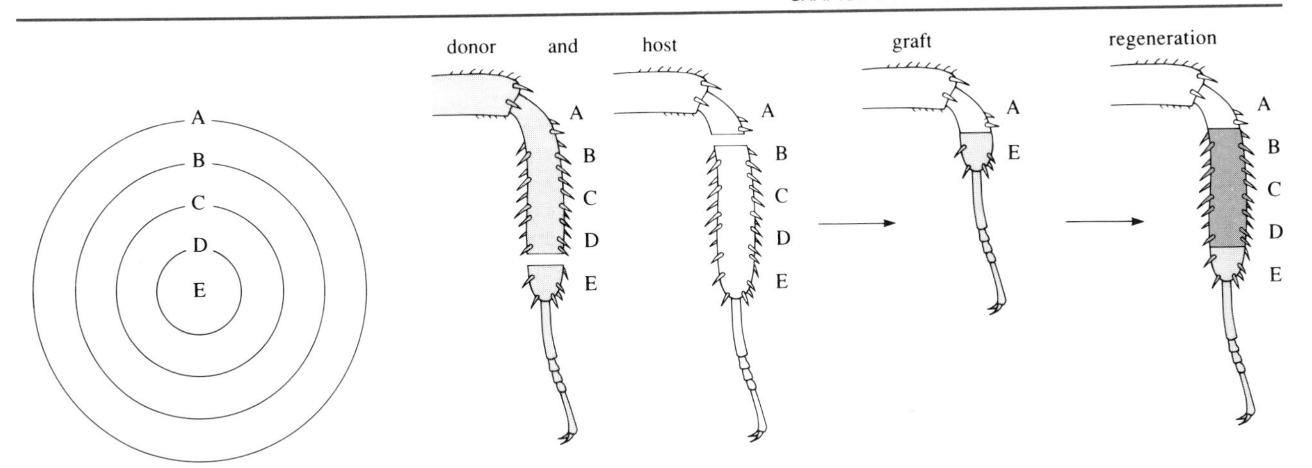

Figure 4.29 Regeneration after grafting a proximal stump to the distal part of a cockroach leg segment.

with the result shown in the figure. In the polar coordinate model, this is a consequence of the rule of intercalation: the morphogenetic field removes the discontinuity by intercalating the missing field values, thus restoring the missing parts. The intercalated tissue comes from both the stump and the graft.

In another such experiment, a proximal tibia level was grafted to a distal stump, as shown in Figure 4.30. Now another discontinuity is produced and the morphogenetic field removes it by intercalation and cell differentiation, producing the structure shown. Notice that the bristles that form on the intercalated tissue continue to point in the proximal–distal direction as determined by the field in the regenerate, as observed experimentally. Here we use the same convention as in Chapter 3 and Figure 4.29: bristles point down the hypothetical gradient, taking A as maximum and E as minimum. The stable retention of field values by stump and graft in these experiments, such as D and B, defining the boundaries of the intercalary region, is often referred to as positional memory.

From a study of intercalary regeneration between different segments, such as grafts between tibia and femur, it emerged that segments are not disting-

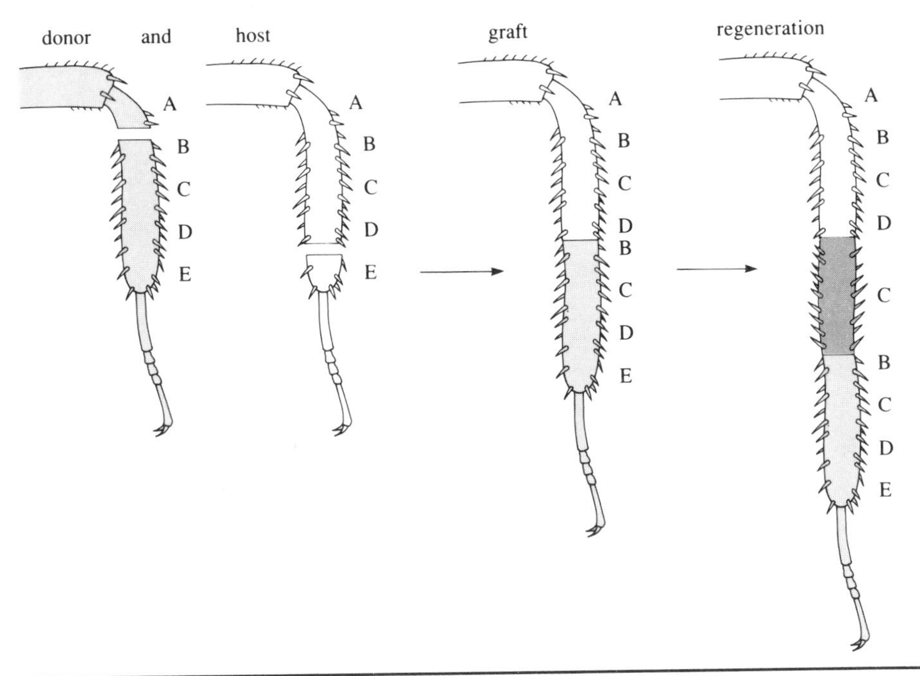

Figure 4.30 Regeneration after grafting a proximal portion of cockroach leg onto a distal stump.

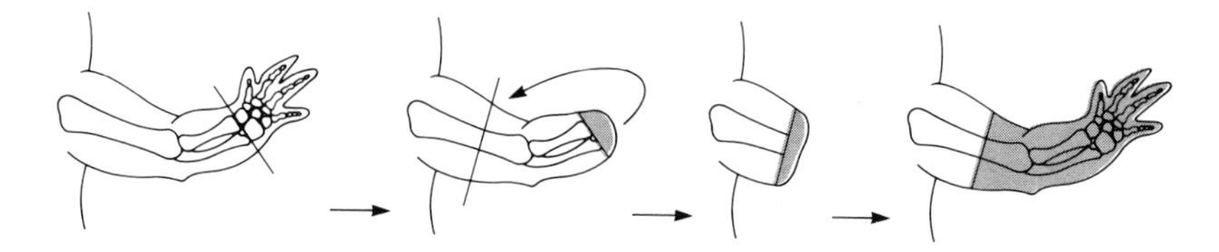

Figure 4.31 Grafting a distal blastema to a proximal level in an amphibian limb. (Compare with Figure 4.29.)

uished from one another. Thus intercalation occurs between tibia and femur tissue as if between tibia and tibia, or femur and femur. As identified by these experiments, the same gradient is therefore repeated in every segment of the limb, as in the body segments described in Chapter 3.

Similar experiments were performed on amphibian limbs, with essentially the same results. However, there are some differences. First, when grafts are carried out to observe the effects of discontinuities along the proximal–distal axis, the tissue that is grafted onto a stump is not part of a limb, as in the cockroach experiments. It is a blastema that is produced by cutting the limb at a particular level and waiting for the blastema to form, then transplanting it. This procedure is followed because differentiated limb tissue does not heal together in amphibians. The equivalent of the cockroach experiment described in Figure 4.29 is shown in Figure 4.31.

◇ Which tissue, host or graft or both, is expected to give rise to the intercalated tissue in this experiment, taking account of the rule of distal transformation?

◆ Since the regenerated tissue between the level of the stump (upper arm) and the graft (lower arm) has values more proximal than those of the graft, this tissue must come from the stump by distal transformation.

When cells were marked so that it was possible to distinguish between regenerated tissue arising from the stump and from the graft, it was found that all the tissue that is intercalated between the level of the stump and that of the graft comes from the stump. The grafted blastema gives rise only to tissue more distal than the level at which it arises, whereas in the cockroach both stump and graft contributed to the intercalation. This is a second difference.

A third difference between cockroach and amphibian limbs is that the graft analogous to that described in Figure 4.30 does not result in intercalary regeneration between stump and blastema. The result is shown in Figure 4.32. In the amphibian limb, a discontinuity remains between a distal stump and the limb tissue produced by a proximal blastema and there is no intercalated region analogous to the region of reversed bristles in Figure 4.30. This result remains an anomaly in relation to the general principle of smoothing in these studies.

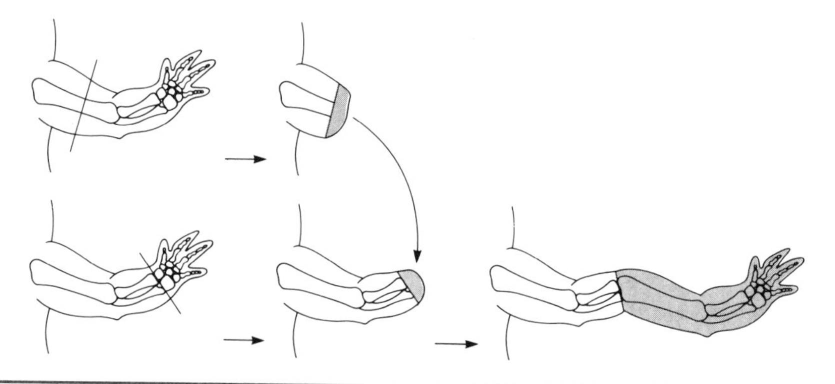

Figure 4.32 Grafting a proximal blastema to a distal level in an amphibian. (Compare with Figure 4.30.)

4.4.2 Supernumerary limbs

We now come to the experimental results on limb regeneration that provided the major stimulus for the construction of the polar coordinate model. Suppose that right and left limbs are cut in mid-tibia and the right limb is grafted to the left stump, as shown in Figure 4.33. There is no discontinuity along the proximal–distal axis, but now dorsal and ventral regions of the limbs are in opposition while the anterior and posterior regions remain in coincidence. How does the morphogenetic field smooth out the discontinuities generated? The experimental result is that two **supernumerary limbs** are produced, one from each of the regions of dorsal–ventral opposition. Furthermore these supernumeraries always have the same handedness as the stump, not the graft. Similar experiments with amphibians produce the same result, as shown in Figure 4.34. In this case the experiment was conducted so that the grafted blastema was orientated on the stump with anterior and posterior regions in opposition, while dorsal and ventral were in coincidence. Two supernumerary limbs arise from the region of opposition, each with the same handedness as the stump.

These results demonstrate quite conclusively that very different types of organism obey the same developmental rules. So it is realistic to formulate universal principles from which the regularities of embryonic development and regeneration can be deduced. This is like the formulation of the rules of

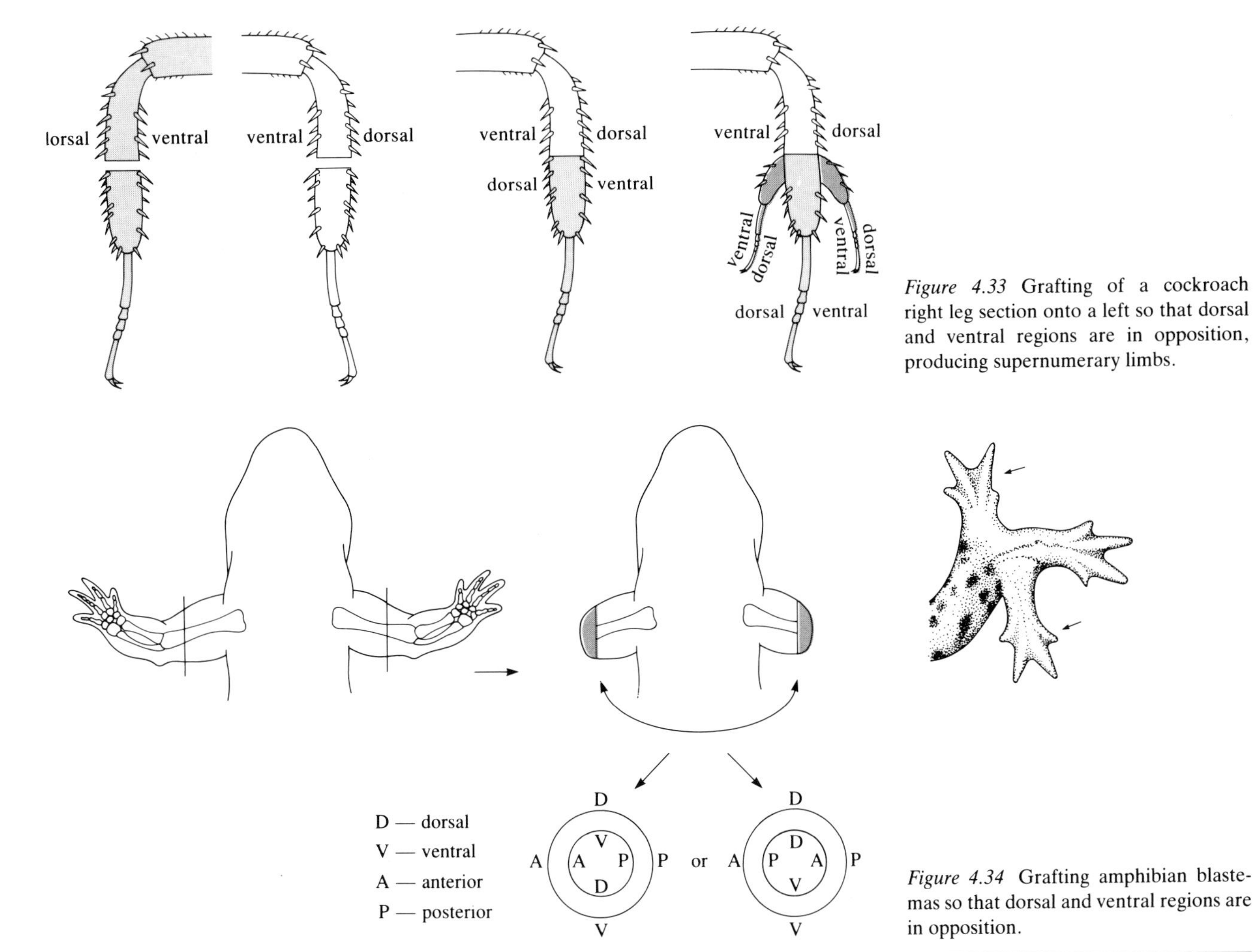

Figure 4.33 Grafting of a cockroach right leg section onto a left so that dorsal and ventral regions are in opposition, producing supernumerary limbs.

Figure 4.34 Grafting amphibian blastemas so that dorsal and ventral regions are in opposition.

Mendelian inheritance which have been extremely significant in making sense of a great diversity of hereditary phenomena — even though there is much that they fail to explain. The polar coordinate model was formulated to explain developmental regularities of the type described in Figures 4.33 and 4.34. How does it do so?

The grafts described in Figures 4.33 and 4.34 involve no discontinuity along the proximal–distal axis, so we can concentrate on the dorsal–ventral discrepancy. The morphogenetic field values around the outer circle of Figure 4.35 describe the circumference of the disc, while the inner circle represents the values around the circumference of the right limb graft. The dorsal–ventral opposition is shown by the 6–12 juxtaposition of field values, while the anterior and posterior values are in coincidence at 9 and 3. The regeneration domain is the space between the circles. It is here that any discrepancies of field values must be resolved by intercalary growth and the filling in of missing values.

If we start from the posterior pole at 3 and work up towards the dorsal stump, then the first discontinuity encountered is between 2 and 4, which can be removed by entering an intercalated 3. Working around the circle in this way, the result is as shown in Figure 4.36. There is now a problem because at the points of dorsal–ventral opposition there are two different ways of filling in the values between 6 and 12: via 3, or via 9. Suppose we follow both routes, via 3 on the posterior side and via 9 on the anterior side. This gives the smoothest possible pattern. The result is shown in Figure 4.37: two complete circles arise at the points of dorsal–ventral opposition each with the same clockwise sequence of number as the stump. According to the complete circle rule of distal transformation, these will each produce a limb. That is, two supernumeraries are predicted, each with the handedness of the stump. So the model generates an elegant and exact description of the double supernumerary result. This is what models are for.

Another value of models is to make predictions and so lead to experimental tests that would not otherwise have been performed. The polar coordinate

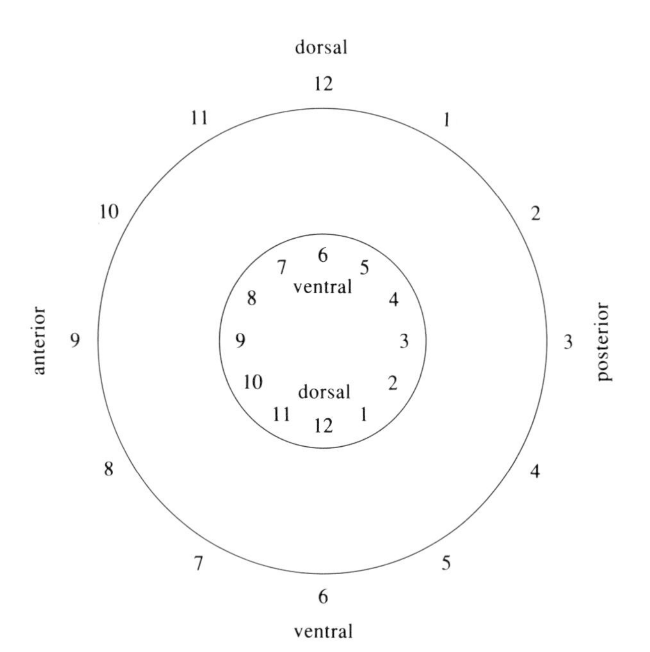

Figure 4.35 Morphogenetic field values in a graft with dorsal and ventral regions in opposition.

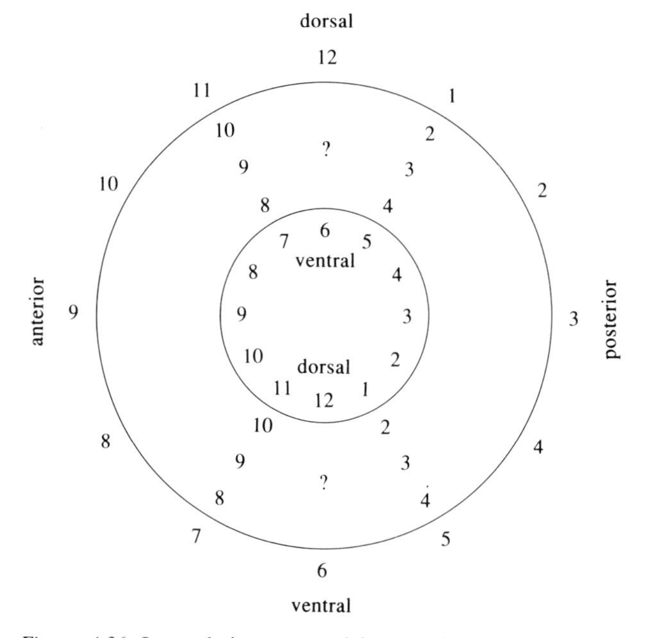

Figure 4.36 Intercalation expected in a graft with dorsal and ventral regions in opposition.

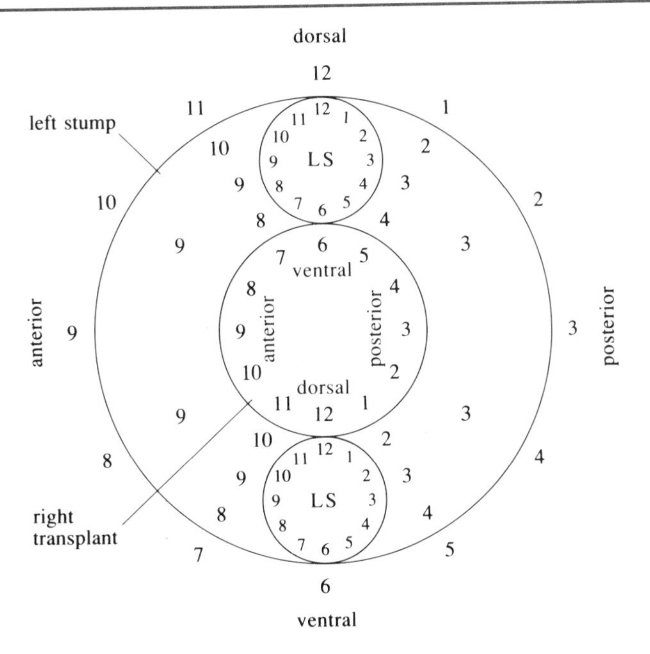

Figure 4.37 Prediction of supernumerary limbs by polar coordinate model.

model did this very effectively, arousing a flurry of research activity in the limb regeneration field. As expected in any scientific study, some results were regarded as anomalies that created problems for virtually all of the interesting theories, for example the failure of amphibian limbs to smooth out the discontinuity resulting from a proximal blastema grafted to a distal stump. But other results challenged central assumptions of the model. The most important of the latter was the observation that limbs with a circumferential pattern that is not described by a complete circle of numbers can nevertheless regenerate. For instance, it is possible to construct limb stumps that have two anterior or two posterior parts by grafting muscle tissue between right and left upper arms, as shown in Figure 4.38a. According to the complete circle rule of distal transformation, regeneration should not occur from stumps produced by cuts through these symmetricized limbs. However, regeneration does occur, the result being a mirror-symmetric limb (Figure 4.38b).

So the model was faced with a serious discrepancy between expectation and reality, leading to something of an impasse. Susan Bryant made various alterations and additions to the rules, and came up with a further set of predictions that something is missing from the analysis, requiring a reappraisal of the phenomena and how they may be modelled.

It seems likely that, just as the one-dimensional gradient model failed and was replaced by a two-dimensional description of morphogenetic field properties, what is now required is a full three-dimensional treatment of limb development and regeneration. We need an approach similar to that which led to the

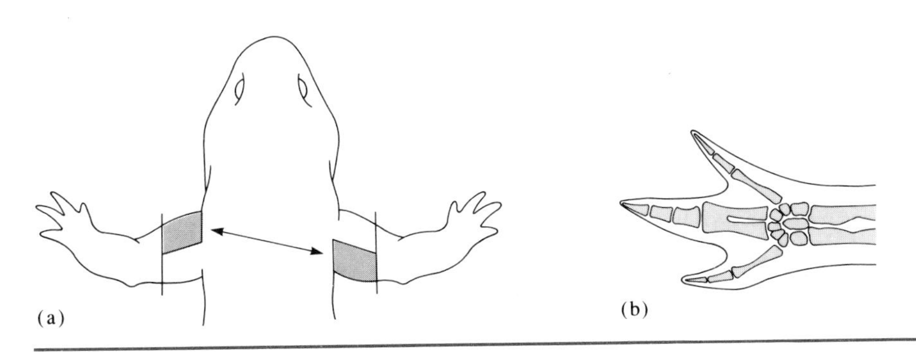

Figure 4.38 The complete circle rule can be proven wrong by creating symmetrical limb stumps. After they are amputated through the symmetrical region, symmetrical limbs regenerate.

description of *Acetabularia* shape formation in Chapter 1 — more physical than the polar coordinate model, taking account of the mechanical as well as the molecular and cellular properties of the morphogenetic fields that organize the segments (insects) and the bones (amphibians). However, the application of the polar coordinate model to regeneration in organisms as taxonomically diverse as *Drosophila*, cockroach and amphibian, has given us a glimpse of the deep regularities within developmental processes. These regularities are encouraging the search for fundamental theories of the type that exist in the more exact sciences.

4.4.3 Changing the boundaries of limb fields

During normal limb regeneration, the parts regenerated replace those that were removed so that the regenerated limb looks just like a normal one. Evidently the regenerative field in the blastema gets information from the stump about the proximal–distal level at which it forms, regenerating structures distal to this level. Such information is called the **boundary value** of a field. Malcolm Maden, working in London in the 1980s, showed that treatment of regenerating newt limbs with high concentrations of retinoic acid (vitamin A) had the remarkable effect of inducing the regeneration of a complete limb from a distal stump, as shown in Figure 4.39, which would normally have produced only structures distal to the wrist.

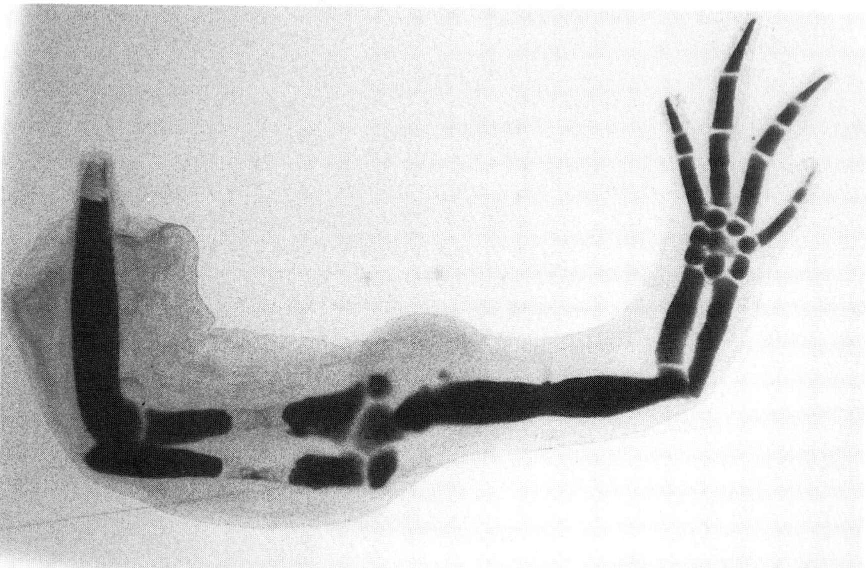

Figure 4.39 Complete newt arm grown from wrist after amputation and treatment with retinoic acid.

The result is like the consequence of grafting a proximal blastema to a distal stump (Figure 4.32). It thus appears that retinoic acid is altering the boundary values of the morphogenetic field in the blastema to a more proximal value. The extent to which the blastema is proximalized depends upon the duration and intensity of the treatment, shorter exposures or lower concentrations resulting in a smaller degree of proximalization of the blastema. (The vitamin A concentrations used in these experiments are about 10 000 times greater than those likely to occur in the human body with the recommended dietary intake of vitamin A. So pregnant women who take vitamin pills need not worry about causing limb abnormalities in their children.) In the USA, David Stocum treated blastemas with retinoic acid and tested them for their tendency to move to appropriate proximal–distal positions when grafted to the stump–blastema junction of a normal regenerating limb cut in the mid-upper region, as described in Chapter 3. The blastemas were found to move to the same level from which they regenerate. Thus the change of boundary value in the blastema is accompanied by a change in affinity or adhesiveness, indicating that the two are correlated.

Retinoic acid has somewhat different effects in different species, changing boundary values in relation to different coordinates of the limb field. In the toad tadpole *Xenopus laevis*, not only does it cause proximalization as in the tailed amphibians, but it also affects pattern along the anterior–posterior axis, resulting in mirror-symmetric structures that can be as complete as two entire limbs, one right and one left, generated from the level of the stump (Figure 4.40). In the chick, on the other hand, retinoic acid affects only the anterior–posterior axis, inducing mirror-symmetric limbs. It has now been shown that retinoic acid is actually present in the low concentrations expected of a pattern modulator in chick limb buds, with a gradient across the anterior axis. This points strongly to its role as an endogenous morphogen involved in regulating pattern generation.

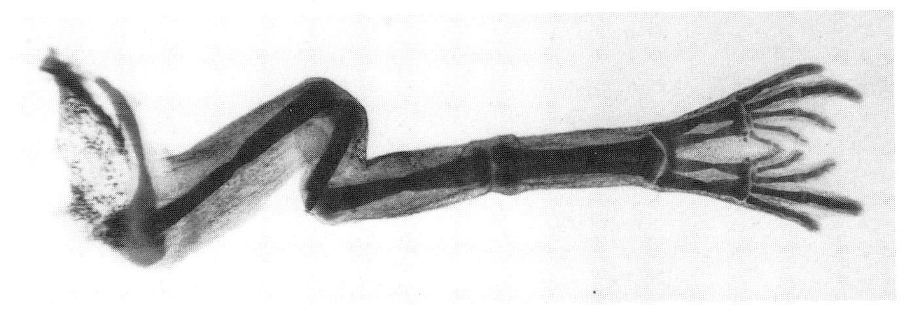

Figure 4.40 Toad double limb regenerated after amputation and treatment with retinoic acid.

Retinoic acid also has other effects on morphogenesis. For example in the developing toad, *Xenopus laevis*, Anthony Durston in Holland has recently shown that a 30-minute exposure of the early embryo at any time prior to gastrulation has the effect of transforming forebrain to midbrain and hindbrain so that the forebrain and eyes fail to form. Again, the vitamin is found in the embryo in low concentration so it appears to be functioning as an endogenous morphogen. The fact that it exerts different effects in different species, and in one species (*Xenopus*) at different times and locations in development, suggests that its action is a rather general modulation of particular properties of morphogenetic fields. The specificity of its effects in different contexts then depends upon the particular competences of the tissue on which it acts, i.e. the details of metabolic state, gene activity, retinoic acid receptors and binding proteins, and whatever gradients are influencing pattern formation in different tissues and species. Exactly how retinoic acid acts is provoking intense research activity among developmental biologists, for it provides a striking example of a relatively simple substance with a diversity of dramatic morphological effects.

The study of pattern formation and morphogenesis has now moved into a stage in which primary focus is on the molecular level of control and the influence of gene products, pursuing lines of enquiry opened up by discoveries such as the effects of retinoic acid and the molecular techniques that reveal patterns of gene activity in *Drosophila* and other species. The next few years could be exceptionally exciting in this field. The directions of research described should bring further insights, and at the same time will require an understanding of the way in which the diversity of molecular processes are integrated into the cellular, tissue, and whole organism levels of morphogenetic coordination. Only within this wider context can the remarkable organization of development be understood, and the general principles that underlie the diversity of morphogenetic processes be revealed.

Summary of Sections 4.3 and 4.4

A two-dimensional model of imaginal disc regeneration, taking account of both radial and circumferential order, explains the behaviour of imaginal discs. This polar coordinate model uses two rules to describe regenerative activity: (1) intercalation, the restoration of missing coordinate values by a smoothing process; (2) the complete circle rule of distal transformation, describing the conditions under which regeneration occurs. The model explains the experimental results on imaginal disc regeneration.

Insect and amphibian limbs regenerate and intercalate missing tissue when grafts are made that produce discontinuities in limb structure. When graft combinations are produced from right and left limbs, the result is two supernumerary limbs, each with the same handedness as that of the stump. The polar coordinate model gives an explanation of these results. However, experiments on mirror-symmetrical limbs showed that they can regenerate, violating the second rule of the polar coordinate model.

The boundary values of limb fields can be altered by retinoic acid, resulting in complete limbs on stumps instead of just the missing parts. Retinoic acid was found to affect different limb axes in different species but always in a similar way, affecting boundary values of the axes. It also alters the boundary of the neural tube in amphibian embryos so that anterior structures are lost. So retinoic acid appears to be an important morphogen, modifying morphogenetic field values.

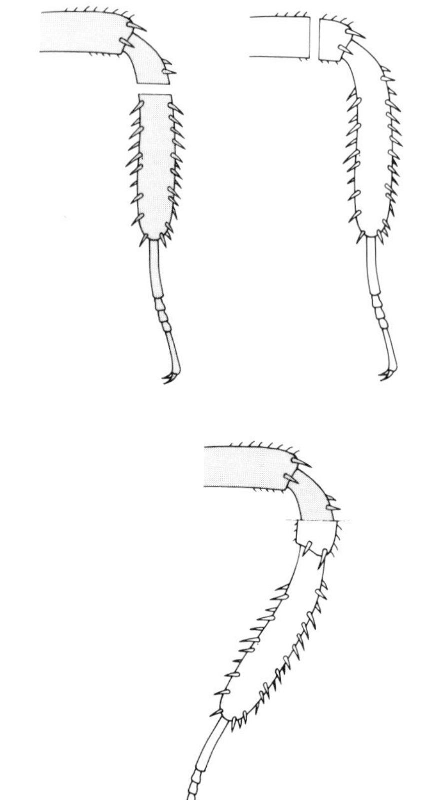

Figure 4.41

Question 2 (*Objectives 4.4–4.9*) The genital disc of *Drosophila* lies in the mid-line of the body (see Figure 4.14) and is bilaterally symmetrical. An experimenter wishing to explore the relative merits of the gradient and the polar coordinate models cuts the disc in half along the line of mirror-symmetry and allows the two halves to grow in an adult abdomen. After transfer to a larva and passage through metamorphosis, it was observed that both halves produced complete structures, i.e. bilaterally symmetrical genitalia. Genital discs were then cut in two at right angles to the line of mirror-symmetry and the result, after *in vivo* culture and metamorphosis, was that each half produced only those parts that it would have anyway, i.e. there was neither regeneration nor duplication. What conclusions do you draw from these observations?

Question 3 (*Objective 4.8*) A cockroach limb is cut at the proximal end of the tibia and to it is grafted a limb cut from the distal end of a femur, as shown in Figure 4.41. Describe the expected result after growth.

Question 4 (*Objective 4.10*) Suppose that the limb treated with retinoic acid, shown in Figure 4.39, is cut at the level of the wrist and the regenerating blastema is treated with retinoic acid. What would you predict to be the structure of the limb?

OBJECTIVES FOR CHAPTER 4

Now that you have completed this chapter, you should be able to:

4.1 Define and use, or recognize definitions and applications of each of the terms printed in **bold** in the text.

4.2 Briefly describe the life cycle of *Drosophila* and its early development, and give reasons for its usefulness in developmental studies. (*Question 1*)

4.3 Give a brief account of the basic molecular techniques that allow the visualization of gene products in developing embryos and present the evidence for a gradient in *bcd* gene product that is related to anterior–posterior pattern formation in *Drosophila*. (*Question 1*)

4.4 Summarize the categories of genes involved in the specification of segments in *Drosophila* and relate their action to the hierarchical nature of pattern formation in developing organisms. (*Questions 1 and 2*)

4.5 Describe the role of imaginal discs in *Drosophila* development, the properties that make them useful for studying pattern formation, and the evidence for a gradient underlying spatial organization. (*Questions 1 and 2*)

4.6 Explain the phenomenon of transdetermination and its significance in relation to commitment and determination of imaginal discs. (*Questions 1 and 2*)

4.7 Show why a single gradient of spatial order in imaginal discs is inadequate to describe the regenerative behaviour. (*Question 2*)

4.8 Describe the rules of the polar coordinate model and apply them to experimental manipulations on imaginal discs and regenerating limbs. (*Questions 2 and 3*)

4.9 Explain the significance of the observation that *Drosophila* discs and regenerating limbs of very different species appear to follow the same rules. (*Question 2*)

4.10 State what is meant by changing the boundaries of limb fields, and present the evidence that retinoic acid has this effect. (*Question 4*)

SEX CHROMOSOMES ◆ CHAPTER 5 ◆

5.1 WHY STUDY THE DEVELOPMENT OF SEX?

An essential feature of all living organisms is their ability to reproduce themselves. An enormous variety of mechanisms of reproduction exist ranging from hermaphroditism in animals and plants, where two kinds of sex organ develop in one individual, to dimorphism where males and females are separate individuals and differ not only in structure but also in behaviour and physiology.

The majority of angiosperm flowers and many animal groups are hermaphrodite and they often show the interesting phenomena of protandry (where male gametes reach maturity first) or protogyny (where female gametes reach maturity before male gametes). Some animal populations have a majority of hermaphrodite individuals and a few pure male and female individuals such as the limpet *Patella coerulea.* Not infrequently in species of plants hermaphrodite individuals occur among a majority of unisexual plants.

Sexual reversal may be a normal occurrence in the life cycle of some species and may involve all members of the species as in the case of the gastropod, *Calyptraea chinensis* and the polychaete worm *Ophryotrocha puerilis* or only one or two individuals in a population as in the case of some fish species. Wrasse, for example, start adult life as females and live in groups of several females and one male. When the male dies, the largest female transforms into a male and the group continues.

This chapter and the following one are confined to the development of separate sexes in a limited number of groups, and to the control of the process of development of their sex differences.

But why study the development of sex differences? Because these differences demonstrate a dramatic change of phenotype — a difference the developmental biologist can use to reveal an understanding of many aspects of development. The emphasis here is on causes, not on descriptions of developmental stages.

At one time all that was known about the control of the development of sex differences was from observations on chromosomes (cytogenetics), and this is where we shall begin in the next section.

5.2 CLASSICAL GENETIC ANALYSIS OF SEX DETERMINATION

The genetic basis of sex determination was first established from classical genetic analysis and forms the basis for modern genetic, developmental and molecular work.

In 1905 Nettie Stevens found that male and female *Tenebrio* (a beetle) can be distinguished cytologically by their chromosomes. Although they have equal numbers of chromosomes in the nuclei, one pair is heteromorphic (of a different shape) in males. One member of this heteromorphic pair is identical to the members of a particular pair in the female; she called this the X chromosome. The other member of the heteromorphic pair is absent in females; she called this the Y chromosome. In summary, *Tenebrio* males have the sex chromosomes XY in addition to the autosomes (non-sex chromosomes) and females have XX (2X) chromosomes.

It is now known that sex chromosomes exist in a large number of species of eukaryotes and furthermore that differences between species occur. For example the heteromorphic pair of chromosomes is not always in the male. In birds this pair is found in females. To distinguish this situation from the XY situation, the heteromorphic chromosomes are called ZW. Yet other groups lack discernible sex chromosomes. The most notable examples are certain reptiles. Some dioecious plant species have a heteromorphic pair of chromosomes, but most do not. These observations are summarized in Table 5.1.

Table 5.1 Sex chromosomes of some species of animals and plants

Organism	Males	Females
Humans, mice and other mammals	XY	XX
Birds	ZZ	ZW
Hemiptera (bugs)	XO	XX
Coleoptera (beetles)	XY	XX
Drosophila (fruit-fly)	XY	XX
Moths	ZZ	ZW
Many reptiles, for example gecko, alligator	no visibly distinguishable sex chromosomes	
A few reptiles, such as snakes	XY	XX
Most dioecious plants, for example holly and dog's mercury	no visibly distinguishable sex chromosomes	
Some dioecious plants such as campion (*Silene alba*), dock and hop	XY	XX

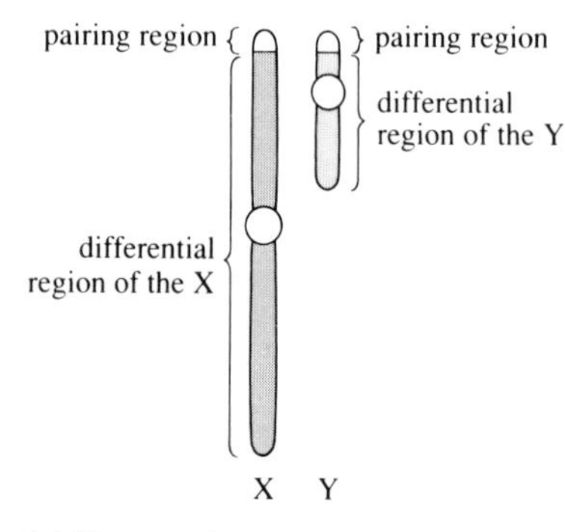

(a) *Homo sapiens*

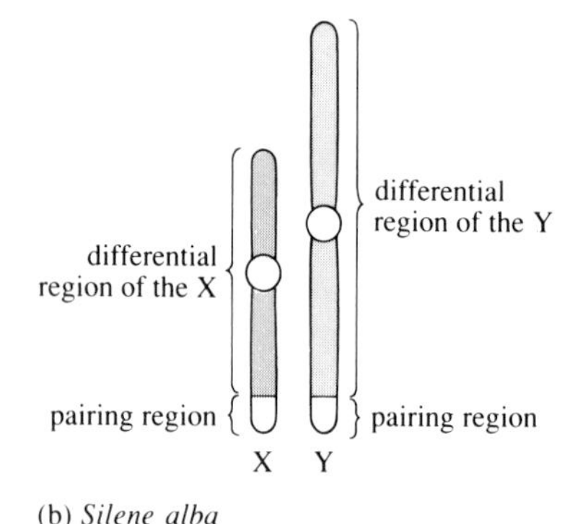

(b) *Silene alba*

Figure 5.1 Differential and pairing regions of sex chromosomes. (a) Humans. (b) The plant *Silene alba*.

With the discovery of sex chromosomes, it was assumed that each embryo had the potential to develop into either one of two possible sexes. Which of these the embryo developed into was assumed to be determined by the genes received by the embryo from the parents. Results of studies in embryology, genetics and physiology have challenged this orthodoxy, a point that will be taken up in the next chapter.

The presence of sex chromosomes in the cells of organisms raises a number of questions about their transmission as well as their involvement in sex-determining mechanisms. For example, how do they behave during the process of meiosis? Importantly, the sex chromosome can be divided into pairing and differential regions (Figure 5.1). These regions have been located chiefly through cytological examination of the heterogametic sex (i.e. the sex having the heteromorphic chromosomes) at meiosis.

Pairing regions

Each species has a characteristic number of chromosomes. In humans the diploid number of chromosomes is 46, consisting of 23 pairs. Each member of a pair of chromosomes is said to be homologous to its partner, i.e. morphologically indistinguishable and each carries the same gene loci in the same order. Sex chromosomes X and Y are the exception since these are morphologically distinguishable and they contain different genes along their differential regions. Very importantly, each gamete contains half the number of chromosomes; in humans this is 23. Thus in meiosis a mechanism is required to ensure that half and only half of the diploid chromosome complement moves to each pole and that each half contains one member of each homologous pair of chromosomes. In the early stages of meiosis, this equal distribution is ensured by the recognition and pairing together of homologous chromosomes so that during division I of meiosis the two homologues separate to opposite ends of the cell. While paired, the close proximity of homologues allows them to exchange segments by means of a process called crossing over. Such crossing over can occur between the pairing regions of the heteromorphic chromosomes in mammals. Pairing is followed by the separation of the two chromatids during division II of meiosis, so that 23 chromosomes (each composed of one chromatid) are pulled to each pole of the cell. The result of the process of meiosis in males is four haploid sperm cells derived from the original cell. A diagram summarizing the stages of meiosis is given in Figure 5.2a.

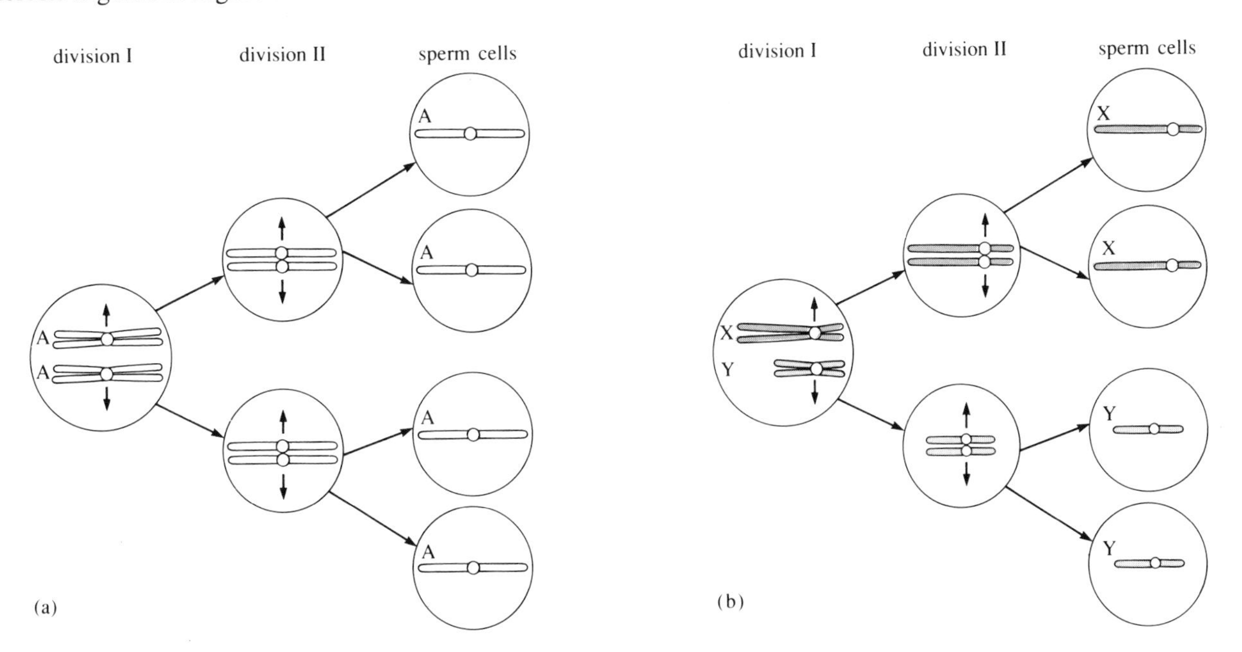

Figure 5.2 Meiotic pairing and segregation. (a) Pair of autosomes (AA). (b) X and Y chromosomes.

Meiosis in human females produces eggs each bearing an X chromosome. In human males, although the X and Y chromosomes are heteromorphic, they pair at meiosis and segregate like homologues in the same manner as a pair of homologous autosomes. This equal distribution of the X and Y chromosomes is ensured by the short pairing regions which are homologous. This is demonstrated in Figure 5.2b. Thus meiosis in the males produces two types of sperm, X-bearing and Y-bearing.

◇ What could the consequences be if the X and Y chromosomes did not pair at meiosis?

◆ Failure of pairing could result in both chromosomes moving to the same pole at division I of meiosis giving rise to sperm either containing both sex chromosomes (24) or containing no sex chromosomes (22).

In some groups, *Hemiptera* for example (Table 5.1), males contain a single X and no Y chromosome in each cell (i.e. XO). Consequently, there is no homologue to pair with during division I of meiosis so that the X chromosome

	♀	♂
	XX	XY
gametes	X or X	X or Y

gametes	X	X
X	XX	XX
Y	XY	XY

Figure 5.3 The pattern of inheritance of sex chromosomes in humans.

moves to one pole along with half the autosomes while the other half of the autosomes travel to the other pole.

If we consider fertilization in humans, union of an egg with an X-bearing sperm will give rise to a female (XX) and union with a Y-bearing sperm will give rise to a male (XY), as shown in Figure 5.3.

◇ What is the striking feature of the progeny resulting from the fertilizations shown in Figure 5.3?

◆ Equal numbers of males and females are expected.

Differential region

The differential region of each sex chromosome contains genes which have no allele on the other kind of sex chromosome, that is, they are hemizygous (half-zygotes) in XY individuals. Genes in the differential region of the X show an inheritance pattern called X-linkage, those in the differential region of the Y show Y-linkage. Many genes not involved in sex determination are known to be X-linked. Some of these known in humans are shown in Figure 5.4.

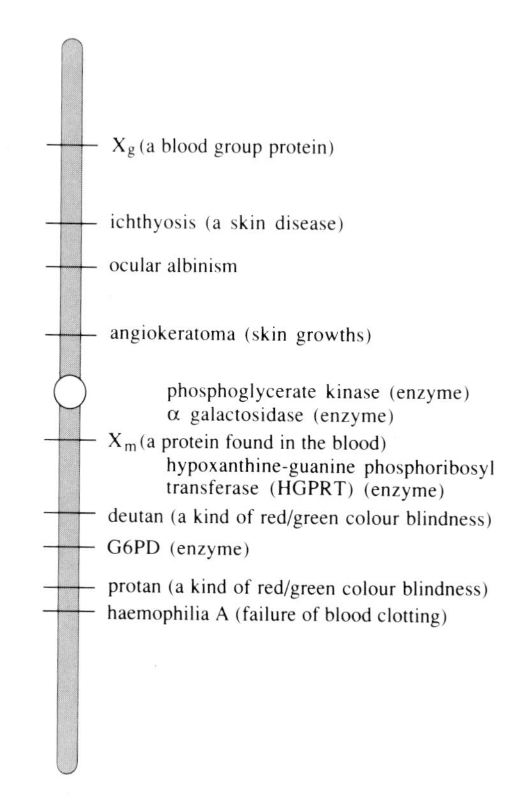

Figure 5.4 A map of the human X chromosome.

X-linked genes segregate like other Mendelian genes. But since only one allele of a gene is present in males, the pattern of their inheritance is modified. In all mammals, including humans, genes on the differential region of the Y chromosome would be inherited only by males, with transmission from father to son. Few genes other than those involved in sex determination and sperm differentiation have been found on the human Y chromosome.

5.2.1 Changes in number and structure of sex chromosomes

Occasionally numerical changes in chromosomes arise, resulting in an individual containing nuclei with one more or one less than the normal number of chromosomes characteristic for the species. The most common means by which such errors arise is that the homologous chromosomes fail to separate at meiosis, a process called non-disjunction. Thus, for example, gametes may arise containing 2X, both an X and Y chromosome or no sex chromosome. At fertilization the zygote contains an abnormal number of sex chromosomes. Such unusual individuals are important for gaining insights into sex determination mechanisms in different species.

◇ What would be the sex chromosome constitution of zygotes following fertilization of an egg containing either 2X chromosomes or no X chromosomes.

◆ Egg XX fertilized by X sperm = XXX
Egg XX fertilized by Y sperm = XXY
Egg O fertilized by X sperm = XO
Egg O fertilized by Y sperm = YO

An XXY human individual is described as having **Klinefelter's syndrome** and an XO as showing **Turner's syndrome**. In such individuals normal differentiation of the ovaries and testes does not occur. Their height and body shape is also affected. In contrast XXX females are fertile, whereas YO zygotes fail to develop.

In addition to numerical changes in chromosomes, structural changes occasionally arise. One way these changes are brought about is by breakage and rejoining. For reasons that are not fully understood, the breakage of chromosomes produces 'sticky' ends with a greater tendency to rejoin than the normal intact ends of chromosomes. These breaks occur spontaneously at low frequency, but their occurrence can be greatly increased by agents such as ionizing radiation and some chemical mutagens.

The loss of part of a chromosome is called a **deletion**. Deletions can occur at the end or within chromosomes (see Figure 5.5). The latter occurs when two breaks in the same chromosome are followed by loss of the intervening

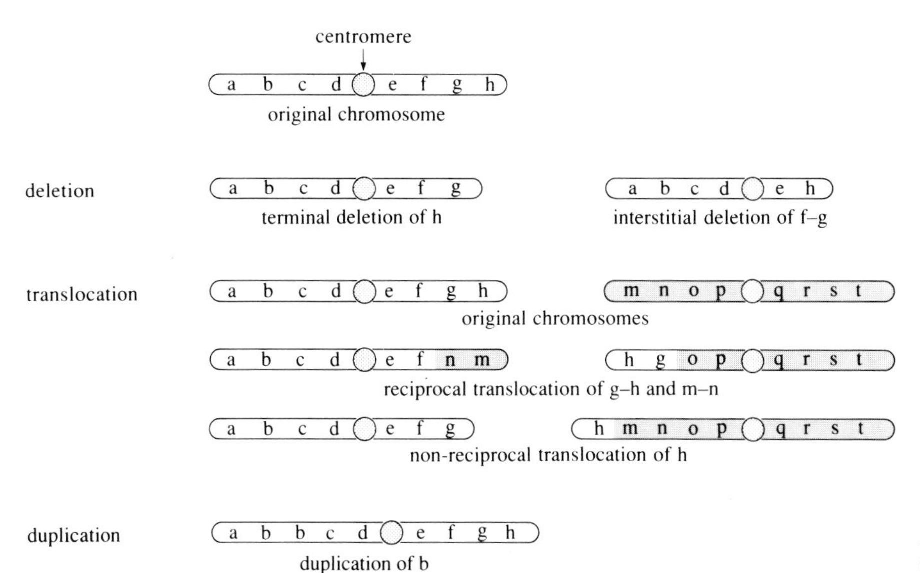

Figure 5.5 The major types of change in chromosome structure.

segment and rejoining of the exposed ends. The size of the lost segment may vary from a single nucleotide to a substantial proportion of the chromosome containing many genes. Smaller deletions often have no visible effect on the phenotype but large deletions usually produce complex phenotypes associated with loss of the genes included in the deletion.

A **duplication** is the presence of an extra copy of a particular chromosome segment. The duplicated segment may be adjacent to the original segment or at a different site within the same chromosome or even within another chromosome (see Figure 5.5).

So far, only breaks within a single chromosome have been described. Consider now the possibility of a single break in each of two non-homologous chromosomes. If the four resulting fragments were to exchange original partners and rejoin, their chromosome segments would be interchanged or translocated. This type of interchange is known as a reciprocal translocation. Non-reciprocal translocations can also occur, when one chromosome loses a segment to another without gaining one in exchange. Both these types of **translocation** are shown in Figure 5.5.

You will learn in the next chapter how numerical and structural changes in chromosomes have proved to be useful tools in elucidating the mechanisms of sex determination in different organisms. In the meantime we are left with the puzzling problem of how male mammals and *Drosophila* overcome the absence of the alleles in the differential region in one of the chromosomes.

5.2.2 Dosage compensation

The presence of genes on the differential region of the X chromosome are required for normal development in both sexes in mammals and in *Drosophila*. If we take an example of genes involved in enzyme production located on the X chromosome, normal females have two alleles for its production whereas males have only one. So how is the difference in allele number between sexes overcome?

In 1965, Ed Grell working in the USA with *Drosophila* eye colour mutants attempted to shed light on this problem. He measured the relative amounts of enzyme activity using two genes that effect the production of the enzyme xanthine dehydrogenase. One gene is the sex-linked gene *maroon-like* (*ma-l*) eye colour whereas the other, *rosy* (*ry*) eye colour, is located on chromosome 3. He constructed stocks wherein some flies had small duplications or had small deletions, thus carrying either additional copies of alleles or fewer copies. Therefore, some flies had only one copy, others had two, and yet others had three copies of the same allele. The results are given in Table 5.2.

Table 5.2 Relative enzyme activity of xanthine dehydrogenase and the number of allele copies

Number of gene copies	*mal-l* X chromosome	*ry* chromosome 3
1	1.0	0.5
2	1.0	1.0
3	1.0	1.5

◇ How would you interpret this data?

◆ The number of copies of alleles of *rosy* is directly reflected in enzyme activity, whereas the activity of the sex-linked gene *maroon-like* remains the same whatever the copy number of alleles. It looks as though in the case of X-linked genes more than one copy is as good as only one copy.

This latter observation is a consequence of **dosage compensation**, a mechanism that compensates for the dosage difference of X-linked genes. In this example of *Drosophila* the amount of enzyme activity for sex-linked genes is about the same in both sexes. Similar observations have been made in mammals, including humans.

◇ Can you suggest two possible mechanisms which could regulate the quantity of gene product for X-linked genes?

◆ 1 Where there is a single allele of a gene, as in males, it produces twice as much gene product as two alleles in a female.
2 Only one of the alleles of a gene in the female is active.

Intriguingly, both these dosage compensation mechanisms have evolved — the first in *Drosophila* and the second in mammals.

If an allele produces twice as much gene product in a male as in a female, it could be because twice as much RNA is produced on the male DNA. The relative transcriptional activity in males and females can be measured — the nucleotide [^{3}H]uridine occurs only in RNA and not in DNA, so measuring its presence in X chromosome RNA gives a measure of transcriptional activity. Results of these measurements, in salivary gland chromosomes of developing larvae of *Drosophila*, show that the single X chromosome in males produces the same amount of RNA as the two Xs in females. So the dosage compensation does occur at the level of transcription. In the next chapter you will learn that in the case of *Drosophila*, there is some evidence that sex-determining genes are intimately involved with the control of dosage compensation.

In mammals, the dosage compensation mechanism is quite different. Here, it is a consequence of one of the X chromosomes being inactivated. But it is not yet known whether the sex-determining genes are involved in this process. The inactive X chromosome becomes 'heterochromatic'; that is, the DNA and the proteins in the chromosome become supercoiled, forming a very tightly packed clump. So again, the control of dosage compensation is at the transcriptional level. But in this case transcription in the tightly coiled X chromosome is prevented altogether.

X-inactivation was first proposed by Mary Lyon in 1960 working in the UK.

Lyon's hypothesis is as follows:

1 Early in embryonic life, when the number of cells in the body of a human female is relatively small, one of the X chromosomes in each cell becomes genetically inactive.

2 Chromosome inactivation occurs at random with respect to maternal or paternal origin of the X.

3 The same X chromosome remains inactivated in all descendants of that cell. Hence the body becomes a mosaic in which some groups of cells have an inactive X chromosome derived from the mother and other groups of cells have an inactive X chromosome derived from the father.

A schematic diagram of somatic cells in a female showing that the female is a mosaic for X-linked genes is shown in Figure 5.6.

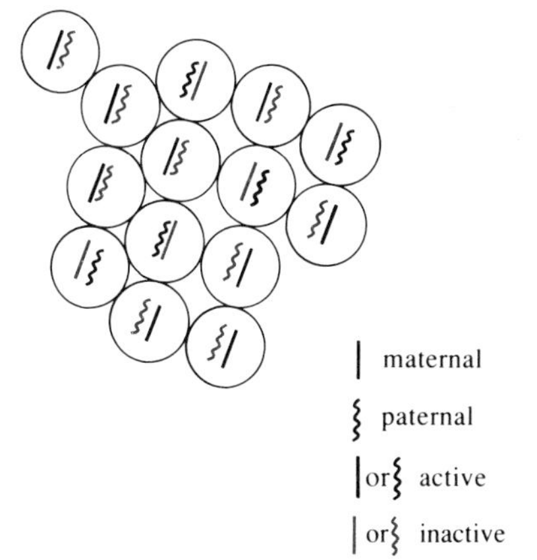

Figure 5.6 A schematic diagram of somatic cells in a female, showing that the female is a mosaic of X-inactivation.

Evidence for X-inactivation

(a) Female mammals heterozygous for X-linked genes provide dramatic evidence for X-inactivation. We shall take the example of anhidrotic ectodermal dysplasia in humans, a condition in which a mutant allele causes the absence of sweat glands. This phenotype can easily be detected by altered electrical resistance of the skin or by the effect of various harmless dyes.

◇ If such a gene were X-linked what would be the expected phenotype of heterozygous females assuming Lyon's hypothesis?

◆ They would have areas of skin with sweat glands and areas without sweat glands, corresponding to regions of cells in which the inactivated X chromosome carries, respectively, either the normal or the mutant allele. Such an individual is described as a mosaic (tissue containing two genetically different cell types).

◇ If inactivation of the X chromosome was random how would the pattern of patches compare in different heterozygous females?

◆ They would show different patterns.

In fact that is precisely what is observed, as shown in Figure 5.7.

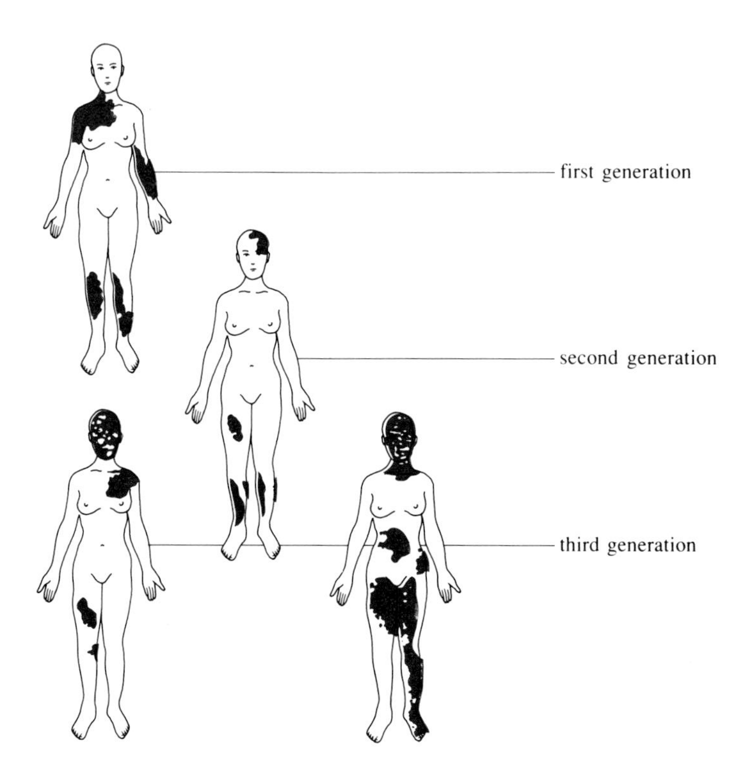

Figure 5.7 Somatic mosaicism in three generations of females heterozygous for sex-linked anhidrotic dysplasia. This abnormality in sweat-gland secretion can be demonstrated with a harmless dye. The location of the mutant tissue is determined by chance, but each female does exhibit the characteristic mosaic expression of a single chromosome.

(b) The undividing cells of normal human females contain a rather mysterious small but stainable clump which is absent from normal males (see Figure

5.8a and b). Such clumps are called Barr bodies after the first person to describe them in detail. Interestingly, women who are XO instead of XX do not have a Barr body, whereas males who are genetically XXY do.

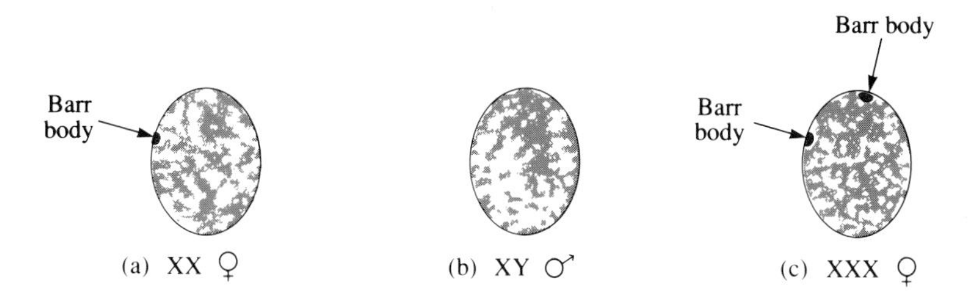

Figure 5.8 Nuclei obtained from cells in the mucous membrane of the human mouth. (a) Nucleus from a female, showing one Barr body. (b) Nucleus from a male, with no Barr body. (c) Nucleus from an XXX female, showing two Barr bodies.

◇ What might be the possible connections between Barr bodies and X inactivation?

◆ A Barr body might be an inactivated X chromosome.

It is now known that the Barr body is a tightly coiled X chromosome that is genetically inactive. It is first observed 5–6 days after fertilization in human embryos, around the time of the blastocyst stage. The chromosome replicates later than its homologue. Instead of uncoiling during the interphase stages of consecutive mitoses, the late-replicating X chromosome remains condensed and appears as a Barr body.

◇ How many Barr bodies would you expect an XXX female to have?

◆ Two Barr bodies per cell and this is indeed what is observed (see Figure 5.8c).

The recognition of sex chromosome transmission and behaviour paved the way for the more recent work of identifying the genes involved in the regulation of sexual development. In Chapter 6 we consider the links between genetics and the development of sex differences.

Summary of Section 5.2

Sex chromosomes are found in many groups of animals, but not all, the notable exception being certain reptiles. They are also found in many dioecious plants.

Sex chromosomes are divided into pairing and differential regions. The equal distribution of sex chromosomes to the progeny cells at meiosis is ensured by the homologous pairing regions.

Sex chromosomes, like other chromosomes, are subject to numerical and structural changes.

The mechanism of dosage compensation regulates the quantity of product for X-linked genes so that it is the same in the two sexes.

Dosage compensation in mammals is a consequence of inactivation of one of the two X chromosomes. The inactivated X chromosome can be identified cytologically as a Barr body.

Question 1 (*Objective 5.2*) Which of the following organisms have (a) XX/XY sex chromosomes, (b) ZW/ZZ sex chromosomes and (c) no sex chromosomes?

(i) Humans

(ii) Alligator

(iii) *Drosophila*

(iv) Birds

(v) Campion

(vi) Mouse

Question 2 (*Objectives 5.1 and 5.3*) What is the function of the pairing regions of heteromorphic sex chromosomes?

Question 3 (*Objectives 5.1 and 5.4*) Define each of the following chromosomal mutations: duplication, deletion and translocation.

Question 4 (*Objective 5.5*) Explain how the difference in allele number between sexes is overcome in (a) *Drosophila* and (b) mammals.

Question 5 (*Objective 5.6*) What are the consequences of X-inactivation in human females?

OBJECTIVES FOR CHAPTER 5

5.1 Define and use, or recognize definitions and applications of each of the terms in **bold** type. (*Questions 2 and 3*)

5.2 Distinguish between different mechanisms of sex determination. (*Question 1*)

5.3 Explain the behaviour of sex chromosomes in meiosis and identify sex linkage. (*Question 2*)

5.4 Define the types of structural and numerical changes that can occur in sex chromosomes. (*Question 3*)

5.5 Describe the two mechanisms of dosage compensation found in *Drosophila* and mammals, respectively. (*Question 4*)

5.6 Describe the inactivation of the X chromosome in female mammals and explain the effect on the expression of X-linked genes. (*Question 5*)

6.1 THE DEVELOPMENT OF SEX DIFFERENCES IN ANIMALS AND PLANTS

This chapter deals with sex-determination mechanisms in certain organisms — specifically in mammals, insects (with particular reference to *Drosophila*), reptiles and plants. We will start with a model which summarizes the sex differentiation in animals. You will find it helpful to bear this in mind when reading about sex determination and sex differentiation in later sections.

6.1.1 A model of the process of development of sex differences

The animals considered here have basically two alternative forms, female and male. Figure 6.1 is a diagram of the development of these sex differences in animals and describes a model which summarizes a number of important features of sexual development. Firstly, the sexual phenotype arises through a sequence of increasingly complex forms which are generated in a hierarchical manner. You will not be surprised by this characteristic of the process since you have learnt about it in earlier chapters of this book. Remember that this involves tissue taking a number of 'decisions' with respect to alternative pathways available to it (Chapter 1). The model indicates that each animal embryo has the remarkable developmental potential to form structures of

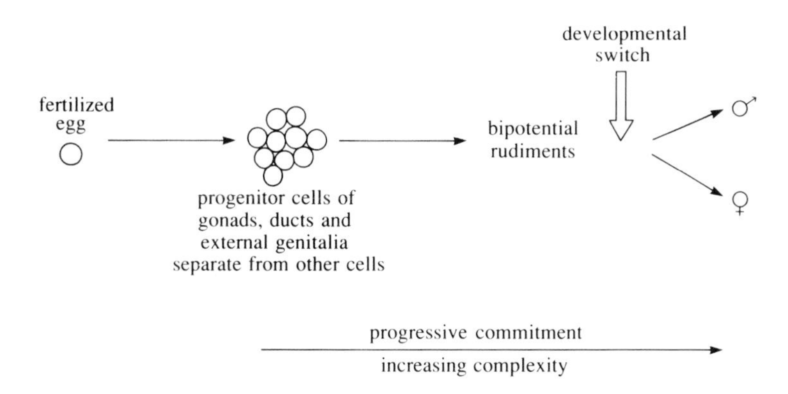

Figure 6.1 Sex development in animals.

either sex from bipotential rudiments. Which sex develops depends upon a developmental switch, triggered either by a genetic or by an environmental influence early in embryonic development. Within these two broad categories of switch mechanism, genetic and environmental, a number of subdivisions have been identified. A study of the nature of this switch will be one of our prime concerns. We shall see that the development of the sexual phenotype introduces another set of decisions to be made. Thus, for example, once cells have become committed to forming a genital ridge that grows into a gonad, a further decision has to be made about whether to develop into ovary or testis.

This takes us to another important feature of the development of the sexual phenotype, that of integration. Differences between the two sexes involves not only morphological features, namely gonads, internal ducts and external genitalia, but also behavioural and physiological aspects. We will concentrate primarily on the development of the morphological sex differences in the embryo.

Many different tissues and organs of the body, with many different functions, are involved in the make-up of a sexually mature individual. The functional coordination of these different structures is essential not only in space but also as the organism develops through time. An individual is shaped from a bipotential set of rudiments that leads to a sexually mature phenotype. The development of more complex forms is coordinated in such a way as to promote 'maleness' or 'femaleness'. Should one tissue make a 'wrong' decision, an infertile intersex or hermaphrodite results. In vertebrates, this integration is achieved by sex hormones. This major concept of development, integration and coordination of processes in space and time is one to which we shall return.

The model illustrated in Figure 6.1, while leaving out much detail, does at least provide a conceptual framework linking disparate signals of the developmental sex switch in different organisms with sexual differentiation. However, one important limitation of this diagram is that the developmental pathway, from the time of the developmental switch to the emergence of the sexually differentiated individual, is shown as a direct one. In fact the process of development is dynamic and progressive so that the mammalian, insect or reptilian embryo does not inevitably develop into one of two discrete categories, male or female. Indeed you will see later that a remarkable range of intermediate morphologies is possible.

In plants, the major expression of the sexual phenotype (in angiosperms) is the flower, sex differences such as size of plant being uncommon. So the functional coordination of different structures in the organism does not apply. Dioecious plants, like animals, also have the potential to develop into either sex.

6.1.2 All genes in all cells

What we are concerned with here is the genetic information available to developing cells and the role that it plays in the development of sex differences. Sex chromosomes, where present, do play a part. But the fact that embryos are bipotential suggests that the two sexes do not have separate sets of genes just as different tissues do not contain different genetic material. John Gurdon and his associates working in Oxford showed that when a nucleus from an adult skin or larval intestine is injected into a fertilized egg of the toad *Xenopus* that had previously had its nucleus destroyed by irradiation, a fully developed tadpole resulted. From such experiments it was concluded that a complete complement of genetic material is present in differentiated cells (although one or two exceptions do exist). If this is the case, then there must be a mechanism which selects genes to be expressed in particular cells, such as haemoglobin in red blood cells. Similarly, certain tissues in females contain proteins (gene products) that are absent from males because in males these genes are not expressed — and vice versa.

One way of bridging the gap between genetics and the embryology of sex differences is by using mutant genes. These are extremely powerful tools for the study of determination and differentiation at both the molecular and phenotypic levels. Mutations of some genes cause one sex to differentiate into the form of the other. Through studying mutants we can explore the effect of particular genes on the development of the sexual phenotype. Furthermore, since genetic mutants identify the specific steps in embryonic development that are subject to genetic control, they can reveal the type of developmental change or decision that normally occurs at these steps.

In the following sections you will learn about some of the developmental decisions and the role that genes play in the development of sex differences. Although it would be a daunting task to describe the actions of all genes and other interacting factors involved in the development of the sexual phenotype, we can attempt to see how differences between males and females arise.

Although there are similarities between the three groups of animals considered here, mammals (with particular reference to mice and humans), insects (with emphasis on *Drosophila*) and reptiles, there are considerable differences in the control of the process of development of sexual form. It is to a study of these similarities and differences in control in these groups and in plants that we now turn, beginning with the major developmental decision, the switch mechanism.

6.1.3 The switch mechanism

There are two broad classes of switch mechanism, environmental and genetic.

A number of examples exist where a single genotype can produce an individual of either sex. In the burrowing marine worm *Bonellia* (see Figure 6.2) sex is determined by the environment to which the immature form is exposed. The worm becomes female with a large bifurcated proboscis if it develops in isolation. However, should the fertilized egg fall onto the proboscis of a female, then it develops into a male. The male is a small ciliated organism parasitic on the female. It lives in her one tube (other than the mouth and anus) which leads to the exterior and which carries both excretory and reproductive products from the coelom.

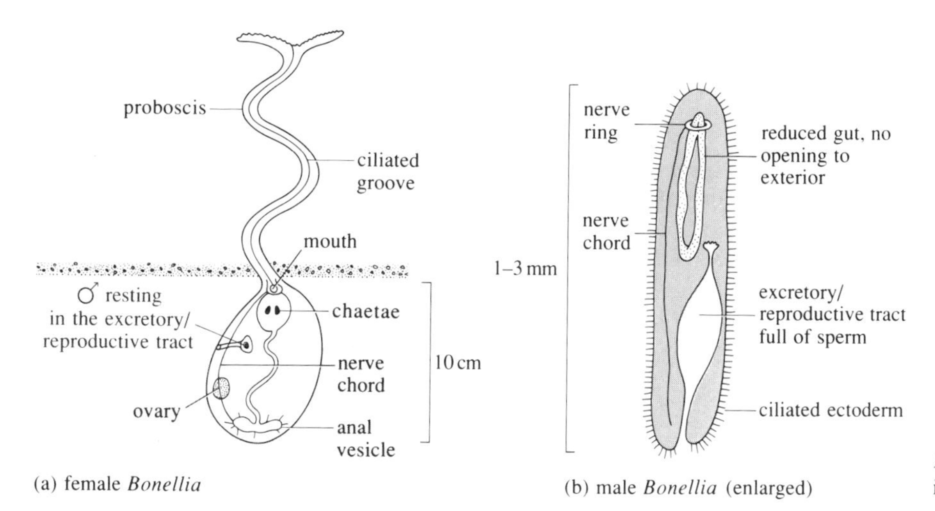

Figure 6.2 Extreme sexual dimorphism in *Bonellia viridis*.

Among fish, wrasse for example, there are examples of sex change during the life of an individual — so these genotypes can also produce individuals of both sexes. Among many species of turtle, and among some lizards and crocodiles, the environmental temperature of egg incubation determines the sex. This is known as temperature sex determination (TSD). Interestingly, there are no apparent differences in the chromosomes between sexes in any of the

Table 6.1 Occurrence of TSD and presence of sex chromosomes in reptiles

Reptile	Family	TSD	Sex chromosomes
lizards	Agamidae	+	−
	Gekkonidae	+	−
	Lacertidae	−	+
snakes	Colubridae	−	+
crocodiles	Alligatoridae	+	−
turtles	Emydidae	+	−
	Testudinidae	+	−
	Cheloniidae	+	−

reptilian species known to have TSD, and in those groups with heteromorphic sex chromosomes TSD is absent (see Table 6.1).

In *Bonellia* and in the reptiles which have TSD, the sex switch is the external environment. However, in another worm, *Dinophilus*, females produce large and small eggs which on fertilization develop into females and males respectively. In this case the determinant appears to be the ratio of cytoplasmic to nuclear volume; that is, the cellular environment rather than the external environment.

We now turn to genetic switch mechanisms which, some biologists argue, have been selected because they are more reliable than environmental control.

Table 6.2 shows the sexual phenotype of individuals with a normal and abnormal number of chromosomes in mammals, humans, *Drosophila* and the plant *Silene*. (A is a haploid set of autosomes, 2A is a diploid set of autosomes and 3A is a triploid set of autosomes. A triploid set is composed of a diploid set plus a haploid set.)

The observations on *Silene* and mammals with unusual numbers of sex chromosomes (including humans) in Table 6.2 give an important clue to understanding the determination of sex in these organisms.

Table 6.2 Chromosome complement and sex determination

Genotype	Sex phenotypes			
	Mammals (other than humans)	Human	*Silene*	*Drosophila*
2A:XX	♀	♀	♀	♀
2A:XY	♂	♂	♂	♂
2A:XO	♀	♀*	—	♂
2A:XXY	♂	♂*	♂	♀
2A:XXX	♀	♀*	♀	♀
3A:XXX	♀	—	♀	♀
3A:XXY	♂	—	♂	intersex

*The sexual phenotype of humans with an unusual chromosome number in general follows the mammalian pattern. However, the phenotypes of individuals with abormal numbers of sex chromosomes are usually sterile with unusual stature. Furthermore, different species of mammals, including humans, differ in the viability associated with a particular chromosomal constitution. For example, over 95% of 2A:XO human embryos spontaneously abort whereas 2A:XO mice are of near normal viability during pregnancy.

◇ What conclusions might you draw about the role of sex chromosomes in sex determination in mammals and *Silene*?

◆ The Y chromosome channels development in the male direction. The number of X chromosomes plays no role in this important decision.

We now know that the Y chromosome in mammals, including humans, acts as a genetically dominant male determinant, switching development away from femaleness and that in *Silene* it has a comparable role.

However, in *Drosophila*, maleness is not triggered by a genetic switch on the Y chromosome. In fact the Y chromosome has no sex-determining function; XO flies are males, although infertile. As early as 1921, Calvin Bridges (one of a group of geneticists working on *Drosophila* in the 'Fly Room' in California) proposed that sex in *Drosophila* depends on a balance between female-determining factors on the X chromosome and male-determining factors on the autosomes. In normal flies an A:X chromosome ratio of 1:1, e.g. 2A:2X is female and the A:X ratio of 1:0.5, e.g. 2A:XY is male (where A is the haploid autosome set, 2A is the diploid set).

◇ If this hypothesis were true, what might be the expected phenotype of a 3A:2X fly (where 3A is the triploid autosome set)?

◆ Such a fly might be expected to have phenotypic characteristics between those of the two sexes, since the A:X ratio is 3:2 or 1:0.67. And indeed flies with this particular chromosome constitution are **intersexes** with structures intermediate in phenotype.

The most striking parallel to *Drosophila* is found in the nematode worm *Caenorhabditis elegans* where primary sex determination depends on the X:A balance of chromosomes. However, in this species the phenotypic end result is quite different since 2A:2X worms are hermaphrodites and 2A:XO worms are males. There are no worms which are only female in this species.

In the next section we will enlarge on sexual differentiation in mammals, the genetic switch mechanism and the integration of the sexual phenotype of the whole organism. The following section compares plants with mammals. Later sections will deal with the genetic control of sexual differentiation in *Drosophila* and the environmental switch mechanism in reptiles. Finally we will return to a discussion of the dynamic process of development in relation to the adult sexual phenotype and then draw some conclusions.

Summary of Section 6.1

Sexual development arises through a hierarchical sequence of forms.

Each embryo has the capacity to develop into either sex from bipotential rudiments.

Development of the complex sexual phenotypes in animals involves the functional coordination of parts of the organism.

Since different parts of an organism do not have different genes, but some parts express sexual characters, there must be mechanisms that select genes to be expressed in particular cells.

There are two broad classes of switch mechanism, environmental and genetic, that switch development in the direction of one particular sex. The genetic mechanism involves either gene determinants or the A:X ratio. The environmental mechanism may involve the cellular or the external environment.

Question 1 (*Objective 6.3*) Give the switch mechanism that determines the sexual phenotype operating in each of the following organisms (a)–(f).

(a) *Drosophila*

(b) *Bonellia*

(c) Mouse

(d) Humans

(e) *Silene*

(f) Alligators

Question 2 (*Objective 6.2*) Select three accurate statements from (a)–(f).

(a) When cells first give rise to vertebrate genital ducts they are undetermined with respect to sexual morphology.

(b) Experiments have shown that male and female mammals at an early stage in development have different genes which code for sex-specific proteins.

(c) The commitment of a gonad to develop into an ovary or testis depends on a developmental switch.

(d) Once a developmental decision to develop into a particular sex has been made then no intermediate form can be produced.

(e) Integration of the sexual phenotype — including phenotype of the genitalia, gonads, behaviour and physiology — is essential in space but not as the organism develops through time.

(f) Development of the sexual phenotype in animals involves a sequence of stages from egg, separation of progenitor cells of sexual structures, development of bipotential rudiments and differentiation of sex-specific structures.

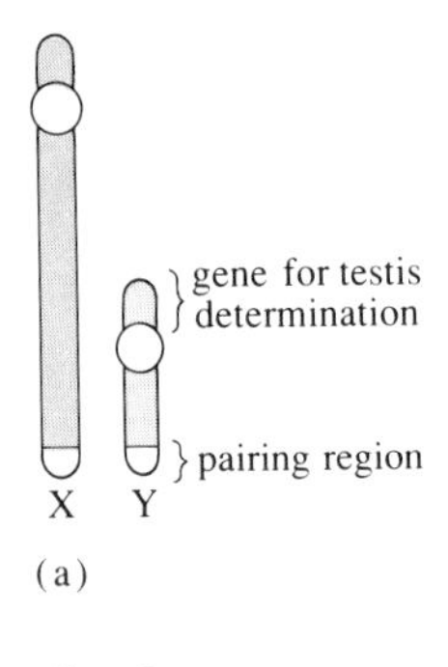

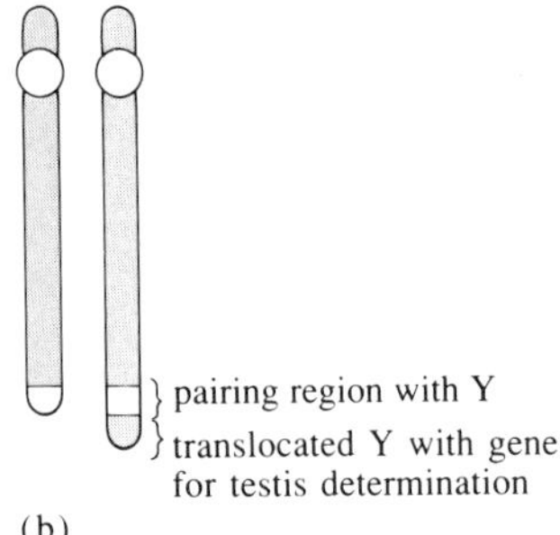

Figure 6.3 Sex reversal in mice. (a) XY chromosomes of a normal male. Note that the pairing regions, in contrast to those in humans, are on the long arms of the X and Y chromosomes and that the X chromosome is quite different in shape from that in humans (Figure 5.1). (b) XX male chromosomes.

6.2 MAMMALIAN SEXUAL DIFFERENTIATION

The embryonic reproductive structures, namely, the gonads, the sets of gonoducts and the structures of the external genitalia, in both sexes in mammals are similar (up to about the seventh week of gestation in humans). The next stage is a response to the developmental switch to become male or female. The undifferentiated gonad is known to develop as an ovary in the absence of any testis-inducing signal, the gene for which was recently located on the short arm of the Y chromosome. The sex hormones produced by the gonads regulate subsequent sexual differentiation and our pursuit of sexual development will depend to a large extent on considering the action of these hormones. We shall begin by considering the genetic switch in more detail, followed by a discussion of the action of hormones that integrate the differentiation of various organs and tissues.

6.2.1 The genetic switch

Until the 1980s, little was known about the link in mammals between the Y chromosome and the development of the testis. A breakthrough came with the study of sex chromosome abnormalities in both people and mice. XX mouse embryos sometimes develop as males! Such sex-reversed mice were cytologically examined and found to have one normal X chromosome and a long X chromosome. This latter chromosome was an X chromosome to which was attached a translocated piece from the Y chromosome, as illustrated in Figure 6.3. Note that in drawing chromosomes, the convention is to show the

Box 6.1 In situ *hybridization*

The position of particular genes on the chromosomes can be observed cytogenetically using the technique of *in situ* hybridization. The procedure is shown in Figure 6.4. Dividing white blood cells (lymphocytes) can be fixed at metaphase of mitosis by the use of the drug colchicine and the chromosomes spread onto microscope slides. Following separation of the strands of DNA, the chromosome spreads are exposed to a radioactive probe. The probe will hybridize (see Figure 3.26) to the complementary sequences in the chromosome *in situ*.

The location of the labelled probes can be detected by autoradiography. By the superimposition of the grains (representing the labelled probes) at a particular site of a chromosome, autoradiography identifies the position of the corresponding genes. The success of these experiments depends on the ability to distinguish between individual mouse or human chromosomes. This can be done by using staining techniques which produce distinctive bands on chromosomes.

Figure 6.4 In situ hybridization on chromosomes.

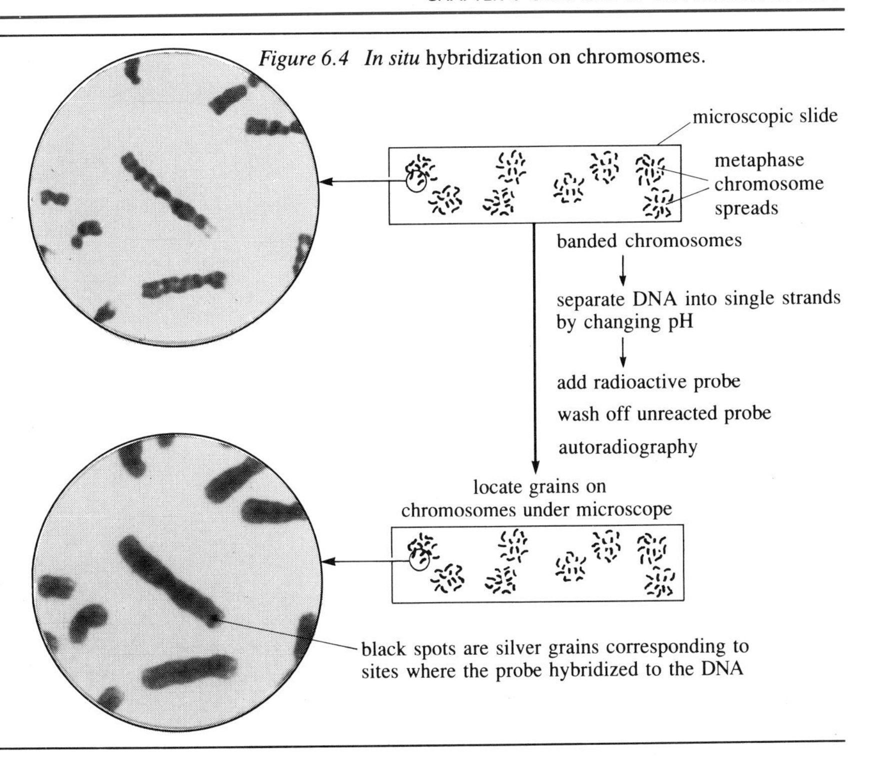

shorter arm above the centromere and the longer one below. In mouse chromosomes the pairing region is on the longer arm and is therefore shown at the bottom. In human chromosomes the pairing region is on the shorter arm (Figure 5.1).

◇ What conclusion can you draw about the genes located on the piece of translocated Y chromosome?

◆ The transposed region of the Y must include the DNA sequences that encode the testis determining factor.

The transposed piece of Y chromosome is termed *Sxr* for sexual reversal. Since it is visible cytologically it must be quite large in terms of the number of nucleotide bases. So it could potentially contain many genes in addition to the testis-determining factor or *Tdy*. It is of interest to know where the *Sxr* region is located in the normal Y chromosome.

To identify which region of the normal Y chromosome is responsible for testis determination, the technique of *in situ* hybridization (Box 6.1) was used on chromosomes. This is similar to the technique described in Chapter 3. However, in this case the positions of the DNA probes are observed microscopically on the chromosomes. In this particular experiment it involved using DNA probes known to hybridize with the *Sxr* region. The results indicated that the *Sxr* region was on the short arm of the Y chromosome.

In humans a different chromosome abnormality was used to locate the gene encoding the testis-determining factor (rather confusingly termed *TDF* instead of *Tdy*). In addition to XX males (which are cytogenetically similar to *Sxr* mice) XY females occur in the population. Such females have a deletion in the short arm of the normal Y chromosome and hence this missing region must contain the *TDF* gene (Figure 6.5).

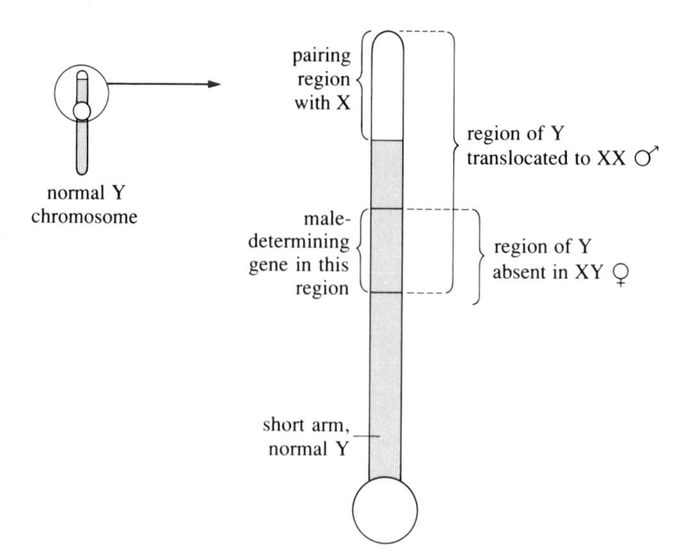

Figure 6.5 The testis-determining region of the human Y chromosome.

◇ How could the *TDF* gene be located in the normal Y chromosome, using information from these XY females?

◆ If deletions are visible cytologically, the missing region of the Y chromosome could be identified.
If the deletions are not visible it would be necessary to use molecular techniques.

The smaller the deletion the more accurately the gene can be localized. Molecular analysis of the genetic deletions of XY females did establish which Y chromosome DNA sequences were absent from these females. This was done by hybridizing extracted DNA with Y chromosome DNA probes on filters — using the **Southern blot technique** (described in Box 6.2) — and then determining the sequence of nucleotides of the appropriate region of the Y chromosome (Figure 6.5). As recently as 1990, *TDF* sequences were precisely identified using powerful molecular techniques, making it possible to determine their structural and functional properties. Using the same probes it was confirmed that these missing DNA sequences were actually present in XX males as a consequence of translocations.

This research on *TDF* and *Tdy* emphasizes the role of genetic information present in the egg and its expression during development as part of an unfolding programme. This tends to reinforce the notion of a preformed plan that guides the embryo, as predicted by the theory of preformation (see Chapter 1). Next we will look at the development of complexity through the interaction of parts, which is identified with the theory of epigenesis.

6.2.2 Integrating the development of the sexual phenotype

To understand how substances encoded in the Y chromosome might be functioning, we need to look at the development of the embryo. We have seen that the mammalian embryo develops into a female in the absence of testis-determining factor. The undifferentiated gonad is made up of two layers, an outer epithelium and an inner medulla, as shown in Figure 6.7a. Sexual differentiation results from the growth and differentiation of one layer and regression of the other. The epithelium forms the ovary and the medulla forms the testis, as shown in Figures 6.7b and c. But which one develops

Box 6.2 Southern blot techique

The Southern blot technique involves a number of stages, the precise details of which are not important here. The DNA is extracted from cells and purified. In humans, DNA is extracted from white blood cells (lymphocytes). Samples of DNA are cut into fragments using restriction endonucleases, enzymes which make cuts in double-stranded DNA molecules at target sites composed of particular nucleotide sequences.

The digested DNA is composed of different-sized fragments which are separated, by analytical gel electrophoresis, into bands according to their size. The size of the DNA fragments can be determined by comparison with the positions of markers of known size (see Figure 6.6a). Following separation of the DNA into single strands, the DNA fragments are transferred on to nitrocellulose paper by capillary action. This process, known as blotting (see Figure 6.6b), results in the paper or blot containing the exact imprint of the gel.

The nucleic acids on the blot can then be hybridized with a radioactive probe. (One way of producing probes is described in Figure 3.26.) The location of the labelled probe can be detected by autoradiography which highlights the fragment corresponding to the gene sequence of interest by giving a positive signal (see Figure 6.6c). If fragments corresponding to the probe are absent from the extracted cellular DNA then no hybridization will occur and no positive signal or band will be observed.

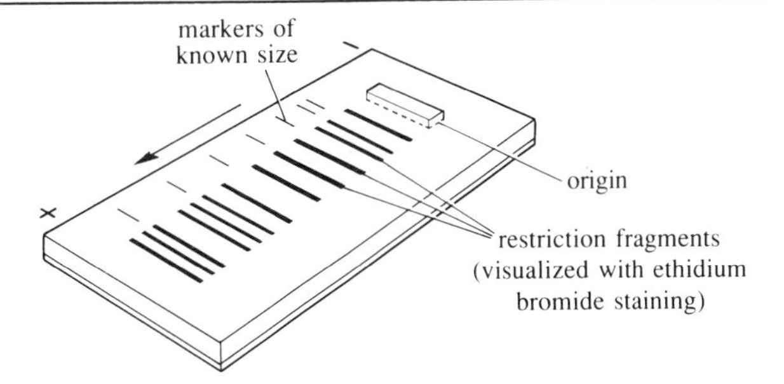

(a) Electrophoresis of restriction fragments on a gel

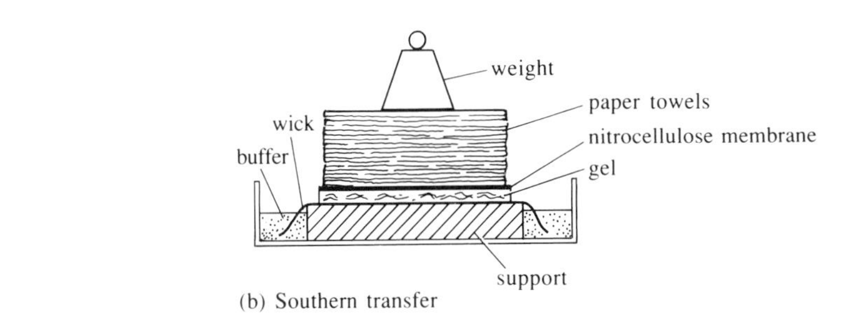

(b) Southern transfer

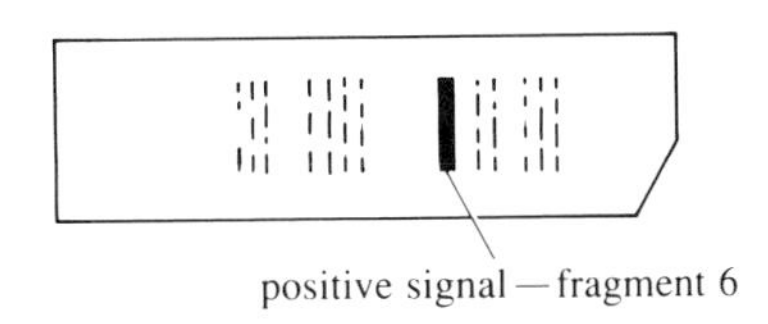

(c) Result of hybridization with radioactive probe and autoradiography

Figure 6.6 Southern blotting and hybridization with radioactive probe.

depends on the genetic switch. The hierarchy of decisions is revealed: first cells are determined to make a gonad but the tissue is still regulative with respect to the kind of gonad. Then the gender of the gonad is determined. The germ cells which will eventually give rise to gametes migrate early in the development to the undifferentiated gonad. Other cells of the gonad secrete sex hormones.

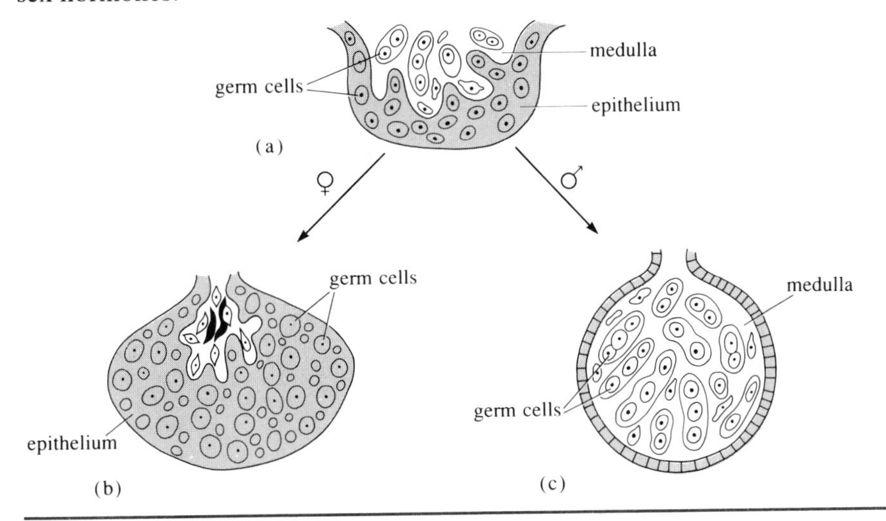

Figure 6.7 Early development of the vertebrate gonad. (a) The undifferentiated gonad. (b) The differentiating female ovary. (c) The differentiating male testis.

The duct systems of both male and female internal genitalia are present in the early embryo (Figure 6.8a). The Müllerian duct in females becomes the fallopian tube, uterus and upper vagina and the Wolffian duct degenerates (Figure 6.8b). The converse occurs in males: the Müllerian duct degenerates and the Wolffian duct differentiates as the epididymis, vas deferens and seminal vesicle (Figure 6.8c). Thus differentiation of the reproductive organs involves the selective elimination of pre-existing structures and growth of surviving structures. This involves the determination of the external genitalia from common structures and their subsequent differentiation. Thus three spatially distinct sets of organs are present in the bipotential embryo (gonads, internal and external genitalia) and their differentiation is temporally separate. We shall now look in some detail at how all these developmental changes both constructive and destructive are coordinated.

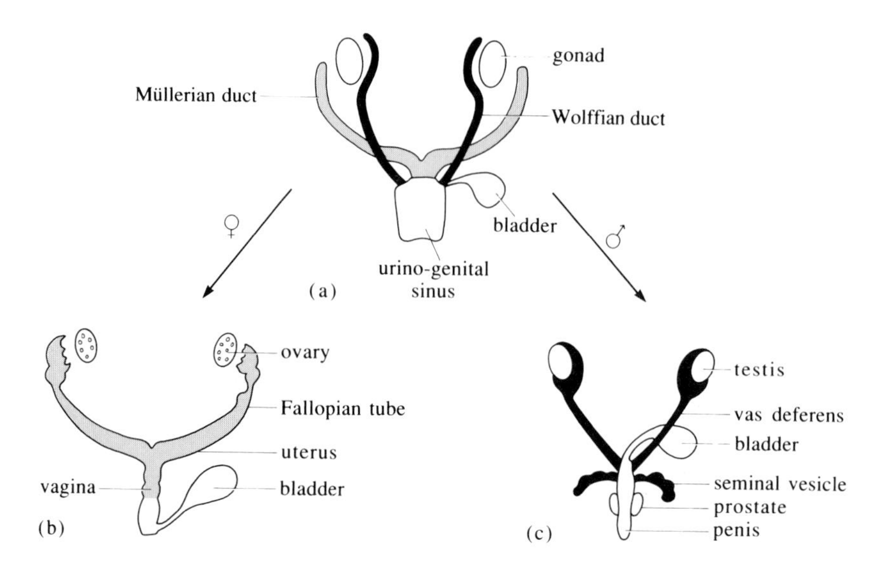

Figure 6.8 The mammalian sex ducts. (a) The bipotential system. (b) Female sex ducts. (c) Male sex ducts.

Male development is established through secretion of three hormones by the testis:

(a) a Müllerian regression factor or anti-Müllerian hormone (AMH) which causes degeneration of the Müllerian ducts.

(b) testosterone which causes the testes and Wolffian ducts to differentiate (as described above)

(c) dihydrotestosterone which induces the masculine pathway of development of the external genitalia.

The sequence of events leading to the formation of the male phenotype in mammals is summarized in Figure 6.9.

The role of the gonad in sexual differentiation was demonstrated by A. Jost in 1965, working in Paris, who removed them from the developing embryo of rabbits and mice whilst at their bipotential stage. Remarkably both sexes developed as females regardless of genetic sex. Now this is surprising because it might have been anticipated that oestrogen, secreted by the ovary, is required for the differentiation of the body in the female direction.

◇ What do these results suggest about the role of oestrogen, secreted by the ovary, for differentiation of the body in the female direction?

◆ The results suggest that oestrogen is not essential for development of female characteristics.

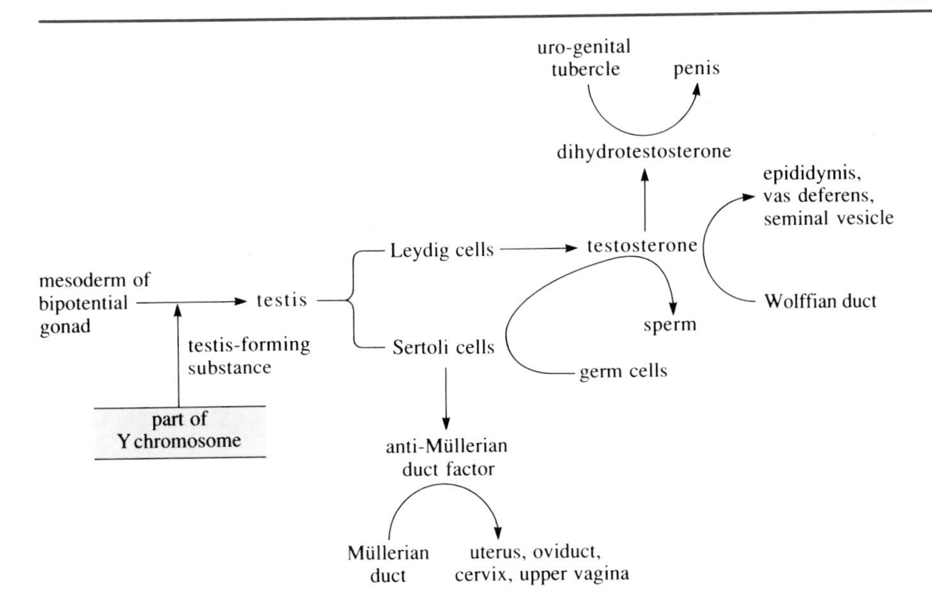

Figure 6.9 Formation of the male phenotype in mammals.

Thus the ovary, the female ducts and genitalia develop if male sex hormones are absent, not because female hormones are present. However oestrogen and other hormones are required for female maturation at puberty.

Clearly sex hormones play a crucial role in initiating the next stage in the sequence of development in males. But they also have a powerful role in integrating the development of different structures with diverse functions in space and, as the organism, develops, in time.

Tissue interactions involving induction, you have learnt, are one of the most important ways in which integrated development occurs in vertebrates. You have already met in earlier chapters examples of groups of cells becoming spatially arranged and so permitting tissue interactions that initiate the next developmental event.

◇Name some of these examples?

◆Formation of the blastodisc in a bird's egg (Figure 3.8). Gastrulation in the sea urchin embryo (Figure 3.9). Gastrulation in amphibians (Figures 3.12–3.14). Embryonic induction by transplanting the dorsal lip of the blastopore into a gastrula.

Another level of integration occurring later in development is achieved by diffusible substances such as hormones and growth regulators. Hormone tissue interactions are strikingly different from the localized nature of tissue interactions during development, since hormones are liberated into the body fluids and presumably reach all tissues of the body. Yet at the same time that genitalia, for example, are responding to these hormones, other cells are not. This specificity is a function of the responding cell population, its competence. An important property that differentiating cells acquire is the ability to be regulated by cues such as hormones. Steroid receptor proteins have been isolated from several adult cell types known to be sensitive to a given steroid. Oestrogen, testosterone and dihydrotestosterone are steroids and therefore suitable receptor proteins must be synthesized in the responding tissue. These receptors are small soluble protein molecules in the cell cytoplasm. The steroid combined with the receptor can enter the cell nucleus and bind to chromosomes (Figure 6.10a) subsequently modifying gene expression. Cell

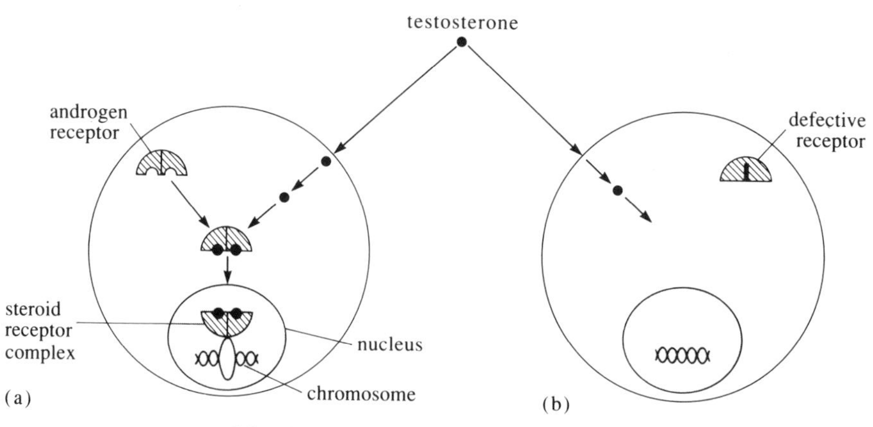

Figure 6.10 The interaction of testosterone and receptor protein. (a) Usual male pathway in an XY individual. (b) Pathway in an XY individual with testicular feminization syndrome.

receptors are covered further in Book 3 of this series*.

The complexity of interactions between hormones and tissues is seen in the development of individuals with one of the rare genetic disorders that reverses the direction of the genetic sex established at fertilization. An X-linked mutant found in both humans and mice, testicular feminization syndrome, renders androgen target organs unresponsive because the receptors in the cells are defective (see Figure 6.10b). So although the testes develop and testosterone is produced in normal amounts, the Wolffian ducts do not differentiate and male genitalia do not develop. Thus the process of communication between testes and other tissues of the body breaks down. Because the cells do not receive the androgen signal they make a wrong decision and proceed to develop along the female pathway.

It is clear that specificity of response to particular hormones is a function of the responding cell population. Is there an underlying sex difference in their response to hormones? We can answer this question by referring to another rare genetic disorder. The adrenal gland (anterior–ventral to the kidney) is important as a secondary source of sex hormones and secretes minute quantities of both oestrogen and androgen in both sexes. In congenital adrenal hyperplasia, the adrenal gland is enlarged and secretes large amounts of testosterone causing masculinization of genitalia and ducts of XX individuals. This occurs because testosterone receptors are found in both male and female target tissue. The important conclusion to draw from these examples of genetic disorders is that the genetic sex of these tissues does not determine their developmental capabilities. It is of interest to compare this situation with that of plants, since the Y chromosome appears to determine maleness both in mammals and plants. So it is to an examination of sex expression in plants that we now turn.

Summary of Section 6.2

The testis-determining DNA sequence in mouse and humans is located in the short arm of the Y chromosome.

The capacity to form the structures of either sex is present in every embryo, regardless of its genetic constitution. The embryonic gonad, the embryonic set of gonoducts and the embryonic structures that give rise to external genitalia, can all form the reproductive organs of either sex.

Sexual development in mammals is in the female direction in the absence of testis-determining factor and male-determining hormones.

*Michael Stewart (ed.) (1991) *Animal Physiology*, Hodder and Stoughton Ltd, in association with the Open University (S203, *Biology: Form and Function*).

Hormones play an important role in tissue interactions and functional coordination.

The study of development of sex differences illustrates how research into molecules and a study of tissue interactions play complementary roles in the understanding of developmental processes.

Question 3 (*Objective 6.4 and Objective 5.2*) If an XX human individual had a piece of Y chromosome containing TDF translocated to an autosome, which one of the following would result?

(a) The individual would be a female because she has two X chromosomes.

(b) The individual would be male because he has the TDF gene.

(c) Neither of the X chromosomes would be inactivated because the individual is male.

Question 4 (*Objective 6.5*) Construct a flow diagram to show the relationships between the genetic switch, and development of testes, male ducts and male genitalia in an XY mammal.

Question 5 (*Objective 6.5*) Construct a flow diagram to show the relationships between the genetic switch and the development of female gonads, ducts and genitalia in an XY mammal with a deletion of the TDF gene.

6.3 SEX EXPRESSION IN PLANTS

Here we will concentrate on angiosperms (flowering plants). Remember that the plant we identify by name is the asexual diploid sporophtye generation. The sexual phase is microscopic and contained largely within the sporophyte.

One of the major differences between plants and animals is the means by which germ cells are derived during development: generally in animals, cells that develop into eggs and sperm are set aside early in development as a separate population of cells or germ-line cells. Very importantly only these cells are destined to give rise to gametes. However, in all plants and some animals such as *Hydra* and urodeles, germ cells are not set aside early as a separate population of cells (see Chapter 1). In animals once gonads develop they remain permanently, whereas plants develop flowers (and gonads) as they need them.

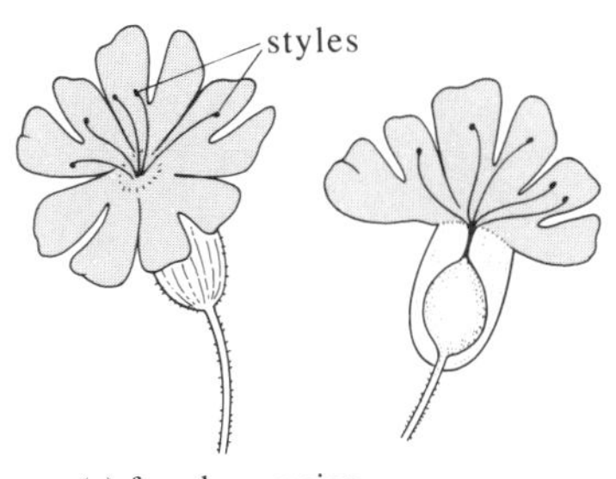

(a) female campion

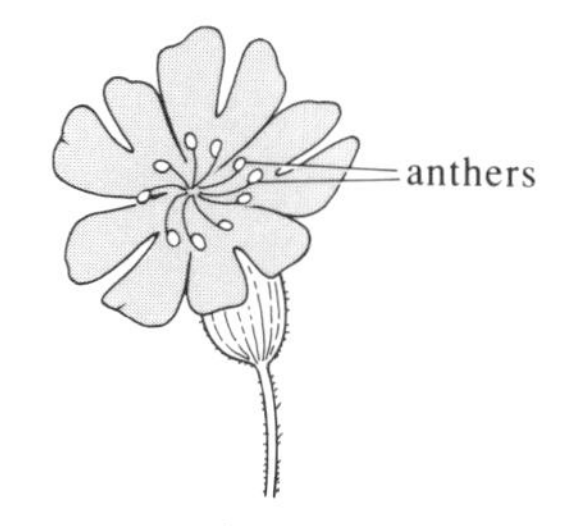

(b) male campion

Figure 6.11 Flowers in the dioecious species *Silene alba* (campion). (a) Female form. (b) Male form.

Flower organization falls into three broad categories which can be arranged in an evolutionary sequence:

1 Primitive flowers are hermaphrodite, with both anthers and ovules, for example, tobacco (*Nicotiana tabacum*).

2 Unisexual flowers with either anthers or ovules, are borne on the same monoecious plant, for example, the monocotyledon maize (*Zea mays*), and the dicotyledons hazel (*Corylus avellana*), beach (*Fagus sylvatica*) and oak (*Quercus pedunculata*).

3 Dioecious species have anthers and ovules in separate flowers on separate plants, see Figure 6.11. Identifiable sex chromosomes are present in some species, but not in all members of this group. Of the species with heteromorphic sex chromosomes, a large proportion have an XY system (see Table 5.1 in Chapter 5). Examples of such heterogametic species are hop (*Humulus lupulus*), dock (*Rumex angiocarpus*) and wild campion (*Silene alba*, formerly *Melandrium album*).

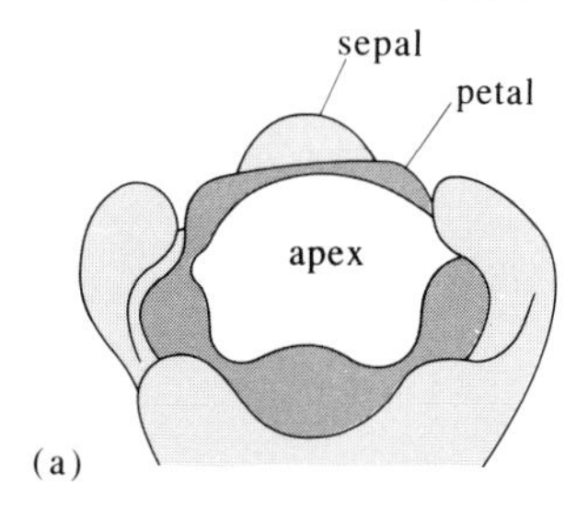

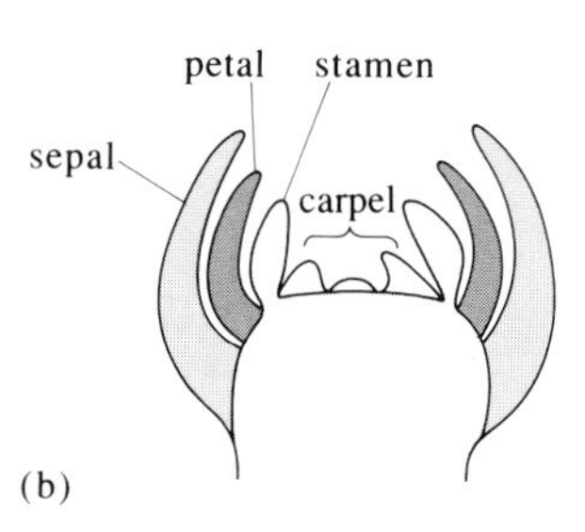

Figure 6.12 Schematic representation of the development of floral organ systems. (a) Floral meristem during the emergence of the petal whorl. (b) Flower longitudinal section after the emergence of the carpel or pistil primordia. The carpel forms at the summit of the floral meristem.

Experimental manipulation of the floral meristem of the hermaphrodite tobacco has shown that the primordia arise sequentially (see Figure 6.12). This is followed by their progressive and irreversible commitment to develop into specific organ systems such as petals or stamens. At early stages of flower development the floral meristem is able to produce any organ of the floral system. Primordia become committed to differentiate into a specific organ as development proceeds, exactly the same as in animals.

Perhaps not surprisingly, unisexual flowers in monoecious plants, like hermaphrodites, have both sets of ovule and anther primordia present in each apex. During early development one or other sex is then suppressed to give either a male or female flower. In some species of dioecious plants both sets of primordia are also present. In others only one set is identifiable, perhaps because suppression occurs very early in development. In terms of the evolution of genetic control, what this means is that only two mutations were required to convert a monoecious plant to a dioecious plant, namely ovule suppression in one set of individuals and anther suppression in another.

The existence of such genes were, in fact, identified in *Silene* by M. Westergaard working in Denmark in 1958, by means of cytogenetic investigations of plants with deletions in the Y chromosome. Figure 6.13 shows the Y deletions observed, along with their associated phenotypes, and the sex chromosomes of normal male and female plants. Part (c) of the figure confirms that each plant has the developmental potential to develop into either or both sexes. Study the whole figure and answer the following questions. The chromosomes are divided into numbered segments for ease of reference.

◇ Since deletion of region 1 of the Y chromosome results in a bisexual phenotype, what gene function must be present in this region?

◆ It must contain a female suppressor gene.

◇ When region 3 of the Y chromosome is deleted the resultant phenotype is a sterile male in which the anthers degenerate. What gene must segment 3 carry?

◆ Segment 3 must carry genes involved in the late stages of anther development.

From comparison of these Y deletions, Westergaard concluded that segment 2 contains genes initiating anther development.

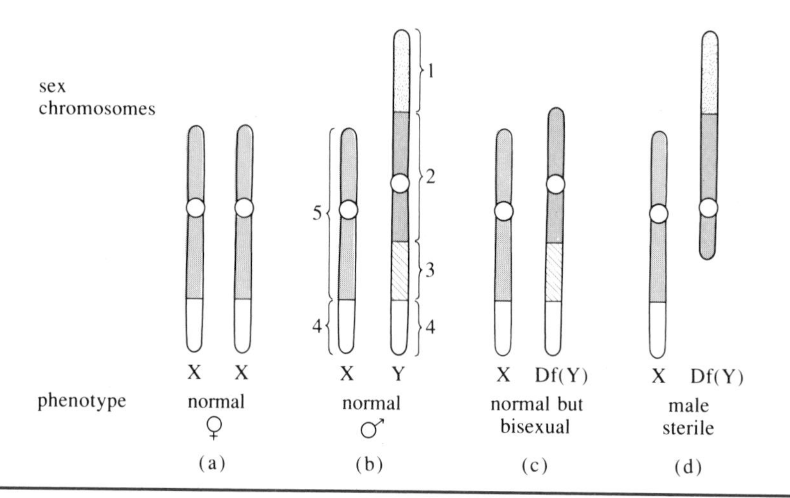

Figure 6.13 Westergaard's (1958) analysis of how Y chromosome deficiencies Df(Y) affect sex in *Silene alba*.

◇ What genes might segment 5, the differential region of the X chromosome, contain?

◆ It presumably carries genes involved in ovule development.

Plants differ very greatly in their sensitivity to external conditions, the initiation of flowering in some species being relatively insensitive whereas in others it is very sensitive. Flower formation can be affected by a range of factors, including temperature, photoperiod, nitrogen supply and mineral supply, and in some cases by exposure to plant growth regulators. At present there is a bewildering mass of facts on the effects of these environmental factors but it is difficult to see how these fit together along with genetic effects to give an overall picture.

Summary of Section 6.3

Germ cells are not set aside in plants as a separate population of cells as they are in most animal species.

The sex mechanisms in angiosperms can be ordered in an evolutionary progression.

Plants, like animals, have the developmental potential to differentiate either (or both) sexual phenotypes.

The genetic sex mechanism in *Silene alba*, a dioecious plant, involves closely linked genes for the activation of one sexual pathway and the inactivation of the other.

Initiation of flowering is affected by many environmental factors.

Question 6 (*Objective 6.6*) List the three steps in the evolutionary sequence of flower organization in angiosperms.

Question 7 (*Objective 6.7*) A particular plant that is normally dioecious has bisexual flowers. Which of the following statements (a)–(c) could explain this observation?

(a) Cytogenetically the plant could be XY with a deletion of part of the Y chromosome containing the female suppressor gene.

(b) Cytogenetically the plant could be XX with deletions for genes for anther suppression.

(c) The plant has the developmental potential to differentiate into either or both sexes.

6.4 SEX DETERMINATION AND DIFFERENTIATION IN *DROSOPHILA*

A comparison of the development of sexual features of *Drosophila* with mammals reveals some striking similarities on the one hand and some dramatic differences on the other. The undifferentiated gonads in *Drosophila*, like mammals, contain both male and female progenitor cells in the two sexes. The internal genital ducts of both males and females in *Drosophila*, again like mammals, are present in every embryo, subsequent development involving the growth of one set and regression of the other. However, sexual development is triggered not by genes on the Y chromosome as in mammals, but by the X:A ratio. Furthermore, their differentiation is regulated not by hormones but by direct gene action.

6.4.1 Sexual dimorphism

Sexual dimorphism in the adult fly is extensive. The external sex differences shown in Figure 6.14 include body size, pigmentation of the posterior abdominal segments, the presence of a row of morphologically unique bristles, the sex comb on the forelegs in males, and external genitalia involving the last few segments of the abdomen. There are also differences in the internal reproductive system and innate behaviour. Most work on the control of sexual development has focused on the events that determine somatic sex differences.

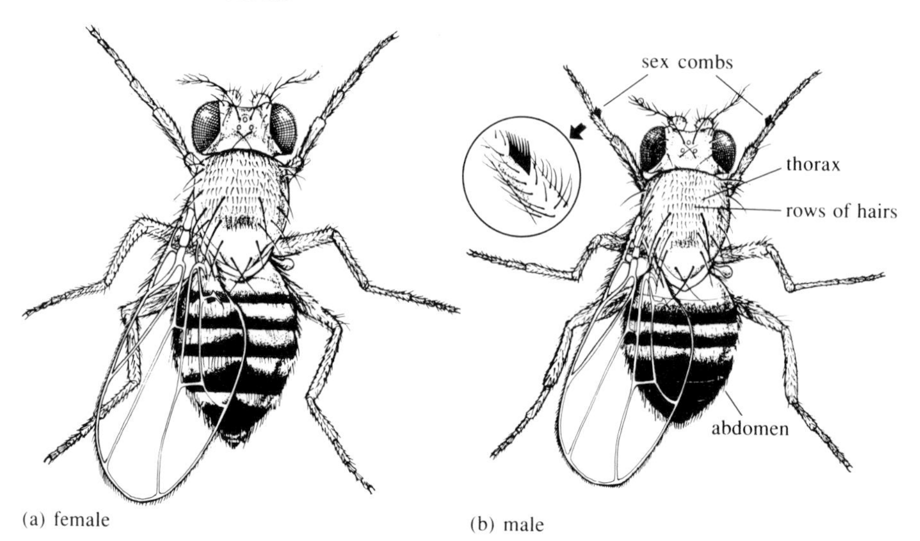

Figure 6.14 Sexual dimorphism in adult *Drosophila*. (a) Female. (b) Male.

The cells that give rise to the somatic structures and gonads of the adult become separated from each other as well as from the progenitor cells of the larval tissues early in embryonic development. Sex dimorphism is not found at the larval stage since all larvae look morphologically identical. During this stage the progenitor cells of the adult are present as undifferentiated dividing cells called imaginal discs (Chapter 4). Not until the pupal stage do these cells stop dividing and undergo differentiation. Nevertheless, it is during embryonic and larval periods that many of the regulatory events occur that determine the later patterns of differentiation including sexual differentiation.

6.4.2 Sex and gynandromorphs

One distinctive feature of sex development in *Drosophila* and other insects, in marked contrast to mammals, is the absence of sex hormones that influence sexual development of the whole fly. Integration of the sexual phenotype is not brought about by sex hormones, but by the genotype of each individual cell. Thus sex (like many characteristics in insects) is a **cell autonomous** phenotype; that is, each cell differentiates according to its own genotype. Consider the following example which demonstrates this phenomenon.

Very rarely individuals arise which contain cells of different genotypes. Such individuals, which can arise as a result of loss of one of the eight chromosomes during development of the embryo, are **genetically mosaic**. Sexual mosaics, part male, part female, arise through the loss of one X chromosome from XX *Drosophila* embryos, giving rise to tissue in the same fly which is either XX (female) or XO (male). Because each cell forms either male or female structures according to the chromosome constitution of the nucleus, phenotypic boundaries between cells of opposite sexual genotype can be identified.

◇ What would be the proportion of male to female tissue if loss of an X chromosome occurred at the first mitotic division of the zygote.

◆ The resultant fly will have half male and half female tissue.

Such sexual mosaics, illustrated in Figure 6.15, are called **gynandromorphs**.

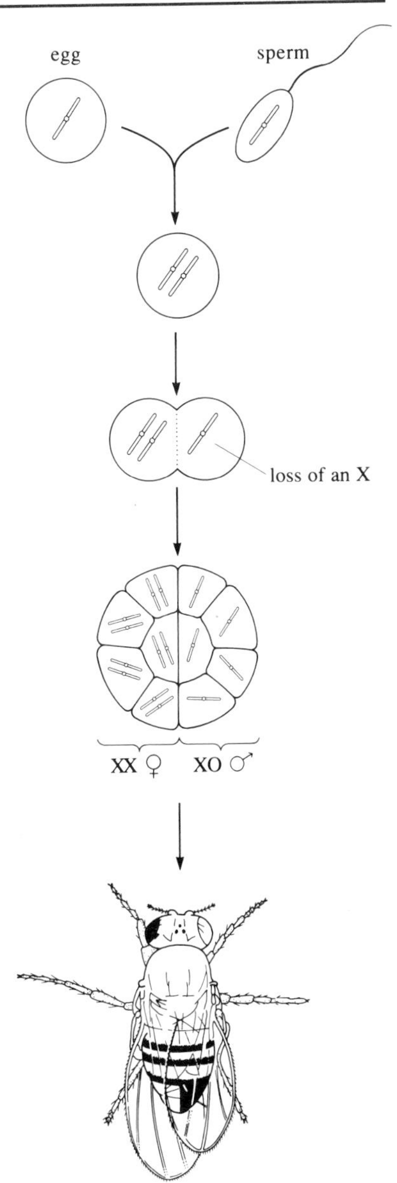

Figure 6.15 Production of a mosaic by the loss of an X chromosome shortly after fertilization in *Drosophila*.

6.4.3 Regulation of sex development in *Drosophila*

As in the case of mammals, mutants have proved to be an indispensable tool for studying sex development in *Drosophila.* When this is combined with molecular and developmental biology, the principles of the control of sexual differentiation are revealed. Much more is known about gene action at the molecular level in *Drosophila* than other animals. As mentioned in Chapter 4, we know a lot about *Drosophila* because it is easily cultured in the laboratory, it produces large numbers of offspring and it has already been studied extensively. However, it is important to note that its mechanism of sex determination is not representative for the group Diptera.

In order to understand the genetic interactions of the sex-determining and sex-differentiation genes, it is important to bear in mind the relationship between DNA, mRNA and proteins. This is illustrated in Figure 6.16.

Genes in eukaryote organisms consist of exons interspersed with introns. The nucleotide sequence of exons are ultimately expressed, while the intervening sequences, the introns do not code for a product. The primary RNA produced by transcription from genes is longer than the mRNA that is ultimately translated. This is because there is a stage between transcription and translation during which the RNA is modified. The transcribed intron RNA is spliced out and the RNA transcribed from exons is joined together. Some genes are quite complex and contain many exons. More than one gene product can be produced from the same gene by splicing together of different combinations of exons (see Figure 6.16). This is true both for some regulatory genes and for some genes which determine structural proteins.

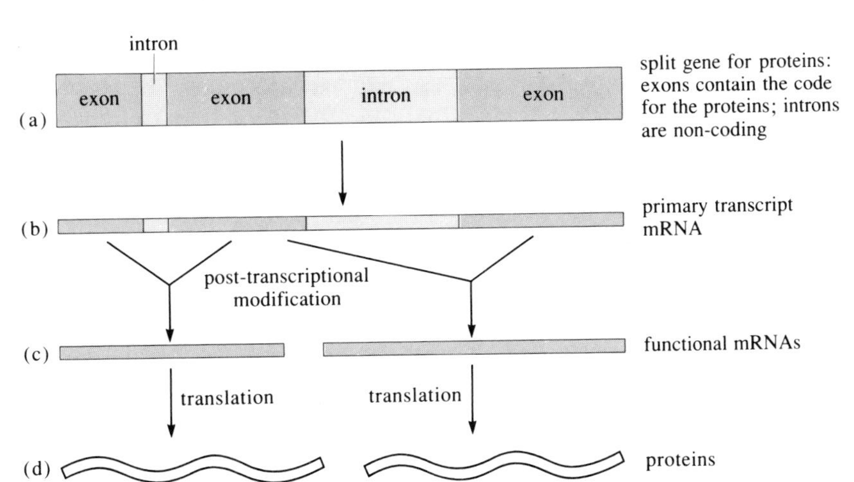

Figure 6.16 The transcription and subsequent translation of a split gene. Note that both exons and introns are initially transcribed. The newly produced mRNA is then modified to the form in which it is translated.

The genetic pathway leading to the development of either the male or the female sexual phenotype in *Drosophila* is sequential, as shown in Figure 6.17. One important feature of the model is that sex development involves genes which regulate other genes. The sex-differentiation or target genes (d_1–d_4) are under the control of the sex-determining or regulatory genes (s_1–s_4). The detail of each step described is not so important as an understanding of the general principles and sequence of events leading to sexual differentiation. We shall explore the action of some of these genes in more detail, beginning with the early-acting genes.

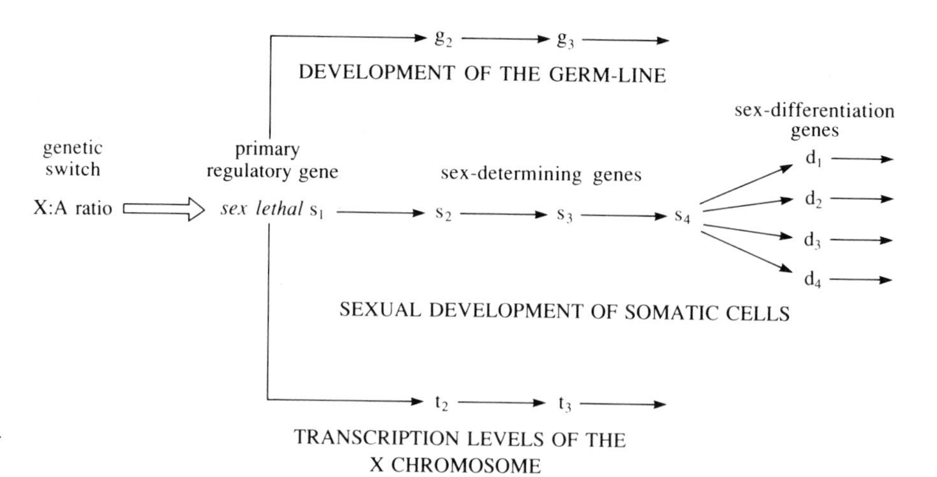

Figure 6.17 The genetic sequence of sexual development in *Drosophila*.

At the top of this hierarchy or sequence is the X:A ratio. This affects sexual development through its action on the gene *sex lethal* (*sxl*) which is the primary regulatory or sex-determining gene. This gene in turn regulates three separate series of regulatory genes, each series of which controls one of the pathways, namely, development of the germ-line, sexual development of somatic cells, or transcription levels of the X chromosomes. Thus these three pathways of development have a common genetic control at an early stage of development.

We focus in this chapter on only one of these pathways, the development of the somatic sexual phenotype.

Sex-determining genes

A number of the sex-determining genes shown in Figure 6.17, have been identified. These sex-determining genes have sex-specific functions. Interestingly, this specificity is the result of alternative splicing of the primary gene products rather than sex-specific transcription. Thus the same primary gene product is differently spliced at the post-transcriptional level to produce different gene products which specify male or female development.

In females, splicing results in a functional transcript which is translated, whereas in males the transcripts are non-coding. This is due to the presence of an additional exon with stop codons which prevent the transcript from being translated.

The last-acting sex-determining gene called *double sex* (*dsx*), shown as s_4 in Figure 6.17, is a pivotal gene in the regulatory pathway. The gene is spliced either into male-specific or female-specific versions of coding transcripts. The specificity of function of *double sex* is regulated by the other sex-determining genes including *sex lethal*. Mutations which result in no *dsx* function convert

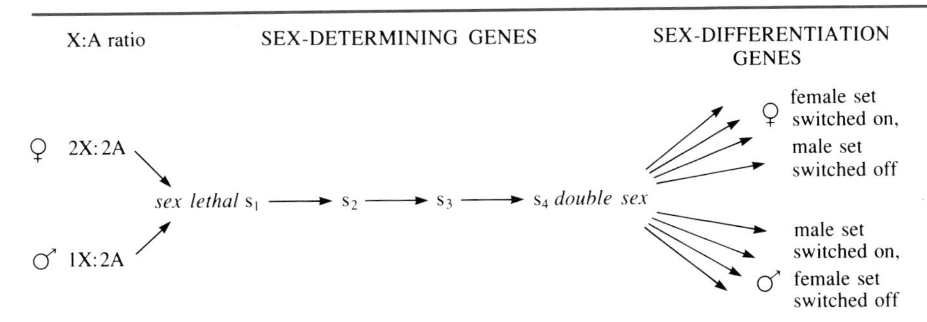

Figure 6.18 Control of somatic sex development in *Drosophila*.

both XX and XY individuals into intersexes with poorly developed male and female genitalia and with both ovaries and testes. It has been suggested that the 'male' activity of the wild-type *dsx* gene represses female development while allowing male development; and that the 'female' activity of the wild-type *dsx* gene acts in the opposite manner to specify the female pathway and repress male-specific differentiation, as illustrated in Figure 6.18.

One important lesson we learn from studying sex-determining mutants is that genes do play a role in developmental decision-making. The failure of any one of the regulatory genes can affect the developmental fate of a tissue, shifting the phenotype away from one sex to the other. This suggests that the products of sex-determining genes regulate the activity of other genes necessary for the differentiation of particular tissues. It is to a study of this regulation that we now turn.

Sex-differentiation genes

In contrast to the sex-determining genes, the sex-differentiation genes produce mRNA transcripts in one sex only — their transcription is switched on or off by the sex-determining genes. When and how the sex-determining genes regulate the expression of the target or sex-differentiation genes has been elucidated by examining mutants at the molecular, tissue and whole organism levels. These studies have revealed two different mechanisms by which the sex differentiation genes are regulated. These are summarized in Figure 6.19.

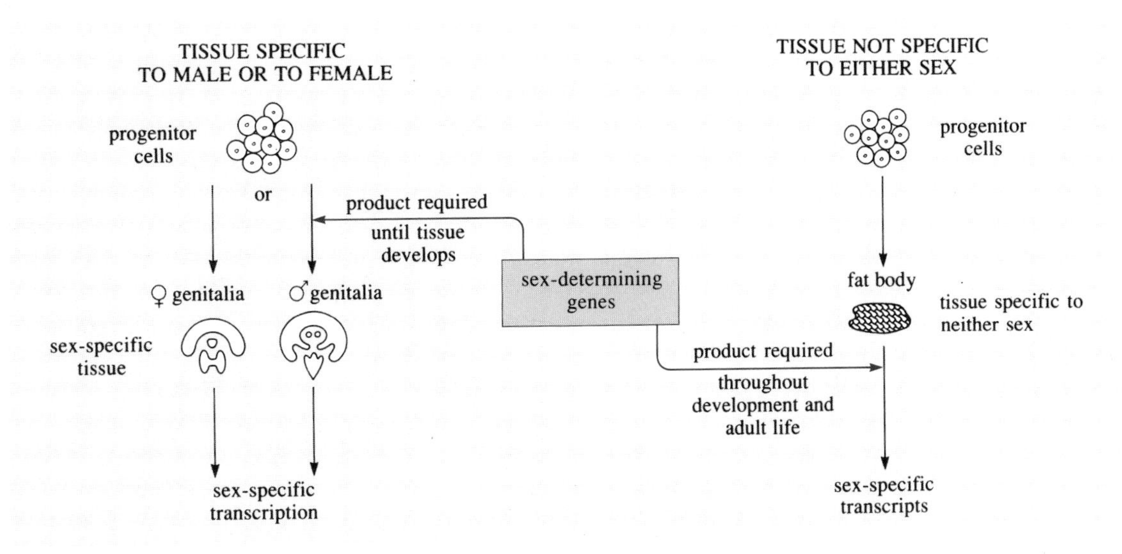

Figure 6.19 Two ways in which sex-determining genes regulate sex-differentiation genes.

1 One way in which sex-determining genes regulate sexual differentiation is by influencing the formation of sex-specific tissues such as genital ducts and glands. Each larva has both male and female genital primordia. Which primordium develops is specified by the sex-determining genes and in particular by *dsx*. Once the decision to develop male genitalia, for example, has been made then continued control by the sex-determining genes is no longer required. When the fly emerges from the pupa, genital development is complete.

2 A second way in which sex-determining genes regulate the sex-differentiation genes occurs in tissue which is not specific to a particular sex, such as fat body. Although this tissue is common to both sexes, there are protein differences. For example, three yolk proteins are synthesized in the fat body of female flies only. In contrast to the regulation in sex-specific tissue, the synthesis of yolk proteins depends on the continuous function of the sex-determining genes not only during the larval stage but throughout adult life.

It is possible that further mechanisms of sex regulation by sex-determining genes will be revealed from detailed study of other sex-differentiation genes. In addition, results from such studies may help to explain how their products affect other characteristics such as behavioural differences between sexes.

6.4.4 Sex-determining mechanisms in other diptera

Since the mechanism of sex determination between mammals and insects is different it is of interest to know whether the X : A ratio of sex determination is common to all dipteran insects. In fact, the mechanism of sex determination in *Drosophila* is found in a very few insect species although cell autonomy of sex expression in general does apply.

◇ Why should sex development as a cell-autonomous phenotype be common to insects?

◆ Sex hormones are generally absent from insects.

Dipteran insects illustrate a somewhat daunting variety of sex-determining mechanisms (as shown in Table 6.3). There is an indication in some species of a male dominant determining factor on the Y chromosome as in the house fly, *Musca domestica*. In the case of the mosquito, *Anopheles*, the male-determining factor is found on different chromosomes in different strains! In contrast to these species, in *Chrysomya rufifacies* sex is determined by a dominant female-determining factor.

Table 6.3 Some examples of sex-determining mechanisms in dipteran insects

Sex-determining mechanism	Genus/species
X : A ratio	*Drosophila*
male-dominant-determining factor on Y	*Musca domestica*
male-dominant-determining factor; position varies from strain to strain	*Anopheles*
female-dominant-determining factor	*Chrysomya rufifacies*
XX–XO system of somatic cells	*Sciara coprophilia*
Environmental sex-determination	
— temperature	*Aedes simulans*
— nutritional state of haemolymph	*Heteropeza*

Other insects have more complex systems. All zygotes of the dipteran fly, *Sciara coprophilia* have two sets of autosomes, one of maternal origin (A^m) and one of paternal origin (A^p), but have three X chromosomes, one being of maternal origin (X^m) and two of paternal origin (X^pX^p). Hence zygotes have the chromosome constitution $A^mA^pX^mX^pX^p$. During development one X, of paternal origin, is eliminated in the germ cells of both sexes.

Therefore the germ-lines are identical in both males and females, that is, $A^mA^pX^mX^p$. The somatic cells of females have the same composition as the germ-line since they also lose an X^p chromosome but the somatic cells of males lose two X^p chromosomes and are thus chromosomally $A^mA^pX^m$. This is summarized diagrammatically in Figure 6.20.

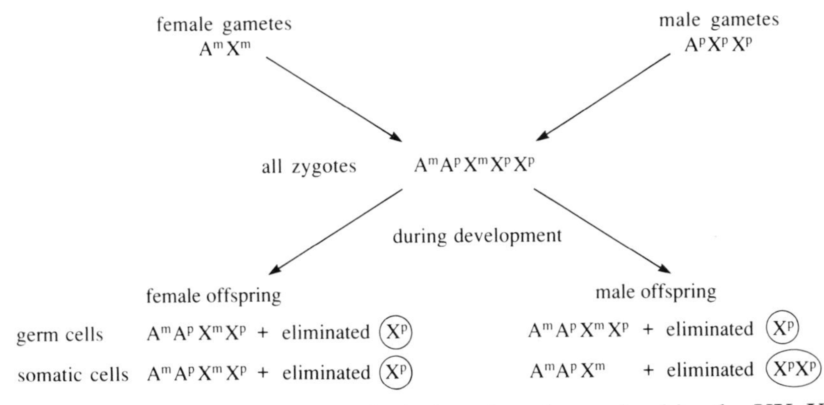

Figure 6.20

Phenotypic sex in *Sciara coprophilia* is therefore determined by the XX–XO system of the somatic cells and gonadal differentiation depends on the amount of gene product produced by the X chromosome of somatic tissue. Therefore in *Sciara*, cell autonomy of sex expression does not apply but instead there is an interaction between the gene product of the somatic cells and the cells of the germ-line!

Environmental sex determination is operative in some Diptera. In the sub-arctic species of mosquito, *Aedes simulans*, sex determination depends on the temperature and the genotype of the developing egg. A system with complete environmental sex determination is found in the gall midge, *Heteropeza*, where the nutritional state of the haemolymph of the female determines the sex of her offspring. However, environmental sex determination is rare in insects but it is characteristic of many species of reptile and it is to a study of this group that we now turn.

Summary of Section 6.4

Sexual dimorphism in the adult *Drosophila* affects the body extensively.

Sex, in *Drosophila*, is a cell-autonomous phenotype. Each cell differentiates according to its own genotype.

Sexual development of the bipotential parts is triggered by the X:A ratio.

The sex-determining genes play a role in developmental decision making and regulate the activity of the sex-differentiation genes.

Products of the sex-determining genes either direct the formation of sex-specific tissues or regulate the activity of sex-differentiation genes in tissue not specific to either sex.

An enormous variety of sex-determining mechanisms exist in dipteran insects, that of the X:A ratio being quite rare.

Question 8 (*Objective 6.8*) A fruit fly has female genitalia, ducts and ovaries but has a male sex comb on the right foreleg (see Figure 6.14). Which of the following statements (a)–(c) could explain this observation.

(a) The fly developed from an XX zygote but lost an X chromosome in the cells that developed into the right foreleg.

(b) The fly developed from an XY zygote but lost a Y chromosome in all cells other than those that developed into the right foreleg.

(c) The fly developed from an XY zygote with a deletion for part of the Y chromosome.

Question 9 (*Objective 6.9*) Construct a flow diagram to show how (a), (b) and (c) are linked in the development of the sexual phenotype of *Drosophila*.

(a) the X:A ratio

(b) the sex-determining genes

(c) the sex-differentiation genes.

6.5 TEMPERATURE AND SEX IN REPTILES: COLD FEMALES AND HOT MALES

A number of studies have collectively suggested that sex can be determined by incubation temperature in many, but not all, reptiles. Laboratory studies backed by field observations indicate that the temperature at which eggs are incubated affects the sex ratio of hatchlings. We discuss some of these studies below. The results show that nest temperature determines sex; giving rise to unusual sex ratios, as well as indicating that genetic sex determination is not operative. In this Section we concentrate on the switch mechanism of temperature sex determination (TSD) and compare it with that of genetic sex determination (GSD) found in mammals and *Drosophila*.

6.5.1 Temperature sex determination

We begin by looking at the relationship between temperature and sex determination in a species of alligator, *Alligator mississippiensis*. Unlike most birds, which sit on their eggs and incubate them, alligators leave their eggs buried in decaying vegetation. In 1982, Mark Ferguson and Ted Joanen removed eggs from wild nests in the Louisiana swamps within 12 hours of egg laying. The eggs were randomly placed in one of six temperature groups (26°C, 28°C, 30°C, 32°C 34°C, 36°C) and incubated in laboratory incubators with nesting conditions simulating those found in the wild.

The sex of hatchlings in each temperature group is given in Table 6.4. The sex of the alligators was resolved at the time of hatching both macroscopically by dissection of the reproductive tract and microscopically by histological examination of the gonads. There was no evidence of structural intersexes or hermaphroditism, indicating that sex is fully determined by the time of hatching. This is particularly interesting in the case of eggs incubated at intermediate temperatures. Note that eggs incubated below 26°C or above 36°C died.

◇ What conclusions about the temperature of egg incubation and sex of hatchlings can you draw?

Table 6.4 Sex and mortality of alligators artificially incubated at various temperatures

	Temperature of egg incubation (±0.2°C)					
	26°C	28°C	30°C	32°C	34°C	36°C
No. of eggs	50	100	100	100	100	50
No. of dead embryos (% of total)	40(80)	4(4)	3(3)	2(2)	6(6)	43(86)
Females (% of surviving eggs)	100	100	100	86.7	0	0
Males (% of surviving eggs)	0	0	0	13.3	100	100

◆ (a) All surviving embryos incubated at ≤30°C developed into females whereas those incubated at 34°C or greater developed into males. Surviving embryos incubated at the intermediate temperature of 32°C had a sex ratio of 86.7%: 13.3%, females to males.

(b) At the extreme temperatures a high percentage of embryos died.

Although the data do not show it, it is reasonable to conclude that these embryos would have developed into all males (36°C) or all females (26°C).

◇ Could differences in embryonic mortality between eggs incubated at 28°C, 30°C, 32°C or 34°C account for the differences of sex ratios between these temperature groups?

◆ Even if all the dead embryos had been of opposite sex to those that were living, the differences between the temperature groups would still be statistically significant.

It might be argued that this association between sex and temperature in the laboratory might be different in the wild where nest temperatures fluctuate depending on the topographical location of the nest, the prevailing weather and, of course, whether the embryos have a genetic predisposition towards the sex. To test this, Ferguson and Joanen placed temperature probes, connected to continuous chart recorders, in different locations within wild alligator nests in order to monitor the temperature throughout the incubation period (Figure 6.21). Temperatures were recorded for 60 days (normal incubation is 65 days) after which the nests were opened and accurate nest maps drawn to indicate the position of eggs and temperature probes. Sex of each animal was correlated with position in the nest and nest temperature. It is interesting to note that where the average nest temperature fell between

Figure 6.21 Map showing the location of eggs and temperature probes (TP) in a dry marsh nest of *A. mississippiensis*.

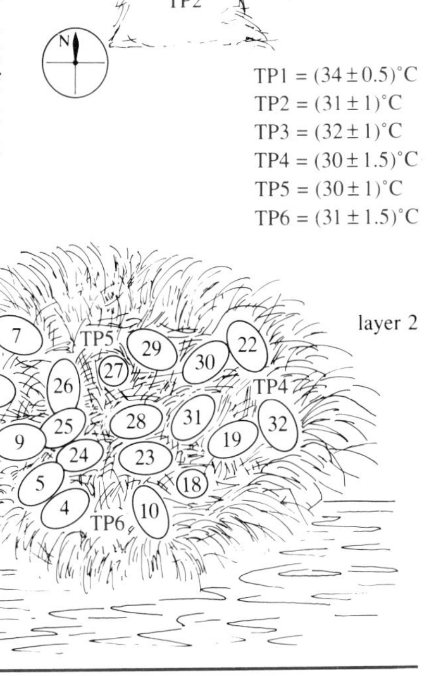

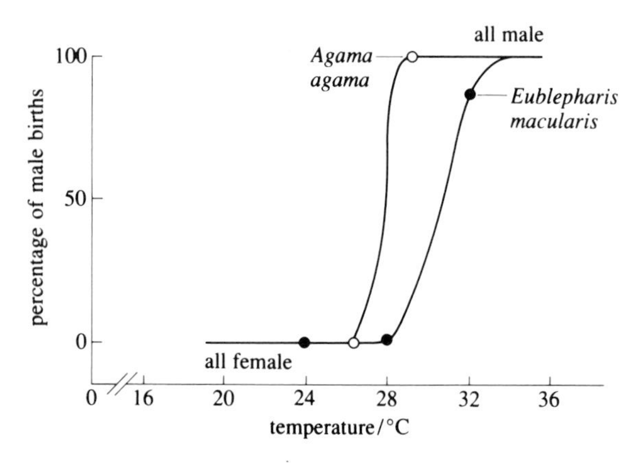

(a) Lizards (2 species)

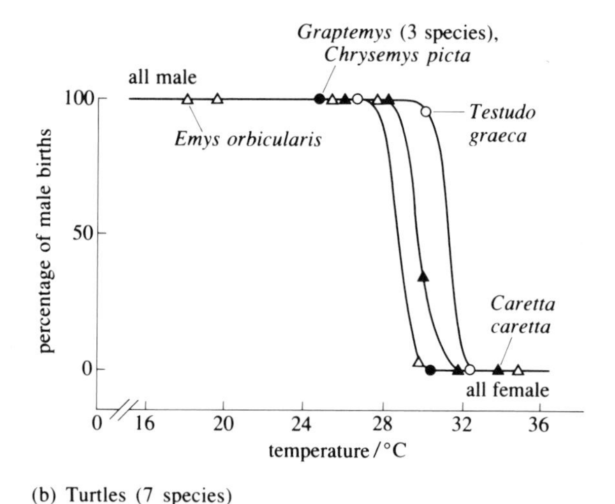

(b) Turtles (7 species)

Figure 6.22 Sex ratio and incubation temperature in some reptile species.

34°C and 35°C, the alligators were male whereas at the average nest temperature of 29°C to 31°C, the alligators were female. Evidently sex is determined by egg incubation temperatures in the wild as well as in laboratory experiments.

In addition, they observed that surviving embryos incubated at intermediate temperatures in the wild develop into either males or females as in laboratory experiments (see Table 6.4 surviving eggs at 32°C). These observations suggest that other environmental factors may also play a role in sex determination in reptiles.

Reptiles in which TSD operates show various relationships between temperature and sex determination. Figure 6.22 gives the responses in sex ratios to incubation temperature in some other species of reptiles. Study this figure and answer the following questions about the relationship between temperature and sex.

◇ Do all reptiles show the same pattern of male producing and female producing temperatures?

◆ No. Lizards show the same pattern as alligators but turtles show the reverse pattern. Low temperatures produce females in lizards and males in turtles.

◇ Is the temperature at which the shift in sex ratio occurs the same in all species?

◆ No, it is not the same. However, in each of the species shown in Figure 6.22a and b it does lie between 27°C and 32°C.

Sensitive period

Since the genetic switch in mammals occurs at a specific developmental stage it is of interest to know whether there is a particular developmental period when the TSD switch mechanism is sensitive to temperature. A few studies have been undertaken to answer this question, using the following procedure. Eggs of alligators are incubated at a male-producing temperature for the first stages of the developmental period and then shifted to a female-producing one for the remainder of the developmental period (or vice versa). The stage at which the shift in temperature is made differs for each sample of eggs. Each group of eggs gives rise to hatchlings which are all male, all female or some of each. When results from different groups of eggs were compared it was found that temperatures up to a certain developmental stage had no effect on sex and those beyond a second stage had no effect on the sex ratio. But between these two stages, sex determination was influenced by temperature. This interval between these two stages of development is defined as the sensitive period. This period is about a week or so in length, occurring in the middle of development, perhaps not surprisingly coinciding with gonad differentiation. This contrasts with mammals where the switch occurs much earlier in development, and during a short specific developmental stage.

Consequences of TSD

A remarkable consequence of TSD is the unequal sex ratio in populations of these species. For a population of *Alligator mississippiensis* in the Wild Life Refuge of Louisiana, the female : male ratio was found to be close to 5 : 1. The sex ratio heavily biased towards females in this species means that classical genotypic mechanisms probably have no role in alligator sex determination.

Why some reptiles have TSD and not genetic sex determination (GSD) is a matter of conjecture, although some scientists argue that this mechanism of sex determination presents a selective advantage by skewing the sex ratio. Moreover, although the sex-determining mechanisms of ancient reptiles are unknown, the selective extinction of dinosaurs might be explained by the relatively sudden change in climate to one that is either hotter or colder.

It would be interesting to know how the mechanism of TSD works at the physiological and biochemical levels to trigger gonad differentiation. A particular nucleotide sequence which is similar to the testis-determining factor in mammals has been identified in alligators. Thus the actual genes involved in sex determination in vertebrates with TSD may be very much like those with GSD. The difference between the two groups is in the regulation of these genes and the time during development at which they function. As in mammals, once gonad differentiation has been triggered by temperature in species with TSD the fetal gonad produces sex hormones which influence further development of the sexual phenotype.

A group of lizards found in Australia, the *Gehyra* lizards, may contribute to our understanding of the relationship between TSD and GSD. Races which are apparently phenotypically identical can be distinguished by their mode of sex determination — some having heteromorphic sex chromosomes and others having TSD.

Recent observations show that the temperature experienced by the embryos of reptiles (with TSD) not only determines gonad differentiation, but directly influences other aspects of sexual phenotype such as reproductive behaviour and endocrine physiology. Thus environment affects various levels of sexual phenotype (morphology, physiology and behaviour). In the next section we go on to examine the complexity of the sexual phenotype and its development in more detail.

Summary of Section 6.5

Egg incubation temperature determines sex in some reptiles; lizards, alligators and turtles. There is a temperature-sensitive period during development which coincides with gonad differentiation.

The pattern of male-producing and female-producing temperatures differs between species.

One consequence of TSD, which cannot be accounted for by differential egg mortality, is the unequal ratio of sexes in a population.

Question 10 (*Objective 6.10*) Which of the following statements (a)–(e) provide evidence for TSD in alligators.

(a) Absence of sex chromosomes in this genus.

(b) Heavily biased female sex ratio.

(c) All eggs incubated ≤30°C develop into females, whereas eggs incubated at ≥34°C develop into males.

(d) There is no evidence of structural intersexes in eggs incubated at intermediate temperatures.

(e) The length of the sensitive period during which the switch mechanism is sensitive to temperature is about a week.

6.6 THE SHIFTING BOUNDARY OF THE SEXUAL PHENOTYPE

So far we have restricted discussion to chromosome composition and to the development of the gonad and morphological characteristics. However, there is more to the adult sexual phenotype than this. Here we examine in more detail the complex nature of the sexual phenotype of the adult organism.

Sexual differentiation is not limited to chromosome composition, gonad structure and morphology. In addition, there exists physiological sex (which refers principally to the nature and pattern of hormone secretion), behavioural sex and, in humans, psychological sex (which refers to gender identity). But is there a consistent relationship between sex chromosomes, body form, gonad sex and behavioural sex?

You have already seen some examples in mammals and *Drosophila* that demonstrate that the relationship between these different features of the sexual phenotype is not always consistent. Having female sex chromosomes, XX, does not necessarily make the human or *Drosophila* body female. Recall, for example, XX *dsx*/*dsx* individuals are intersexes with a body form intermediate between that of male and female. Nor is there a consistent relationship between gonad sex and body form.

◇ Recall an example where male gonad sex can be separated from male morphological sex in mammals.

◆ In testicular feminization syndrome (Section 6.2.2) in mice and humans, XY individuals with androgen-secreting testes develop female secondary sex characteristics due to defective cellular androgen receptors.

There is evidence that sex hormone levels early in development exert a powerful influence on sexual physiology at maturity. There are three times in the life history of a human male where there are peaks of testosterone synthesis: six weeks after fertilization, around the time of birth and at puberty, see Table 6.5. The perinatal testosterone production directs the hypothalamus and pituitary (endocrine glands) to release their hormones in a 'male' pattern as a continuous secretion at puberty. In the absence of perinatal testosterone these endocrine glands secrete hormones in a cyclical 'female' pattern (see Table 6.5).

Table 6.5 The effects of hormones on reproductive development in male and female mammals

Stages of development	Male	Female
early in fetal life	testosterone, dihydrotestosterone and AMH establish male development	absence of these hormones establishes female development
perinatal (around time of birth)	testosterone imprints endocrine glands to release hormones in a 'male' pattern as a continuous secretion at puberty	absence of testosterone imprints endocrine glands to secrete hormones in a cyclical 'female' pattern
at puberty	increase in endocrine gland hormones leads to growth of testes and testosterone output → male secondary characteristics	increase in endocrine gland hormones leads to growth of ovaries and oestrogen output → female secondary characteristics and ovulation

It is clear that sex hormones also exert a significant influence on sexual behaviour in both mammals with chromosomal sex determination and in reptiles with TSD. This was dramatically demonstrated in a classic experiment in which newborn morphologically female rats were injected with testosterone. At maturity these rats exhibited male copulatory behaviour, possibly because part of the brain had been rendered structurally similar to that of morphological males. Thus hormones influence many organs which are all parts of the developing phenotype.

The same is true for species with TSD. A shift in reproductive behaviour away from that typical of the morphological sex has been found in the leopard gecko (*Eublepharis macularis*). W. H. N. Gutzke and David Crews working in the USA recently (1988) showed that not only is morphological sex determined by the incubation temperature but so also are other features of the sexual phenotype including endocrine physiology and reproductive behaviour. Embryos of this species incubated at high temperatures develop into morphological males whereas those incubated at low temperatures developed into morphological females. However, none of the females from 'hot' incubation temperatures, i.e. the highest temperatures producing a female anatomy, mated or laid eggs, and many responded to the courtship of males as if they themselves were males. It would appear that the primary determinant of sex, in this case temperature, also directly affects non-gonadal aspects of sexual development, possibly as in mammals by 'setting' the hypothalamus.

Similarly in *Drosophila*, which does not have sex hormones, individuals can have behaviour patterns typical of the opposite sex. *Drosophila* predominantly exhibit reproductive behaviour characteristic of their own gonadal sex which involves clearly defined courtship patterns for both males and females. In 1975, Seymour Benzer, working at the California Institute of Technology, reported a remarkable mutant of *Drosophila* called 'fruity'. Chromosomally male individuals homozygous for this mutation are morphologically male. However, 'fruity' flies behave like females in that they stimulate other males to court them and instead of rejecting courting males they show 'female' acceptance response to mounting.

In these examples, what we see is that sexual differentiation is a dynamic phenomenon involving the whole organism as it develops through time. Disturbances, genetic, hormonal or environmental, can cause changes in the developmental pathways followed by an organism, resulting in the transformation of part of the phenotype, such as secondary sexual features or behavioural sex, or of the whole organism as in the case of XY mammals with a deletion for *TDF* and in reptiles with TSD.

Summary of Section 6.6

Sexual differentiation is a dynamic phenomenon involving the whole organism as it develops through time. The result is that the relationship between the various features of the sexual phenotype is not consistent. Chromosomal, gonadal, morphological, behavioural and physiological sex can have different expressions within an individual.

Question 11 (*Objective 6.11*) Freemartins are sterile XX calves whose gonads have been completely or partially masculinized. These individuals are only found when the affected cow is one of a pair of twins, one male and one female, and the placental blood circulation has united the two fetuses. Suggest the possible cause of freemartins.

6.7 CONCLUSIONS

The examples reviewed here illustrate some of the variety of sex mechanisms found in eukaryotes. Both male and female primordia are present together in all sexually dimorphic organisms studied, and development of gonadal sex in animals and male or female flowers in plants, depends on the growth of one of these and suppression of the other. The decision as to which will grow depends either on an environmental switch or on a genetic switch.

What the presence or absence of such gene products does is to produce conditions that favour the development of one or other of the two sexes. An environmental stimulus does the same; it produces conditions that favour the development of a particular sex. If indeed the same male-determining DNA sequences are present in both mammals and reptiles, then what is different is the way in which they are regulated.

OBJECTIVES FOR CHAPTER 6

6.1 Define, recognize or place in the right context all the words in **bold** type.

6.2 Describe in words and diagrams the basic model of the process of development of sex differences in vertebrates. (*Question 2*)

6.3 Describe the main classes of switch mechanism that direct development towards a particular sexual phenotype. (*Question 1*)

6.4 Provide genetic evidence for the location of testis-determining factor (TDF) to a particular region of the Y chromosome in mammals. (*Question 3*)

6.5 Describe the roles of the genetic switch, gonad development and hormone production and explain the inter-relationship of these three factors in sexual differentiation in mammals. (*Questions 4 and 5*)

6.6 Describe the evolutionary sequence of sex mechanisms in angiosperms. (*Question 6*)

6.7 Explain the genetic control of sex expression in a plant such as *Silene*. (*Question 7*)

6.8 Explain how sex in insects is a cell autonomous phenotype with reference to gynandromorphs. (*Question 8*)

6.9 Describe in words and diagrams the chromosomal and genetic control of sexual differentiation in *Drosophila*. (*Question 9*)

6.10 Provide evidence for temperature sex determination (TSD) in reptiles. (*Question 10*)

6.11 Describe variations among mammals, reptiles and *Drosophila* in the way that gonadal, morphological, behavioural and physiological sex are related to one another within an individual. (*Question 11*)

DEVELOPMENTAL DEFECTS AND HEALING: TWO HUMAN CASE STUDIES

◆ CHAPTER 7 ◆

The last chapter of this book has two major aims. The first is to show how some of the concepts learned from developmental biology can help our understanding of normal and abnormal development; the second is to review the information you learned in earlier chapters of this book.

To achieve these aims two different examples are considered: the developmental causes of neural tube defects, with a focus on **spina bifida**; and the structure, maintenance and repair of the human skin. These examples contrast with one another for a number of reasons. Neural tube defects arise as congenital malformations in early embryogenesis whereas the properties of the human skin are to be understood in terms of processes characteristic of the embryo which persist throughout adult life. Also neural tube defects, having occurred, cannot be developmentally corrected unlike skin which, once damaged, has remarkable powers of healing.

7.1 NEURAL TUBE DEVELOPMENT AND MALFORMATIONS

7.1.1 Causes of neural tube defects

The precise cause or aetiology of neural tube defects is unknown. However, certain environmental and genetic factors are known to contribute to an increased risk of a human infant developing a neural tube defect. These factors, which implicate particular developmental processes, are outlined in this short section.

Neural tube defects are a group of clinically important human developmental abnormalities. Partial failure of the neural tube to close during embryogenesis presents at birth as spina bifida or **anencephaly** (where the forebrain tissue becomes exposed to the surface and degenerates). Such abnormalities occur with an overall world frequency of about 2 in 1000 births. A woman can be diagnosed as carrying an affected infant by detection of alphafeto-protein in the amniotic fluid which surrounds the developing fetus. The production of this protein, however, is not confined to fetuses with neural tube defects but may also be produced by those with some other developmental or genetic abnormality.

Environmental factors

A number of population studies have been undertaken to determine possible causes of neural tube defects. These studies have revealed no single cause but have provided evidence that a number of environmental factors may contribute to an increased risk. These include:

geography
month of conception
maternal age
birth order
socio-economic class
maternal diet
maternal illness or disease.

Although these population studies suggest a role of environmental agents in increasing the risk, they have not established these factors as the primary causes of neural tube defects. This is because such studies compare frequencies of neural tube defects between populations, and no study has revealed a frequency of zero. So genetic factors may also play an important role.

Genetic factors

A number of studies of families with a history of neural tube defects have been carried out. The pattern of inheritance fits neither that of single-gene inheritance nor a pattern of multiple or many gene inheritance. But these studies have shown that offspring with either an affected mother or father have an increased risk of developing the trait of about three per cent.

Without going into detail, what all these studies indicate is that no single factor influences the development of neural tube defects, but that its expression is the result of many different and interacting environmental and genetic factors.

7.1.2 Experimental approaches

One experimental approach to an understanding of the basic causes of neural tube defects involves the *in vitro* culture of mammalian embryos. This has helped to elucidate the cellular mechanisms affecting the formation of the neural tube in vertebrates. In this section the morphological events of neural tube formation are reviewed, then the changes at the cellular level are examined, and finally the cellular processes involved in neurulation are explored.

Morphological events in neural tube formation

An understanding of the normal process of neuroembryological development enables us to gain insights into the pathogenesis of neural tube defects. The mechanism and timing of neurulation and its relationship to other developmental processes was described in Chapter 3.

◇Recall the basic principles involved in the process of neurulation in the gastrula of vertebrates.

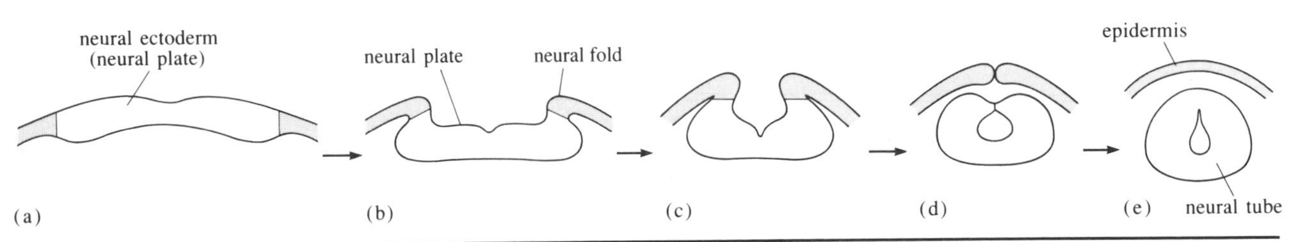

Figure 7.1 Sequence of events in neural tube formation.

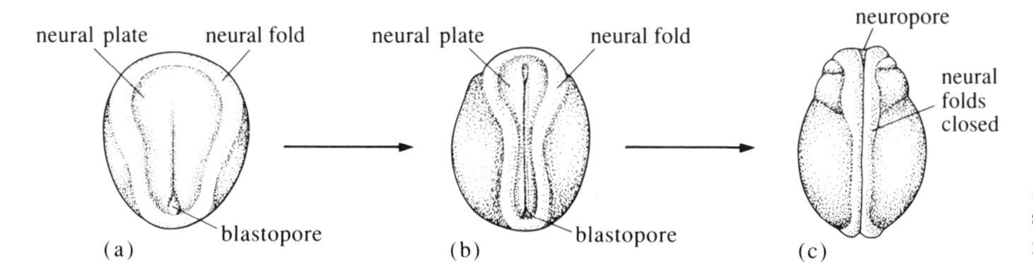

Figure 7.2 Dorsal view of neurulation stages. (a) Early neurula. (b) Middle neurula. (c) Late neurula.

◆ Dorsal mesoderm in the gastrula induces the overlying ectoderm to form the hollow neural tube which will eventually differentiate into the brain and spinal cord.

Here we are concerned with the response by various ectodermal tissues to the mechanism of primary embryonic induction. The events of neurulation are shown in Figures 7.1 and 7.2. The original ectoderm is divided into two regions, namely the neural ectoderm (called the **neural plate**) which thickens and folds to form the neural tube, and the epidermal ectoderm which thins and forms the epidermis. The lateral borders of the neural ectoderm elevate to form **neural folds**, (Figure 7.1b and c and Figure 7.2a and b), so that the neural plate is now concave in cross-section. These folds meet at the dorsal midline (Figures 7.1d and 7.2c) and then fuse to form the neural tube beneath the overlying epidermis (Figure 7.1e).

Closure of the neural tube does not occur simultaneously along its length. It begins at the cephalic region and progresses to the caudal region. This leaves two open ends to the neural tube (Figure 7.3a and b) called the anterior and posterior **neuropores** respectively. Failure of the former to close results in the lethal condition, anencephaly. Failure to close the posterior neuropore results in spina bifida, the severity of expression depending upon how much of the spinal cord remains open.

These tissue movements involve changes at the cellular level and these are examined in the next section.

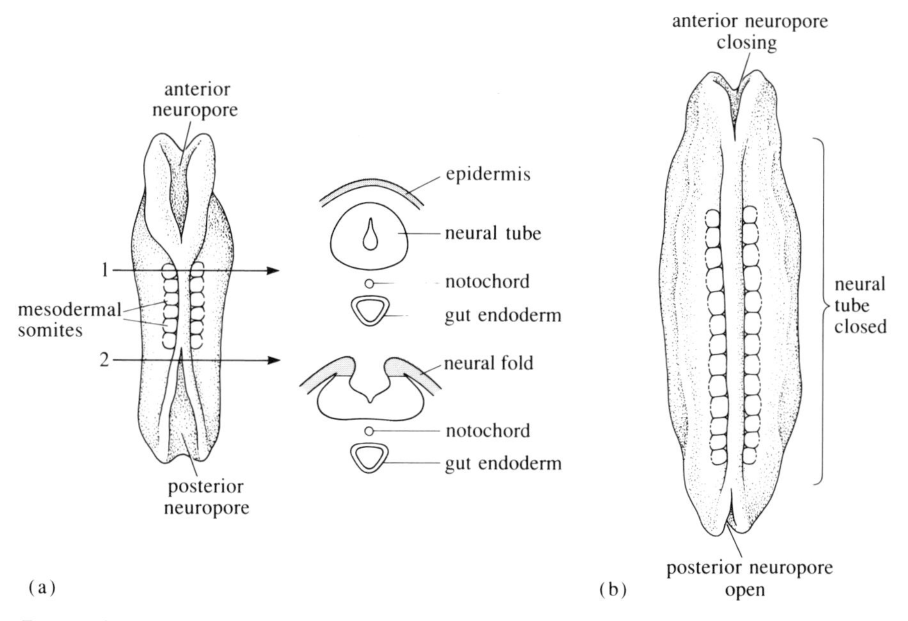

Figure 7.3 Formation and closure of the neuropores. (a) Initiation of neurulation; dorsal view and transverse sections. (b) Late neurula; dorsal view.

Mechanisms of neural tube formation

As early as 1885, Wilhelm Roux suggested that active changes in cell shape could bring about tissue folding. Since that time, documented evidence has accumulated on the cellular changes occurring in the **neural ectoderm** during neural tube formation. This can be summarized as follows.

(a) Throughout neurulation the cells of the neural ectoderm are undergoing mitosis. However, the increase in cell number contributes at this time to longitudinal growth and not to the cross-sectional area of the neural ectoderm, which remains constant.

(b) Histological observations of rodent embryos have shown that neural tube formation is intimately linked to changes in shape of the cells of the neural ectoderm.

(i) At the gastrula stage the cells are columnar (as shown in Figure 7.4a), but as the neural plate broadens the cells elongate thereby increasing in height (see Figure 7.4b).

(ii) During the elevation of the neural folds, the apices of the cells constrict resulting in a 'purse-string' effect, the neural tube being comprised of cells with narrow necks (see Figure 7.4c).

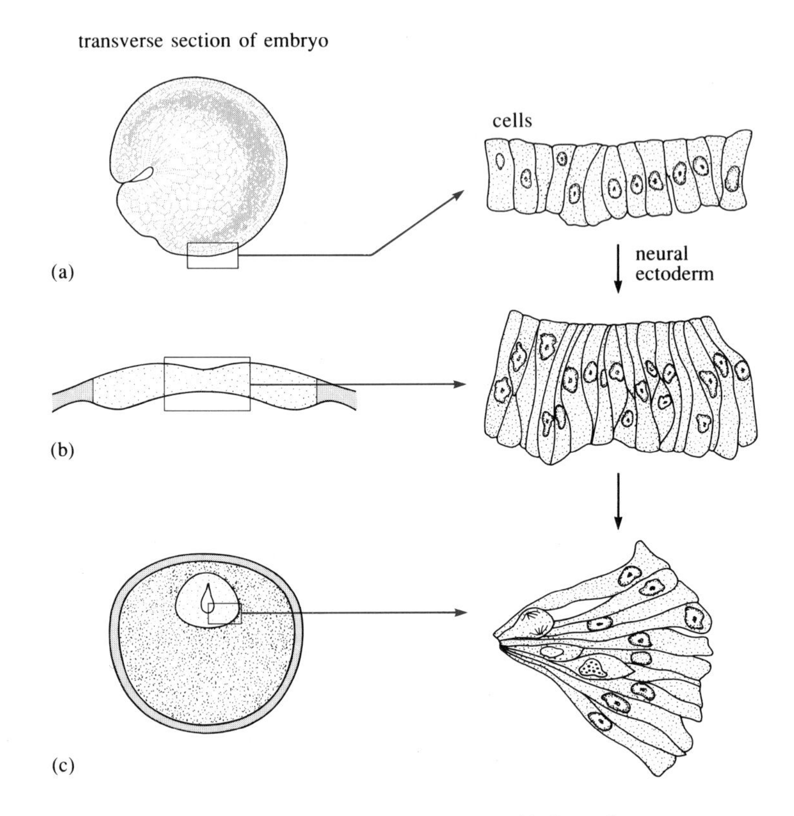

Figure 7.4 Changes in neural ectoderm cell shape. (a) Gastrula stage. (b) Neural plate stage. (c) Neural tube stage.

G. M. Odell and colleagues in 1981 demonstrated by means of computer simulation that a simple model of cell shape change based on the contractile properties of actin microfilaments is sufficient to transform a circle of cells first into a flattened sheet and then into a hollow tube (as summarized in Figure 7.5). This result supports the conclusions of developmental biologists concerning these morphogenetic movements.

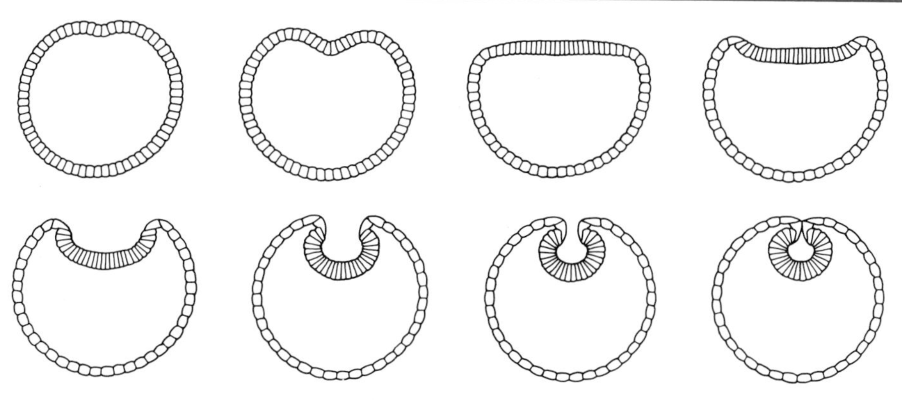

Figure 7.5 Computer-generated model of neural tube formation in amphibians.

The driving forces for change in cell shape

If the change in cell shape is indeed sufficient to bring about the remarkable process of neural fold elevation then it would be of interest to know how the change in cell shape is achieved. You may not be surprised to learn that the cytoskeleton (see Chapter 1, Section 1.1.2) plays an important role. Evidence for this comes from two sources, electron microscopy which enables microtubules and microfilaments to be clearly visualized, and studies using inhibitors.

As the neural ectoderm cells elongate, randomly arranged microtubules align themselves parallel to the lengthening axis, as shown in Figure 7.6. Significantly, this stage of neural tube formation can be blocked by **colchicine** which inhibits microtubule function by binding to the constituent protein, tubulin. It is interesting to note that mouse embryos cultured *in vitro* in serum containing colchicine over the period of neural tube formation, show neural tube defects.

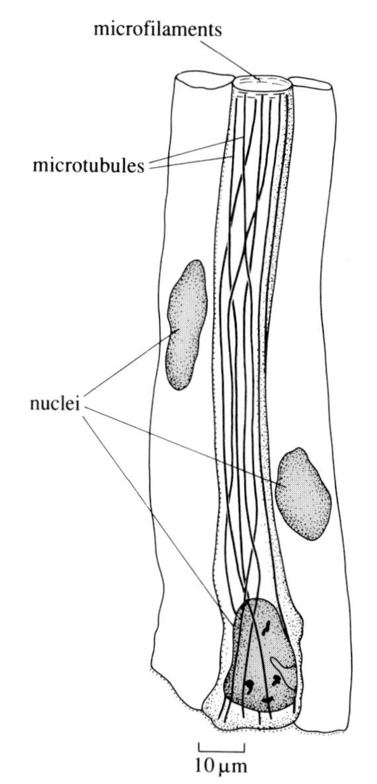

Figure 7.6 Microtubule and microfilament orientation in neural plate cells. Diagram showing three cells.

The second change in cell shape which results in a narrow neck at the apical end of the cells is directed by a ring of contractile microfilaments, as shown in Figure 7.6. It is significant that in embryos cultured *in vitro* in the presence of **cytochalasin**, an agent known to disrupt microfilament functioning, the neural ectoderm cells can elongate but cannot constrict.

◇ Would you expect the neural folds to elevate in such embryos?

◆ No, because the process of elevation of the neural folds is brought about by the apical constriction of the cells.

As you might expect, mammalian embryos exposed to cytochalasin *in vitro* show neural tube defects.

Thus it appears that the thickening and subsequent elevation of the neural ectoderm in mammalian embryos is driven, at least in part, by changes in the shapes of the neural ectoderm cells. Thus disturbance of the cytoskeleton might be the cause of some neural tube defects in human embryos. The next section considers the role of calcium in the function of microfilaments.

The role of calcium in neural tube elevation

A number of experiments have demonstrated that there is a calcium requirement for neural tube elevation.

(a) Extracellular calcium is required for neural fold elevation since removal or reduction of calcium in the medium in which mammalian embryos are cultured causes the folds to flatten.

(b) By using radioactive calcium the movement of calcium can be traced. During neurulation calcium moves from the medium into the cells of the neural ectoderm.

(c) Addition of the drug papaverine, which alters calcium fluxes, inhibits elevation of the neural folds.

Since contraction of microfilaments is controlled by changes in the concentration of intracellular free calcium, it would appear that the changes in cell shape are dependent on calcium movement. The replacement of calcium with magnesium or strontium in the medium in which mammalian embryos are cultured does not maintain the elevation of the neural folds, suggesting that the requirement for calcium is specific. These observations raise the possibility that disturbance of the calcium balance at a critical time in human development may result in incomplete closure of the neural tube.

7.1.3 Genetic models of neural tube defects

You have learnt how experiments have elucidated some of the cellular mechanisms which bring about normal neurulation in mammalian embryos. Clues as to how the mechanisms involved may be disturbed have come from a completely different experimental approach, that of studying mammalian mutant strains. Of those studied the most interesting is the homozygous recessive *curly-tail* (*ct*) mutant in mouse since it has the greatest phenotypic resemblance to the human condition. About half of homozygous *ct/ct* mice develop spina bifida, the rest are normal.

The formation of spina-bifida-type abnormalities in these mice is associated with a delay in closure of the posterior neuropore. The extent to which neuropore closure is delayed determines the type of neural tube defect that results; embryos with severe delay develop open spina bifida plus a bent tail whereas embryos with moderate delay develop bent tails only. Recently (1988) Andrew Copp and his colleagues working at Oxford carried out investigations to try to determine the factors which affect this delay.

One factor they investigated was the role of cell proliferation during mouse neurulation. To do this they compared **mitotic indices** (the percentage of cells undergoing mitosis) in different tissues in *curly-tail* embryos developing spinal neural tube defects and in their normally developing littermates. They also compared two other cell cycle parameters in these two sets of embryos namely, the uptake of ^{3}H thymidine (which is incorporated into DNA) and the length of S-phase (so-called because at this phase in the cell cycle DNA is synthesized).

These three measures of cell proliferation rate were determined in the neural ectoderm, surface ectoderm, mesoderm, notochord and gut endoderm (see Figure 7.3 for the relationship of these tissues to each other in the developing mammal at the time of neurulation). Their results indicated a reduced rate of proliferation of gut endoderm and notochord cells in the neuropore regions of embryos developing spinal neural tube defects compared with normally developing controls.

In order to understand the effect on the embryo it is important to bear in mind that it is developing in three dimensions. Section 7.1.2 was mainly concerned with changes in the dorsal–ventral axis. Here we are concerned primarily with development on the longitudinal or anterior–posterior axis.

Cell division during neurulation contributes to the longitudinal growth of the embryo. The imbalance in growth due to a cell proliferation defect in the ventral structures in *ct/ct* embryos results in stress forces within the embryonic tailbud such that ventral curvature occurs in this region (Figure 7.7). Such curvature would counteract the forces in the neural ectoderm that lead to neuropore closure.

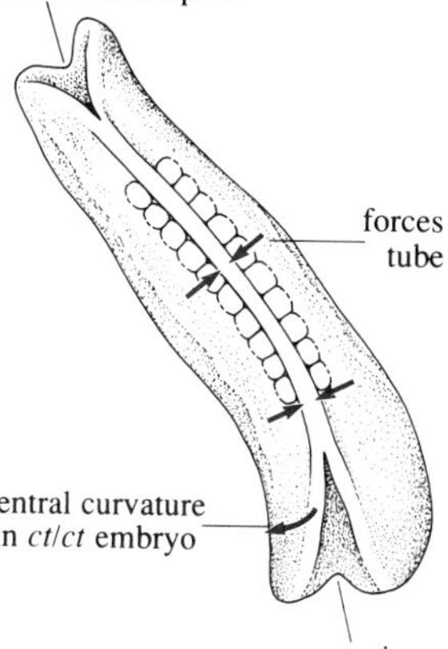

Interestingly, the closure of the neuropore in *ct/ct* embryos is influenced by a number of environmental factors such as temperature and administration of vitamins, some of which are known to contribute to an increased risk of neural tube defects in humans.

Copp and his colleagues found that *in vitro* culture of *ct/ct* embryos at 40.5°C instead of 38°C for a 24 hour period before posterior neural tube closure led to normalization of this process. They observed that the slight increase in temperature diminishes cell proliferation to a greater extent in the neural ectoderm than in any other cells. Thus it appears that the reduced rate of cell division in this tissue balances the rate of cell division in the gut endoderm/notochord in *ct/ct* embryos so that opposing stresses in the tailbud are no longer produced. It is significant that a number of other agents, **mitomycin** or **hydroxyurea** which inhibit cell division, can also greatly reduce the expression of the *curly-tail* gene.

Figure 7.7 Diagram of a *curly-tail* mouse embryo showing tube closure forces and opposing stresses in tailbud.

In humans there is circumstantial evidence that early embryonic growth retardation may also be associated with a reduced prevalence of spina bifida. For example, twin pregnancies are associated with growth retardation early in gestation and show a reduced prevalence compared with single-fetus pregnancies.

Another environmental factor known to affect the rate of cell division in the neural ectoderm of rodent cells is vitamin A (retinoic acid, see Chapter 4). It is not possible to give the details here, but depending on the time in development when it is administered, it either increases or reduces the percentage of affected individuals. This observation has found application in the preventative treatment of women at increased risk of having an affected infant, who are now given vitamin supplementation around the time of conception. However, whether the preventative effect of vitamins is mediated through an effect on cell division in humans or on a process of the type described in relation to limb formation in Chapter 4, is still unknown.

7.1.4 Conclusions

You have learnt how experiments with non-human mammalian embryos have increased our understanding of the cellular mechanisms which bring about the morphogenesis of the neural tube. In addition, animal models allow the testing of clues from human data, such as the possibility of certain environmental agents affecting neural tube development. Most importantly these models also provide information about the process of normal neurulation.

The process of neurulation is very complex involving embryonic induction, tissue movement, changes in cell shape and cell division. So it is not surprising that it is affected by many genetic and environmental factors.

Summary of Section 7.1

Failure of neural tube closure can lead to neural tube defects.

No single agent influences the development of neural tube defects but its expression is the result of interacting genetic and environmental factors.

Neural tube closure can be affected by agents each of which has a different mechanism of action.

Neural tube defects may be produced by several different mechanisms including defective cell shape or abnormal rate of cell division.

A single agent such as Vitamin A can either increase or decrease the risk of neural tube defects depending on the time of administration.

Question 1 (*Objective 7.3*) Each of the following six statements describes a stage in neurulation. Put them in the correct developmental sequence.

(a) Neural ectoderm elevates to form neural folds.

(b) Neural folds fuse at dorsal midline.

(c) Posterior neuropore closes.

(d) Anterior neuropore closes.

(e) Cells of the neural ectoderm increase in height.

(f) Apices of the cells of the neural ectoderm constrict.

Question 2 (*Objectives 7.3 and 7.4*) Each statement (a)–(c) describes the cellular effect of a drug and each statement (i)–(iii) describes a stage in the development of the neural tube which may be interfered with by one of the drugs. For each drug (a)–(c), select a developmental stage which it would affect.

(a) Colchicine inhibits microtubule function.

(b) Cytochalasin disrupts microfilament functioning.

(c) Papaverine alters calcium fluxes in and out of cells.

(i) Neural ectoderm cells elongate.

(ii) Neural folds elevate.

(iii) Neural folds fuse at dorsal midline.

Question 3 (*Objective 7.5*) The administration of mitomycin to the medium supporting the *in vitro* development of *curly-tail* mouse embryos reduces the percentage of individuals with spina bifida. Explain how this might be brought about.

Question 4 (*Objectives 7.2, 7.4 and 7.5*) Explain why women who have previously given birth to an infant with a neural tube defect might reduce the risk of later siblings being affected if they take vitamin supplements at about the time of conception.

7.2 DEVELOPMENT AND HEALING: THE SKIN

The properties and behaviour of the skin provide one of the most dramatic, as well as the most familiar, illustrations of the dynamic state of the adult organism, whose form is maintained by continuous loss and replacement of materials. Cells sloughed off through wear and tear at the surface of the skin are replaced by new cells moving up from lower layers, where cell divisions exactly balance cell loss. Skin cells proceed through a precise pattern of differentiation as they move towards the surface, maintaining a well defined spatial pattern in the tissue. When this pattern is disrupted by a wound, cells behave in characteristic ways that initiate the healing process and restore the original spatial order of the tissue. Here we encounter another example of the type of process that was studied in Chapter 1, where *Hydra* was used to examine the processes involved in the maintenance and regeneration of the adult form as a dynamic flow pattern. As remarked there, these processes characterize development in general, regeneration being a persistence of developmental or morphogenetic field properties in the adult. And we shall now see that, once again, dynamic balance is dependent upon the influence of activators and inhibitors, differentially distributed in the tissue in accordance with the distribution of cell types.

7.2.1 Histology of the skin

Skin develops from embryonic ectoderm, providing a protective boundary over the whole body surface. The **epidermis** is described as stratified squamous epithelium (Figure 7.8) because of the layered structure of the nucleated cells seen above the basement membrane. Notice the hexagonal (6-sided) pattern of the surface cells.

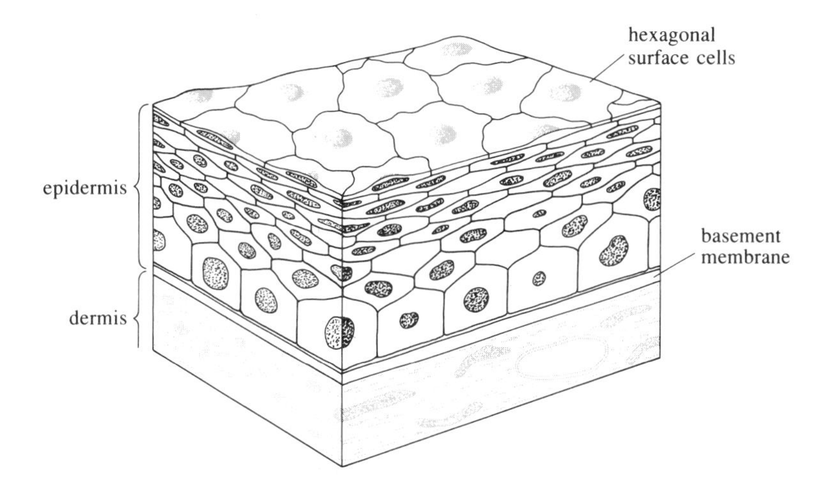

Figure 7.8 Stratified squamous epithelium.

◇ Why do you think that cells take this hexagonal rather than a circular or some other form?

◆ A surface can be covered smoothly (without overlap) by hexagons, which define the minimal energy structure for closely packed units on a surface, so cells naturally assume this shape. A surface cannot be covered without overlap by circles.

The basement membrane is a thin noncellular layer rich in proteins and polysaccharides. Beneath it is the **dermis**, made up largely of connective tissue which contains parallel arrays of collagen fibres, collagen being an inert protein with great mechanical strength. These fibres form bundles that run in characteristic directions in different parts of the body. You can determine their direction by pinching your skin together and finding out which way it folds most easily into wrinkles. These are known as cleavage lines. They are important to the surgeon because an incision that separates parallel bundles of collagen fibres without rupturing them heals with a fine line, whereas an incision severing and disrupting them produces a broad scar.

◇ Which way do the collagen fibres run on your torso?

◆ Transversely.

◇ So which is the best direction for an incision to remove the appendix?

◆ A transverse incision.

Figure 7.8 is an idealized drawing of stratified squamous epithelium. The skin has this general structure but is more complex, as shown in Figure 7.9. The basal layer of cells in the epidermis is folded over projections (papillae) from the dermis that give the whole structure mechanical resistance to shearing stresses that would otherwise strip off the epidermis. The dermis is copiously supplied with blood vessels and nerves. The skin also contains a number of appendages such as nails, hairs, sebaceous glands and sweat glands, all derived from the epidermis.

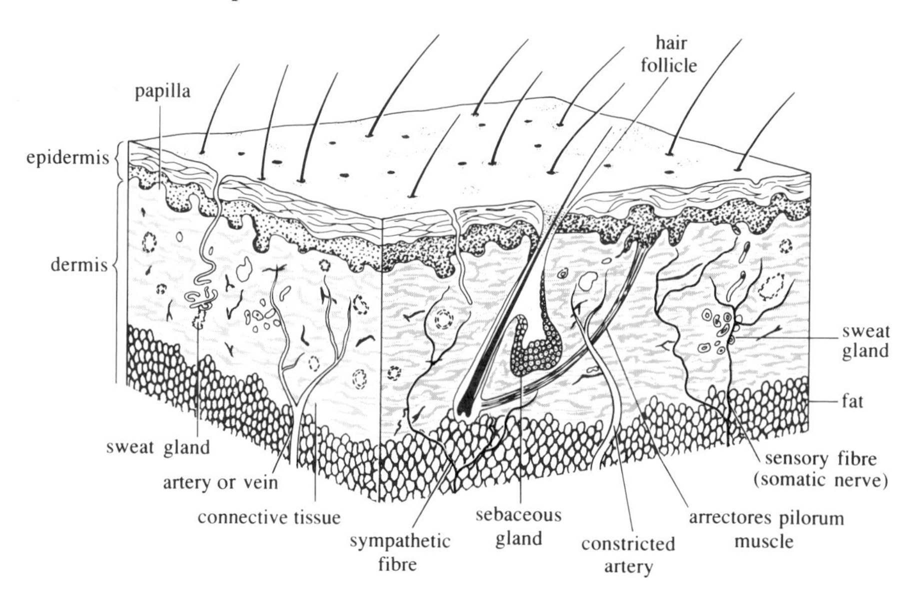

Figure 7.9 The general structure of the skin, showing epidermis, dermis, epidermal derivatives, nerves and blood vessels.

7.2.2 Maintaining the spatial pattern

A closer look at the structure of the epidermis (Figure 7.10) shows how cells alter their shape and their state as they move away from the basement membrane towards the surface, where they get sloughed off as hardened, dessicated squames, flat sheets of dried skin. The cells of the basal layer adhere strongly to the basement membrane. Among them are cells that divide on average once every 4–5 days. Daughter cells move away from the basement membrane and begin the process of differentiation, changing their state as they move, becoming progressively flattened and filled with the protein, **keratin**. The cells are joined strongly together by structures called **desmosomes** (shown as parallel lines). The different layers, spinous, granular, and cornified, are distinguished simply by the states of differentiation of the cells within them. In the cornified layer, which can be very thick in parts of the body such as the soles of the feet or in calloused areas, the cells have lost their nuclei and cytoplasmic organelles. They are flattened sheets packed with keratin and surrounded by an insoluble protein envelope closely apposed to the plasmalemma. Keratin and the protein envelope give the epidermis its mechanical toughness as well as its water-repellent properties.

The hexagonal pattern of the surface cells shown in Figure 7.8 extends down to the basal layer to give a basic structural element, the epidermal proliferative unit (EPU). A number of these are shown in Figure 7.11, each consisting

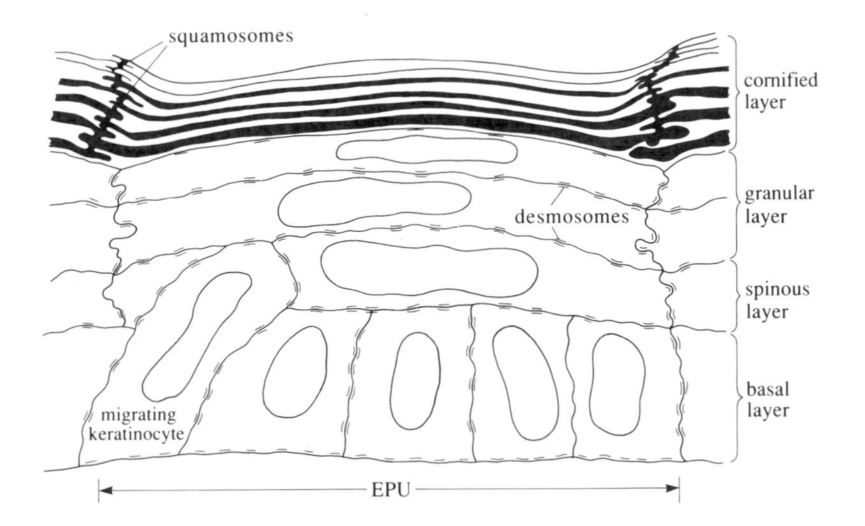

Figure 7.10 A schematic structure through an epidermal proliferative unit (EPU), showing the layered structure and the pattern of cell differentiation with dividing cells in the basal layer.

of a stack of flattened hexagonal cells and several cuboidal cells in the basal layer. Cell divisions are normally restricted to this layer. The overall organization of the skin is reminiscent of that in *Hydra* where populations of dividing cells (interstitial cells), located next to a membrane, the mesogloea, replace those lost from tentacles, into buds, and from the foot. The details of the cellular flow pattern and states of cell differentiation are clearly different, but the principles of dynamic maintenance and flow are the same.

◇ Recalling the factors involved in head and foot formation in *Hydra*, what might you expect to find in the epidermis that influences cell states and their transitions?

◆ Activators and inhibitors of cell division and differentiation are expected to be involved.

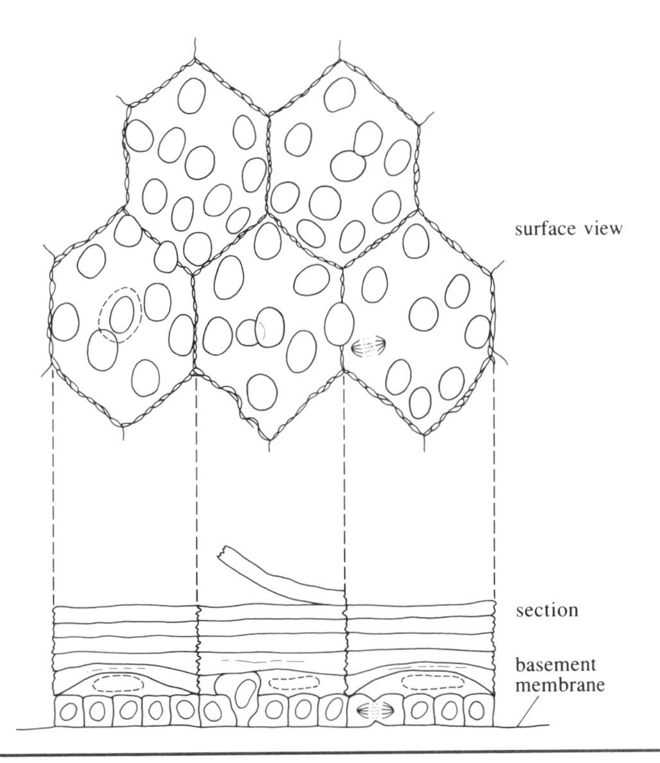

Figure 7.11 Surface and section views of an EPU, showing the hexagonal structure of the differentiating cells and distribution of nuclei in cells of the basal layer.

The existence of regulatory factors influencing rates of cell division in the epidermis can be demonstrated by wounding the skin. If superficial wounds are given to hairless mice by applying a bit of sticky tape and stripping it off, thus removing some of the surface layers of cells, the mitotic index (fraction of cells in mitosis) in the basal layer increases. Some kind of signal is therefore received by these basal cells as a result of the stimulus of wounding.

◇ What are the basic categories of control signal that could produce such a response?

◆ A stimulator of cell division could be released from damaged surface cells and diffuse to the basal cells; or an inhibitor of cell division could be removed with the surface cells, resulting in a reduced level of inhibition in the basal region.

7.2.3 Control of cell division

Study of the factors that control human epidermal cell proliferation and differentiation is greatly facilitated by using *in vitro* cultures of **keratinocytes**, the name used for epidermal cells because they produce keratin. The source of cells for such cultures is skin removed during breast reduction and other forms of plastic surgery, bald patches of skin excised from the scalp, and foreskin. There is always a question in such work about the resemblance between cultures of cells grown in artificial media and those in the epidermis itself. The evidence is that the cultured cells retain the basic characteristics of epidermal cells. They become stratified, with proliferation confined to the basal layers. The cells contain keratin filaments and have abundant desmosomes. They also develop cornified envelopes containing **involucrin**, the major precursor protein of these structures in the cornified layers of human skin. In culture the onset of expression of this differentiation product is immediately above the basal layer (spinous), whereas *in vivo* involucrin synthesis usually begins in the next layer up.

To discover the best conditions for growth of cells in culture, it is necessary to examine how various extracts from blood serum and tissue affect rates of cell proliferation. **Epidermal growth factor (EGF)** is a polypeptide that was found this way. It promotes proliferation of cultured keratinocytes by stimulating lateral movement of the peripheral zone of dividing cells, encouraging expansion and growth of colonies. Another polypeptide, named **transforming growth factor alpha (TGFα)** stimulates cell division in human keratinocyte cultures. Both dividing and differentiating keratinocytes in the epidermis itself produce mRNA that codes for TGFα, demonstrating the activity of the gene in these cells. Furthermore, adding EGF or TGFα to the culture medium stimulates the production and secretion of TGFα by keratinocytes. So it appears that there is a positive feedback loop operating here, resulting in auto-induction of growth and growth factor expression in keratinocytes.

◇ What would be the expected consequence of such a positive feedback effect on the epidermis if it were the only stimulus acting?

◆ The epidermis would enter a condition of unstable, unrestricted growth and behave like a skin tumour.

To achieve the precise balance of cell production and loss that occurs in the adult epidermis, there must also be a negative feedback system at work. One of the substances found to exert this influence is the polypeptide called **transforming growth factor beta (TGFβ)**, which has an inhibitory effect on keratinocyte proliferation. Cultured keratinocytes secrete TGFβ-like molecules into the medium and have TGFβ cell surface receptors. There is also another inhibitor, purified from mouse epidermis, which reversibly inhibits keratinocyte proliferation both *in vivo* and *in vitro*. It is a pentapeptide with the amino acid sequence glutamate-glutamate-aspartate-serine-glycine. This inhibitor also acts on human keratinocytes in culture. It is called epidermal pentapeptide (EPP).

7.2.4 Control of cell differentiation

What induces the basal cells of the epidermis to stop dividing, move away from the basement membrane, and start differentiating into the flat, keratin-containing cells characteristic of the epidermis? Growth activators and inhibitors themselves have little effect on the proportion of terminally differentiating keratinocytes in culture. However, if the cells are removed from the petri dish surface, to which they normally attach and spread along, and are held in suspension by stirring, within 24 hours there is irreversible inhibition of division. By 3 days the cells are surrounded by cornified envelopes. So loss of cell–substrate and cell–cell contact can stimulate differentiation, as measured by involucrin production. It has also been observed that keratinocytes containing involucrin adhere and spread more slowly than those without. Reduced adhesiveness appears to be correlated with the beginning of keratinocyte differentiation, possibly accounting for the upward migration of cells as they enter the pathway of terminal change. Involucrin synthesis and decreased adhesive properties may therefore be components of a positive feedback loop that results in cells starting to differentiate and leaving the basal layer. This process must also be linked to the number of cells proliferating in the basal layer, to maintain constant proportions of cells in different states throughout the epidermis. Cells communicate with one another via gap junctions, whereby small molecular weight substances pass from cell to cell. Change in rates of production of growth activators and inhibitors as a function of cell state can therefore lead to a dynamic balance within the different levels of the epidermis that maintains a stable spatial distribution of cell types, as in *Hydra*.

◇ Where would you expect to find the highest concentrations of TGFα in the epidermis?

◆ In the basal layer, since these cells produce TGFα, which stimulates further production by basal layer cells.

7.2.5 Wound response and the healing process

The adult skin, in response to wounding, undergoes a characteristic sequence of changes that restores the original organization. These are conventionally divided into three stages: inflammation (blood clot formation, influx of leucocytes, decontamination), wound closure, and matrix remodelling. The end result of this process is an intact skin with a scar. The first stage of inflammation involves blood platelet adhesion and blood coagulation to provide a provisional extracellular matrix which is invaded by blood cells, such as neutrophils and macrophages, that clear contaminating bacteria. This is followed by the migration of epidermal cells that spread across the wound. Platelets release a protein called **platelet-derived growth factor (PDGF)** which stimulates mitosis in blood cells involved in tissue repair, and in epidermal and dermal cells. If only the epidermis is removed, wound closure is achieved entirely by epidermal migration from the edges of the wound. However in a full-depth wound, involving removal of both dermis and epidermis, the exposed area is closed by a combination of contraction in which the edges of the wound move together, and epidermal migration.

Wounds to the epidermis

The cells of the epidermis, whether dividing or differentiating, are normally non-motile: that is, they are unable to migrate laterally within the tissue. Their only movement is displacement towards the surface layer of the skin. However at the margin of a wound, epidermal cells are mobilized within a few hours and begin to move towards the centre of the wound through the matrix created by coagulated blood. The spreading sheet of flattened keratinocytes is only one or two cells deep at the advancing margin. As they move across the wound surface they deposit a protein called **fibronectin** which facilitates the migration of more epidermal cells. Other substances are also secreted by migrating cells, particularly polysaccharides called glycosaminoglycans (GAGs). These give qualities of flexibility and resilience to the non-cellular environment surrounding the cells, which is known as the **extracellular matrix (ECM)**. Once the wound has been covered, the epidermal cells revert to their normal phenotype. The basement membrane is reconstructed from the margin of the wound inwards, in a zipper-like fashion. The cells adhere to one another by demosomes and collagen fibrils are laid down as the basement membrane forms. Thus the epidermis is reconstituted.

Full-depth wounds

Reconstruction of the dermis in full-depth wounds depends upon the migration of another cell type, the fibroblast. This does not start till several days after wounding, when epidermal reformation is well advanced. These cells, constituents of normal dermis, undergo transformation to a migrating phenotype, as do epidermal cells, and invade the wound area beneath the epidermis. As they migrate they diversify to collagen-synthesizing cells and myofibroblasts, cells characterized by actin bundles that give them contractile properties. These two types reconstitute the dermis by the cooperative activities of wound contraction and collagen deposition.

In the final stage of tissue remodelling, collagen bundles grow, increasing the tensile strength of the new tissue, and proteoglycans appear, giving it resilience. Apart from collagen and proteoglycans, a host of different protein and polysaccharide components of the extracellular matrix (ECM) have been identified. What emerges from these studies is the dynamic reciprocity between cell state and the extracellular environment, the ECM. Cells create the matrix by their secretions, and these in turn act on the cells, modulating their state. The whole is a beautifully orchestrated pattern of reciprocal influences that maintains a dynamic balance over the skin, and restores this even after the extensive disturbance of severe wounds.

There are degrees of wounding from which the body is unable to recover because the damaged areas are too large for the wound repair mechanisms to be effective. Extensive third degree burning in which dermis as well as epidermis is destroyed is an example of this critical condition. In these cases the use of cultured epidermal cells can be crucial to saving lives. However, because of immunological rejection, the cells must come from the patients. The procedure involves taking a thin layer of healthy epidermis from the patient, which causes no stress, culturing the cells and then grafting the cell sheets onto the exposed surface. This has proved to be an extremely effective treatment.

7.2.6 Embryonic wound healing

Whereas wound healing in adult animals is always accompanied by the formation of scar tissue, embryonic wounds heal without trace. If we understood the causes of these differences, we might be able to suggest a way of treating adult wounds so as to reduce or eliminate scarring.

Studies of fetal wound healing in rabbits and sheep have shown that the process occurs with substantially less inflammation than in the adult, and considerably faster. Somehow the embryo is able to do a better job more rapidly, which points to differences in the way the basic structures are regenerated. Since, as we have seen, the healing process depends upon a dynamic interaction between cell state and the extracellular matrix throughout the reconstitution of the damaged tissue, this interaction is where to look for differences in the quality of healing. It was discovered that the fetal extracellular matrix (ECM) in the early stages of recovery from wounding has a significantly different composition from that in the adult. It is much richer in glycosaminoglycans (GAGs), the materials that give the ECM its resilience and plasticity. A particularly abundant constituent is the polymer **hyaluronic acid (HA)**. This substance is found wherever there is rapid cell proliferation, regeneration and repair. It inhibits cell differentiation and promotes cell division. Fetal wound fluid stimulates and sustains hyaluronic acid synthesis, and a glycoprotein has been discovered in fetal serum which stimulates the production of hyaluronic acid. This substance persists in fetal wounds for prolonged periods. Invading cells are thus induced to secrete copious amounts of HA. Collagen deposition proceeds rapidly within this environment and it is assembled into fibres and bundles without the puckering and deformations that result in scarring. The collagen in fetal wounds is also different from that in the adult, having initially a lower breaking strength though it later strengthens. Finally, there is substantially less contraction of the wound in the fetus as compared with the adult. The result is a remodelling of the damaged area by a process that has the characteristics of speed and a highly plastic mode of molecular organization in which the ECM and the cells that produce it develop into a structure that is a virtually perfect reconstruction of the original tissue organization.

◇ What treatment of adult wounds does this suggest as a possible means of improving the quality and speed of healing?

◆ Applying hyaluronic acid and other glycosaminoglycans to the wound immediately after injury.

7.2.7 Cell shape and cell state

Cell shape, cell adhesiveness and cell state are evidently connected in some way and are related to the conditions for cell division or differentiation. We saw earlier that proliferating populations of cells adhere to surfaces such as the basement membranes (*in vivo*) or to plastic (*in vitro*). Differentiating cells in the skin are less adhesive and leave the basement membrane. However, in suspension culture, normal keratinocytes do not divide but undergo differentiation. During wound healing, cells change their phenotypes as they perform different roles: migration, division, and differentiation. What is now emerging from recent studies on a variety of cell types is a fundamental relationship between cell shape and the control of cell state. This occurs through interactions between the cytoskeleton and the extracellular matrix. Understanding this relationship is helping to clarify many different aspects of development.

Mouse fibroblast cells in suspension culture assume a rounded morphology and their growth is arrested. If small glass fibres or beads are added and kept in suspension with the cells, they attach to and spread on the glass surfaces and start dividing. However, they do so only if PDGF and EGF are present. PDGF is described as a competence factor because it makes cells responsive to growth factors such as EGF and insulin, another general growth stimulator. It appears to do so via effects on the cytoskeleton. Within 2–3 minutes of adding PDGF to the culture, the spatial pattern of distribution of the cytoskeletal proteins, actin and vinculin, changes from association with adhesion plaques at the cell surface to sites of aggregation near the nucleus. Only then are the cells sensitive to EGF.

A possible pathway of response of these competent cells to EGF and insulin is via the second messenger signalling system involving calcium (see Chapter 1, Figure 1.22). Calmodulin is a protein that responds to Ca^{2+} and activates many cellular processes such as protein phosphorylation and microtubule assembly. It is associated with cytoskeletal structures, predominantly microfilaments. It is also involved in the Ca^{2+}-directed depolymerization of microtubules. The binding of growth factors to receptors could mobilize calmodulin from microfilaments and so activate a series of Ca^{2+}-calmodulin dependent processes, including the reorganization of microtubules. The action of EGF is to be understood within the context of this state of arousal of the cell.

Three genes are known to be activated as a result of this process. Two of them, known as *c-fos* and *c-myc*, code for proteins that perform a regulatory function in the nucleus, binding to DNA and controlling the activity of other genes. These strange names reflect the history of their discovery. The first was found in the virus feline osteosarcoma and was called *v-fos*. Then a homologue was discovered in normal cells and labelled *c-fos*. A similar history explains *c-myc*, first found in the leukemia virus avian myelcytomatosis and then identified in normal cells. The third protein in this group is the gene for actin, the major cytoskeletal protein which is found in the nucleus as well as the cytoplasm. What role could this play as a primary regulator of gene action? A recent idea, suggested by Sheldon Penman working at the Massachusetts Institute of Technology, is that mRNA attaches to components of the cytoskeleton when it is actively translated. So changes in cell shape, reflected in the state of the cytoskeleton, can directly influence rates of protein synthesis in cells. This is supported by an increasing amount of evidence. Cells of different types can be induced to increase the synthesis of specific proteins by agents that disturb the cytoskeleton and induce a redistribution of microfilaments. For example chondrocytes, a cell type involved in the initial stages of bone formation, can be induced to synthesize collagen by exposure to the drug cytochalasin, which disrupts bundles of actin filaments called stress fibres and causes them to assume a distributed pattern in the cell.

It is also proposed that these changes of cell shape propagate to the nucleus, where genes become either more or less accessible to transcription depending upon their spatial state, determined by the condition of the nuclear cytoskeleton which has connections with the cytoskeleton in the cytoplasm. The latter is also connected with the extracellular matrix and other cells in normal tissues, providing a continuous channel of influence, both chemical and mechanical, from the environment to the nucleus. These observations emphasize the intimacy of the relationships between form and function at a variety of levels of biological organization, from the activity of a gene in the nucleus to the context of a cell in a tissue, where structure arises from and acts back upon the shape and state of its constituent cells.

◇ Recall the evidence linking calcium, cell shape, and cell division rates to neural tube closure and its failure, resulting in spina bifida.

◆ Reduced calcium at crucial stages in neural tube formation results in spina bifida, presumably preventing normal changes in cell shape that underlie the process; and in *ct/ct* mice cell division rates are reduced in endoderm and notochord cells in the neuropore regions of embryos developing spina bifida.

7.2.8 Skin cancer

If the adult organism had no dividing cells there would be no cancer; it is a disease associated with the strategy of maintaining the adult form by continuous cell renewal. Machines are free of cancer; they just wear out. The skin, first defence against the external environment, encounters a host of potential invaders. One of these is ultraviolet irradiation from the sun. Most ultraviolet radiation is absorbed or reflected by the heavily keratinized surface layers. But some gets to the basal layers where it can disrupt chemical bonds and result in mutations in the DNA. This is one of the primary causes of skin cancer, though there is also a significant incidence of spontaneous mutations that lead to tumorous growths. The vast majority of skin tumours are **benign** — localized lumps of tissue that grow to a certain size and then reach a stable balance between cell growth and cell death. **Malignant tumours** consist of cells that have the ability to invade normal tissue and set up secondary colonies.

The underlying cause of skin cancer is a change in the responsiveness of epidermal or dermal cells to the normal system of checks and balances that maintain the skin in a dynamically stable state. We have encountered a few of these in the form of growth activators and inhibitors, each with a specific cell surface receptor. Mutations which enhance the effectiveness of the growth factors, or which reduce the influence of cell division inhibitors, will increase the rate of cell division and produce a lump of tissue. This is usually a benign tumour, but occasionally it will be malignant.

◇ Given that loss of sensitivity to normal inhibitors is a significant contributory factor to cell malignancy, why not simply treat skin cancers with high levels of inhibitors such as the epidermal pentapeptide?

◆ Such treatment could indeed reduce the rate of malignant cell growth, and possibly even stop it. But since the malignant cells are likely to be less sensitive to inhibition than normal cells, concentrations that arrest growth in the former will certainly stop cell divisions in the latter as well. Therefore unless secondary factors are brought into play that selectively destroy the tumours, treatment with normal inhibitors will simply reduce or stop all normal epidermal cell divisions during the treatment, without necessarily removing the tumour.

There are two experimental procedures for investigating the specific events that initiate transformation of epidermal cells to a cancerous state. One approach is to establish cell lines from human epidermal carcinomas and compare their properties, such as responsiveness to growth and differentiation signals, with those of normal cells. The other approach is to start with normal keratinocytes and to find out what changes they undergo when treated with agents that cause conversion to the malignant condition.

Viral agents

Among the agents that cause malignancy under experimental conditions is a specific group of viruses that induce loss of control over cell growth. After infection, cells acquire the capacity to grow without the presence of external growth factors. This can be due to the production by the transformed cells of large quantities of growth factor themselves. In some cases, the virus-induced growth factors are similar to those produced by normal cells, so only the rate of production is increased. In other cases the growth factors are unrelated to any normal ones previously identified. These provide instances in which malignancy results from excessive self-stimulation of growth by the transformed cells — the positive feedback loop runs out of control.

A particular example is infection of human keratinocytes with simian sarcoma virus, after which the cells produce a protein identified as $P28^{sis}$. This is almost identical to a part of the platelet-derived growth factor (PDGF), which as we have seen stimulates division in epidermal as well as blood and dermal cells. PDGF consists of two polypeptide chains designated A and B. An amino acid sequence in $P28^{sis}$ is virtually identical to that found at a terminal of the B chain in PDGF. Transformation of cells by this virus appears, then, to be due to the production of PDGF-like molecule which stimulates cells to grow and proliferate in a runaway fashion.

Experiments with another gene involved in malignant cell transformation have revealed a close similarity between its protein product and part of the cell surface receptor for EGF. The gene is labelled *erb-B*, first identified in a virus that causes a form of leukaemia in chickens. The region of the EGF receptor molecule which is structurally similar to the *erb-B* protein is that which spans the plasma membrane and lies inside the cell. How this causes malignancy is not clear, but presumably it enhances the responsiveness of cells to EGF and so results in increased proliferative rates.

Mutant genetic agents

Oncogenes are normal genes which, when mutated, make the cell lose its responsiveness to growth controls and lead to cancer. More than 20 have now been identified. These are genes that play a crucial role in the processes regulating cell division. Two of them, *c-fos* and *c-myc*, are already identified as major genes involved in the primary response of cells to PDGF and the initiation of cell division. The normal pattern of cell behaviour — sensitivity to external signals via the second messenger system and changes of cell shape involving the cytoskeleton — can be lost when the gene mutates, so that the cell becomes self-activating and autonomous in the cell division mode. The consequence is a benign tumour. With the accumulation of further mutations affecting adhesiveness and mobility, transformation to invasive malignancy occurs and the cells escape from all the influences that regulate tissue organization. The result is that the organism ceases to function as an integrated whole and so cannot sustain itself.

Summary of Section 7.2

The structure of the skin is maintained by a flow of cells from the basal layer, where divisions occur, to the upper layer, from which differentiated cells are lost through wear and tear, cell differentiation accompanying cell movement and resulting in a constant tissue structure. This pattern of cell division and differentiation involves the action of regulatory substances that influence the rates of cell differentiation by both positive and negative effects. TGFα has a positive feedback effect on dividing cells, which also produce it, whereas TGFβ and EPP, produced by differentiating keratinocytes, inhibit cell division.

The restoration of the skin after wounding depends upon the normal regulatory mechanisms as well as a set of cell responses to damage that result in the stages of inflammation, wound closure, and remodelling that characterize wound healing. Cells interact with and modify their extracellular environments in this process, producing a variety of ECM materials such as fibronectin, glycosaminoglycans, and collagen, as well as regulatory substances to which they respond, restoring the tissue fabric and structure. Cell shape is one of the variables involved in control of cell state, linking form and function. Embryonic wounds heal without scarring whereas in the adult the processes of ECM construction and tissue contraction are such that the remodelling is imperfect, leaving a scar.

Cancer is a result of the failure of normal control processes that maintain the harmonious balance between cell division and cell differentiation in the dynamic maintenance of tissues. A number of genes (oncogenes) whose mutant alleles result in uncontrolled cell growth have been identified and the molecular structure of their products has been determined. These give significant insight into the molecular level of cell state changes associated with cancer.

Question 5 (*Objectives 7.6, 7.8 and 7.10*) Which of the following statements are true and which are false?

(a) All the cells of stratified squamous epithelia of human skin are nucleated.

(b) Cell division in the epidermis is normally restricted to the basal layer.

(c) Dermis and epidermis contain blood vessels and nerves.

(d) Epidermal cell growth inhibitors also induce cell differentiation.

(e) PDGF affects the state of the cytoskeleton in epidermal cells.

(f) Hyaluronic acid enhances cell division and inhibits cell differentiation, and is particularly abundant in fetal wounds.

(g) Oncogenes are genes whose products are primarily involved in the control of cell differentiation.

(h) Cancer necessarily involves a change in the amount of normal growth control substance in the transformed cell.

Question 6 (*Objective 7.7*)

(a) Draw the gradients that you would expect to find in the epidermis for the following substances.

(i) TGFα

(ii) TGFβ

(iii) keratin

Figure 7.12 gives you a coordinate system on which to base your graphs, with the different epidermal layers shown as in Figure 7.10 along the X-axis and concentrations of the substances up the Y-axis. The graphs are intended to show qualitative features only.

(b) On Figure 7.13 draw the expected distributions of the same substances as those in (a), but now in an epidermal skin tumour caused by a mutation that over-produces TGFα. The X-axis now represents position in the tumour measured as distance from the dermis, and the Y-axis measures concentration, as before.

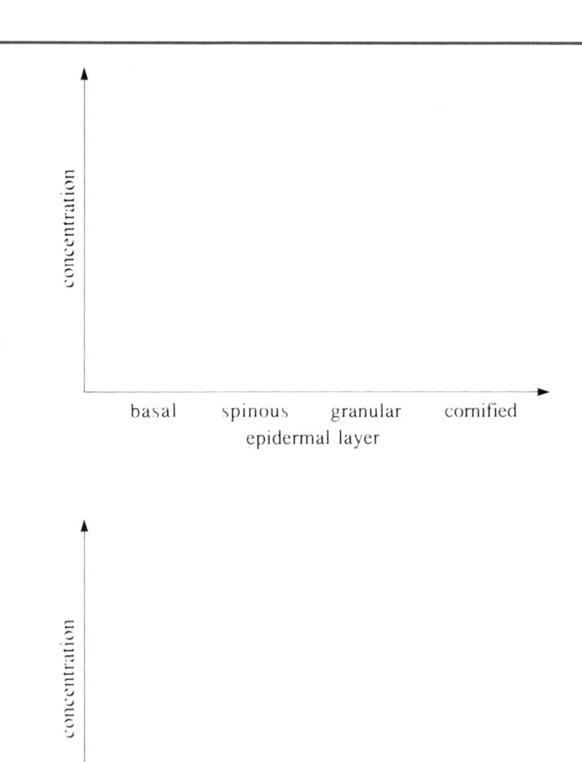

Figure 7.12.

Figure 7.13.

Question 7 (*Objective 7.9*) Each of the following statements describes a stage in epidermal wound healing. Put them in the correct order:

(a) Stimulation of cell division by PDGF.

(b) Deposition of fibronectin, glycosaminoglycans, and other components of the ECM.

(c) Migration of epidermal cells across the wound surface.

(d) Invasion of the provisional matrix by blood cells.

(e) Blood platelet adhesion and blood coagulation.

(f) Formation of desmosomes and cell differentiation.

7.3 CONCLUSIONS

Developmental biology is concerned both with the processes of embryonic development and with regeneration. The intimate relationships between these have been stressed throughout this book, particularly in Chapters 1 and 4. In this chapter it has become evident how a knowledge of these processes provides a basis for understanding the irreversible changes that give rise to birth defects and the healing properties of the skin. Irreversible developmental abnormalities are determined states arising from disturbances that interfere with normal morphogenetic processes sufficiently to deflect them from their normal path, resulting in anomalies such as spina bifida. The great value of developmental studies in such instances is in providing not only descriptions of the processes involved and how they can suffer perturbation, but also indications of their possible prevention. We have seen that both calcium and vitamin A, substances whose morphogenetic influences have been documented elsewhere in this book (Chapters 1, 3 and 4), are important candidates for such preventative treatment. It is also worth bearing in mind that embryonic states classified as determined in one context can be transformed in another context, as in the case of transdetermination of *Drosophila* imaginal discs under the stimulus of persistent growth (Chapter 4). So irreversible defects could become reversible under the appropriate conditions.

Another aspect of development, stressed throughout the course and highlighted in our first case study of spina bifida, is the equivalence of particular genetic and environmental stimuli: either a genetic mutation (such as *curly-tail* in mice) or an environmental disturbance (reduced calcium) to a genetically normal individual can result in the same morphological abnormality. Turning this observation around we can conclude that, in principle, no genetic defect need be absolutely irreversible. The reason is that genes exert their influence on morphogenesis through the limited repertoire of cellular mechanisms described in Chapter 3, so an appropriate environmental stimulus counteracting the genetic influence is always possible. Vitamin A treatment to reduce the incidence of spina bifida is a typical example of this. However, such treatment must always remain within physiological bounds, since as seen in Chapter 4, excessive amounts of vitamin A can cause limb abnormalities. The equivalent effects of genetic and environmental influences were also encountered in Chapter 6, where it was observed that intersexes can result from either genetic or environmental perturbation. An extension of this to comparisons between species gives the fascinating result that sex determination can be either genetic or environmental, as discussed in Chapter 6.

Our second case study, the human skin, emphasizes developmental processes complementary to those that result in irreversible effects: the regulative qualities of embryos that define their most fundamental properties, described in this book in terms of morphogenetic fields (Chapter 1). The skin maintains itself as a dynamic form by virtue of self-organizing patterns of cell state and of extracellular materials, the cells both producing their environment and responding to it. These properties are the basis of health and healing, which is dynamic balance in a self-regulating system. We encountered this in most striking form in *Hydra* regeneration (Chapter 1), where the spatially distributed substances produced by cells and to which they respond were called morphogens. This term could equally well be applied to substances such as TGFα, TGFβ, EGF, etc., and only convention limits its use in this context, mainly because most work on these substances are carried out on cell cultures where gradients related to spatial organization are not studied. But the same principles apply to both, as has been stressed.

Cell adhesion, movement, and shape are also highly significant variables in the maintenance of the skin and during wound healing, as we have also seen in the context of morphogenetic movements in Chapter 3 and in the study of spina bifida. The detailed study of the effects of PDGF on the state of the cytoskeleton in fibroblasts, resulting in responsiveness to inducers such as EGF and other growth factors, points to the importance of structural aspects of the cell, linking form and function and providing a molecular model for the processes of competence and induction described in Chapter 3. The same themes keep recurring, varying in expression in the different cells and tissues of an organism of one species, and between species. These observations take us back to the theme of the introduction to Chapter 1: developmental processes provide the basis for understanding both the diversity of organisms and their unity. It is where all aspects of biological form and function come together in the diverse harmony of life.

OBJECTIVES FOR CHAPTER 7

Now that you have completed this chapter you should be able to:

7.1 Define and use or recognize definitions and applications of the terms in **bold** in the text.

7.2 Outline the known causes of neural tube defects in humans. (*Question 4*)

7.3 Describe the mechanism of neural tube formation at the tissue and cell levels including the importance of changes in cell shape. (*Questions 1 and 2*)

7.4 Explain the roles of the cytoskeleton and calcium in neural tube formation and in control of cell division. (*Questions 2 and 4*)

7.5 Describe how mammalian mutants have contributed to an understanding of the development of neural tube defects. (*Questions 3 and 4*)

7.6 Describe the basic histology of the skin and the dynamic pattern of cell change that maintains it. (*Question 5*)

7.7 Outline the experimental procedures for studying the control of cell division and differentiation in the epidermis, describe the factors that have been identified and how they act to maintain epidermal structure. (*Question 6*)

7.8 Give a description of the way in which skin cells both modify and respond to their extracellular environments, and how cell shape may influence cell state. (*Question 5*)

7.9 Describe the steps involved in wound healing and the behaviour of epidermal and dermal cells in the process of repair. (*Question 7*)

7.10 Explain why cancer can arise in adult tissues and what factors are involved in producing this condition. (*Question 5*)

ANSWERS TO QUESTIONS

CHAPTER I

Question 1 The zygote first breaks spherical symmetry, resulting in a primary axis. One end of this axis branches progressively to form a rhizoid while the other grows to produce the stalk. At the tip of the stalk whorls are produced periodically. These consist of circular rings of hairs which are simple at first, then branch. Finally the cap forms. This is more complex than the whorls in that it has more components, but it has the same circular symmetry.

Question 2 The polarity of regenerating stalk segments of *Acetabularia* can be controlled by an external electric field, the cap regenerating at the end towards the positive pole. By changing the calcium concentration in the medium, the morphological complexity of the developing alga can be controlled. At 3 mmol l^{-1}, tips and whorls are produced but not caps; at 2 mmol l^{-1} there are only growing tips and neither whorls nor caps will form; while at 1 mmol l^{-1}, no growth or change in complexity occurs.

These results suggest that electrical currents are involved in polarity, and calcium is involved in the production of tips, whorls, and caps. But we cannot conclude that normal control of polarity is primarily due to ionic currents, or that calcium concentration is the direct determinant of morphological complexity. These are important variables in the morphogenetic fields that determine these morphological structures, but there are many others besides.

Question 3

(a) False. Regeneration of a cap on an enucleate stalk can be explained by the presence of stable mRNA in the cytoplasm.

(b) False. The presence of stable mRNA in the cytoplasm of a stalk of species 1 will contribute to the formation of a cap and so influence its morphology even in the presence of a nucleus of species 2.

(c) False. Since ribonuclease will break down both pre-existing stable mRNA and newly synthesized mRNA, as well as other types of RNA in the cytoplasm required for protein synthesis, it is not possible to draw any conclusion about the importance of translational control in *Acetabularia* development.

(d) True.

Question 4 Regeneration in *Hydra* involves both movement of cells to new positions and the production of new types of cell from those in the regenerate. The latter depends upon the presence of multipotent cells, capable of forming a variety of differentiated cell types.

Question 5 Totipotent: cell a.
Multipotent: cells b_1, b_2, c_1.
Most multipotent: cell b_1.

Question 6 A graft of sub-hypostomal tissue to a sub-hypostomal site in a host after 4.5 hours in freshwater is expected to result in a lower frequency of head induction than when such tissue is grafted to the mid-gastric region. This is because the level of head inhibition is higher near the host head than further away.

Question 7 Wherever a head begins to form, the inhibitor concentration will rise. The immediate neighbourhood of a developing head will therefore be subject to higher inhibitor concentrations and so head formation will be discouraged. However, at some distance from a head the concentration will be sufficiently reduced for another head to form. So multiple heads are expected. DAG could either reduce the effective inhibitor concentration or increase the tendency of tissue to make heads, thus reducing the normal distance between heads. The result is many heads separated by a shorter distance than normal.

Question 8 In *Acetabularia*, axis formation occurs after gamete fusion with the appearance of a rhizoid–stalk polarity. The axis then develops by the growth of the stalk and the formation of the rhizoid. In an amphibian the egg already has a polarity that defines the animal–vegetal axis. Fertilization then establishes a plane of bilateral symmetry, defined by the animal–vegetal axis and the sperm entry point. The anterior–posterior and the dorsal–ventral axes form within this plane.

Question 9 Tissue from the blastopore region of a stage 10 gastrula is determined relative to the graft performed, so it is expected to induce the formation of a secondary embryo as described in 1.3.1. The removal of this tissue from the blastopore region does not stop the formation of the primary axis. Evidently the other cells in this region continue with this process. However, the fact that the two embryos develop equally suggests that the primary axis is delayed in its formation by the removal of the blastopore cells. This gives the graft time to get established and to organize into a second embryo an equivalent amount of neighbouring tissue, compared with the primary axis.

Question 10 The reason for the development of half embryos from separation of the first blastomeres may have been (a) because cleavage occurred at an angle to the plane of bilateral symmetry, or (b) because the species is mosaic already at first cleavage and fails to regulate irrespective of the plane of first cleavage. To test these possibilities it is necessary to examine separated blastomeres after cleavage along a variety of planes, including that of bilateral symmetry, to see if regulative capacity varies with the cleavage plane, as in the newt.

Question 11

(a) False. The first evidence is soon after fertilization, with the formation of the grey crescent.

(b) True.

(c) True.

(d) False. The example of frog and newt mouth parts provided an example that the difference arises from the responding ectoderm, not the inducing stimulus.

(e) False. Hierarchical development refers to the gradual emergence of spatial detail, organized over the whole organism. The gene–cell–tissue–organism series is another hierarchy that is believed by some to reflect the control sequence of development. However, this must be complemented by reverse influences from the organism to the genes, as in the case of morphogenetic gradients, in order to represent the complete cycle of influences in development.

CHAPTER 2

Question 1 Your diagram should be similar to Figure 2.1. The ovum has half the chromosome numbers of the oogonium.

Since the oogonia are formed by the time of birth, the process in this mammal could take as long as 5 years.

Question 2 Apart from the basic observation that sperm are much smaller than ova, the key differences are that there are many more divisions of the primordial germ cells during sperm production and each spermatocyte gives rise to four sperm, whereas each oocyte gives rise to only one ovum.

Question 3 Since the role of the nurse cells is to produce RNA and that of the follicle cells is to build up food reserves in the oocyte, you would expect to find cells with high levels of metabolic activity.

Question 4 You should have predicted that by the 8-cell stage the factors that influence gut and tuft formation have already been localized by mitotic processes. Thus you would not expect the animal region to produce any gut tissue. Gut tissue would only be found in the vegetal region (see Figure 2.7).

Question 5 Species-specific sperm attractants, possibly small peptides produced by the egg will attract sperm of the correct species. Species-specific sperm activation will generally only allow the acrosome reaction, and thus a pathway for the sperm to the egg membrane, for sperm of the same species as the egg jelly. Changes in the potential of the egg membrane once a single sperm has fused with the egg membrane prevent the fusion of further sperm. The situation in mammals is more complicated, but the question only asked for possible mechanisms in aquatic species which shed eggs and sperm into water.

Question 6 During fertilization there are cytoplasmic changes in a *Styela* egg leading to a definite pattern of distribution. If this is disrupted by centrifugation an abnormal embryo is formed with the tissues in the wrong position relative to each other.

CHAPTER 3

Question 1

(a) False.

(b) True.

(c) True.

(d) False. The inherently precise machine is used to explain fibroblast orientation.

Question 2 (c) and (d). The main features of primary invagination are the migration of vegetal cells into the blastocoel and the infolding or invagination of the vegetal cell sheet. Gustafson and Wolpert suggested that changes in the adhesiveness between vegetal cells cause archenteron formation by making the tissue sheet bend inwards.

Question 3 The rate of cell division is slower in the yolky regions of the egg and thus cells of different size are formed in the amphibian egg. In the sea urchin egg, where there is little yolk and this is evenly distributed throughout the egg, the cells which result from cleavage are similar in size.

Question 4 The change of shape and inward movement of either the cells which will form the bottle cells, or the cells which underlie the surface cells. Some spatial pattern in the early embryo determines which region will form the blastopore.

Question 5 It decreases as the archenteron increases in size and so gradually closes the entrance to the enlarging archenteron.

Question 6 Explanting endoderm tissue and putting a graft of presumptive bottle cells onto it. The presumptive blastopore cells sink into the endoderm forming a groove, and because they are darkly pigmented ectodermal cells, their progress can be readily followed. The ectodermal cells can also be stained to aid observation.

Question 7 Removing the bottle cells once they are formed does not stop invagination. This suggests that the bottle cells may not have a major role. However removal of the sub-surface cells, but not the bottle cells themselves, does stop invagination.

Question 8 The presence of yolk will influence cell movements but otherwise similar processes occur in both cases. However the invagination step is slightly different. In amphibians, specialized cells (either bottle cells or the underlying surface cells) are involved, while in sea urchin gastrulation pseudopodia 'pull' the archenteron into place.

Question 9 Mesoderm cells.

Question 10 B cells will form a ball inside A cells, because the stickiest cells take the central position. Because B is stickier than C, and C is stickier than A, B cells must be stickier than A cells.

Question 11 The amputated limb of the salamander produces a mass of undifferentiated cells (a blastema). These then differentiate to replace just that part of the limb that was lost. Blastemas from different regions, when placed together in tissue culture, will 'sort out' with cells from proximal regions surrounding distal ones. That is, they show different adhesiveness. Blastemas from the forelimb, when grafted to regenerating hindlimbs, will migrate to take up the correct relative position in the new limb. Migration of cells to specific regions is likely to be due to differential cell adhesion.

Question 12 Incubating lung tissue with antibodies to CAMs results in disordered lung development. Thus CAMs appear to be important in influencing cell movements and lung morphogenesis.

Question 13
(a) False.
(b) True.
(c) False.
(d) True.
(e) True.

Question 14
(a) Normal.
(b) Abnormal.
(c) Normal.
(d) Abnormal.
(e) Non-inducer.

Question 15 The unheated bone-marrow extract produces only mesodermal inductions (e.g. muscles). So either mesodermalizing factor is present alone, or more likely, the effect of neuralizing factor is not evident. With increasing length of heat treatment, the induced structures tend more and more to show forebrain and/or hindbrain characteristics. So, with increasing heat treatment, the effect of neuralizing factor becomes evident. Probably this is due to the fact that vegetalizing factor is destroyed by heat while neuralizing factor is not. The results in column B of Table 3.1 suggest that hindbrain characteristics are induced when the relative proportions of neuralizing and vegetalizing factors reach a particular level. These results are in accordance with Saxén and Toivonen's hypothesis, which suggests that the type of structure induced depends on the relative proportions of two inducer substances, which differ in sensitivity to heat. Notice that prolonged heat treatment (150 seconds) appears also to inactivate neuralizing factor.

Question 16
(a) True.
(b) False.
(c) True.
(d) False.
(e) True.

Question 17 In a gradient model, pattern formation depends on both the level of the morphogen and the threshold value, either of which could be altered by a mutation. If the mutation affects gradient level, then it is to be expected that in the mutant *Anabaena* the heterocysts produce more inhibitor than in the wild-type, resulting in a higher concentration near the heterocyst and therefore an increased distance along the filament to the point where the inhibitor level drops below the threshold for pro-heterocyst formation. If the threshold level of response to the inhibitor is altered, then the prediction is that this is decreased so that again a greater distance between heterocysts is required to reach the new threshold for pro-heterocyst formation in the mutant.

Question 18 A distortion of the ripple pattern occurs in (b), (c), (e) and (g). Recall that a disturbed ripple pattern is thought to arise when cells at different gradient levels are adjacent. So, if integument (cuticle) pieces are placed at an identical level in the same segment in the same orientation (a and d), the adult ripple pattern is normal. Similarly, transplantation to a site at the same level in an adjacent segment produces a normal ripple pattern (f). If the pieces are rotated or are transplanted to a different level in a segment, the ripple pattern is abnormal (b and c). Pieces that are rotated and then transplanted to the same level in an adjacent segment cause an abnormal ripple pattern (e). Pieces transplanted into integument at a different level in an adjacent segment also produce a disturbed ripple pattern (g).

CHAPTER 4

Question 1

(a) False. In the newt, nuclear divisions are followed by cell divisions, whereas in *Drosophila* mitoses occur without the formation of cell boundaries, resulting in a syncytium.

(b) True.

(c) True.

(d) False. The level of *bcd* protein can vary over more than a two-fold range without any effect on the morphology of the hatched larva.

(e) False. Maternal effect gene products are made in the nurse cells and transferred to the growing oocyte.

(f) False. Segmentation gene products of the gap, pair-rule and segment polarity categories are distributed in bands, not gradients.

(g) True.

(h) True.

(i) True.

(j) False. Discs change their fate during transdetermination.

Question 2 The first experiment can be explained equally well by both models. If it is assumed that the maximum of the gradient lies somewhere along the mid-line then each half should regenerate the other, which restores the bilaterally symmetric disc. The polar coordinate explanation is given in Figure 4.42a.

However, the second experiment cannot be explained by the gradient model, since the high point of the gradient should be in one or other of the halves, or in both if the cut passed through the maximum. The polar coordinate model gives an explanation of this result, however, as shown in Figure 4.42b.

Question 3 Since the proximal–distal gradient is repeated in every segment of the insect limb, regeneration will occur just as if a distal tibia had been grafted on: the missing limb parts between the level of the stump and that of the graft will be regenerated by the stump and the graft, as in Figure 4.30.

Question 4 The blastema would be proximalized by the retinoic acid so that a whole limb would be regenerated from the wrist. The resulting limb would then have three tandem repeats from shoulder to wrist, the last one ending in a normal hand.

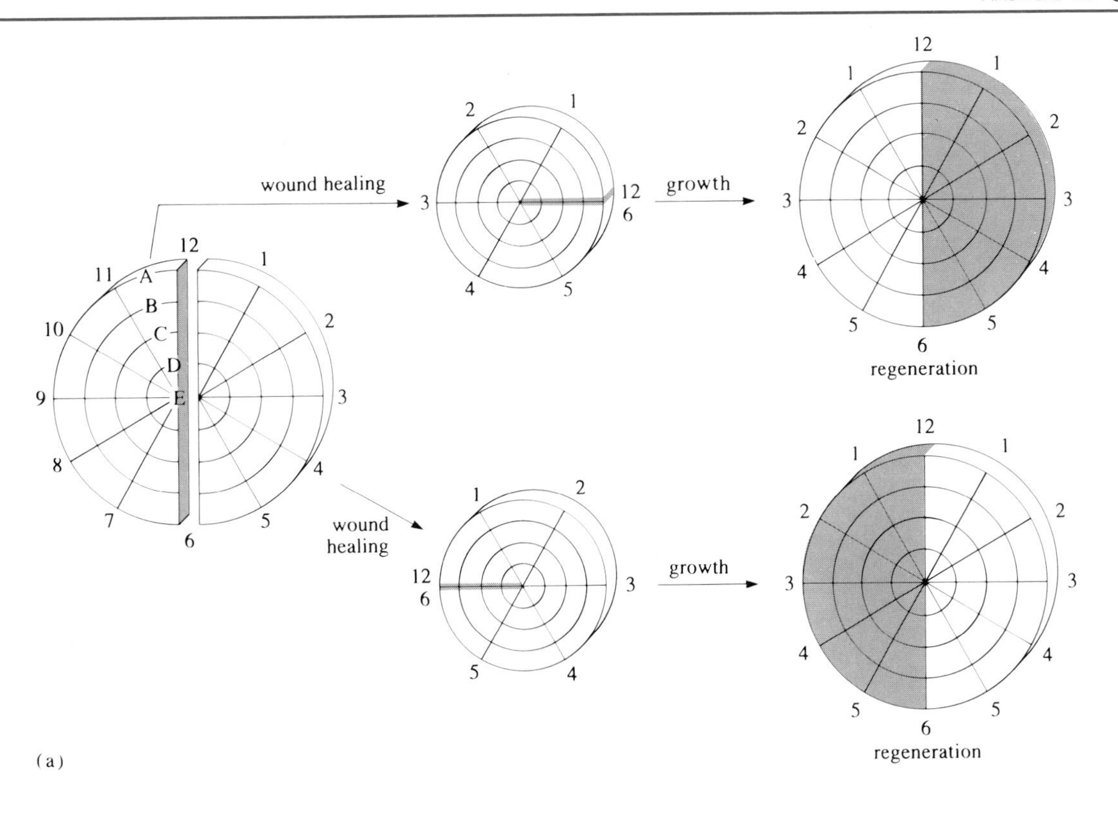

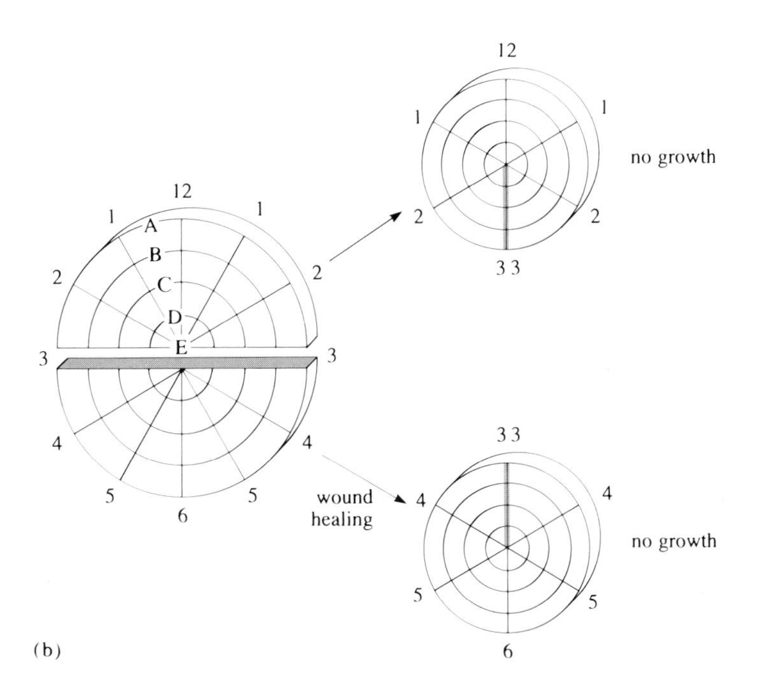

Figure 4.42 (a) Cuts parallel to the axis of symmetry result in regeneration or duplication, and when the cut is medial, regeneration is indistinguishable from duplication. (b) Cuts perpendicular to the axis of symmetry do not lead to the confrontation of different positional values, so neither regeneration nor duplication occurs.

CHAPTER 5

Question 1

(a) All mammals including humans (i) and mouse (vi) have XX/XY sex chromosomes, as has *Drosophila* (iii) and many dioecious plants such as campion (v).

(b) Birds (iv) have ZW/ZZ sex chromosomes.

(c) Alligators (ii), like many reptiles, have no sex chromosomes.

Question 2 The heteromorphic sex chromosomes pair together at meiosis I, thus ensuring that only one chromosome moves to each pole. Consequently, each gamete contains only one sex chromosome.

Question 3 A duplication is an extra copy of a chromosome segment. A deletion is the loss of a chromosome segment. A translocation is the exchange of a chromosome segment to another chromosome.

Question 4

(a) The difference in allele number between sexes in *Drosophila* is overcome by males producing twice as much gene product as females.

(b) In mammals only one of the two alleles in females is genetically active.

Question 5 There are two main consequences:

(i) It equalizes the number of active copies of X-linked alleles with that of males.

(ii) Females are mosaic for X-linked genes since the X-chromosome that is genetically active may be different in different cell lineages.

CHAPTER 6

Question 1

(a) Genetic — X : A ratio.

(b) Environment.

(c) Genetic — the Y chromosome acts as a dominant male determinant.

(d) As (c), as is true for all mammals.

(e) Genetic — the Y chromosome has a similar role to that in mammals.

(f) Environment — temperature.

Question 2 The correct statements are (a), (c) and (f).

(b) Rather than different genes between sexes, mechanisms exist that select genes to be expressed in cells of one sex and not the other.

(d) The process of development is a dynamic one with each cell and tissue responding to the conditions prevailing at that time in development.

(e) Integration of the sexual phenotype is essential in space and as the organism develops through time. This is a general developmental phenomenon that has been stressed in earlier parts of this book.

Question 3 The correct prediction is (b) because TDF is present.

(a) TDF is dominant to any number of X chromosomes (see Chapter 5).

(c) Only one X chromosome is active whether the individual is male or female (see Chapter 5).

Question 4 Your flow diagram should look something like Figure 6.23.

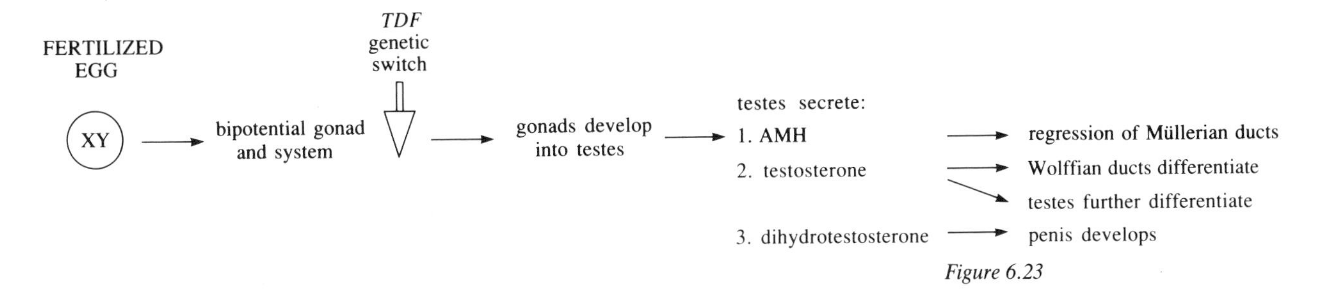

Figure 6.23

Question 5 Your flow diagram should look something like Figure 6.24.

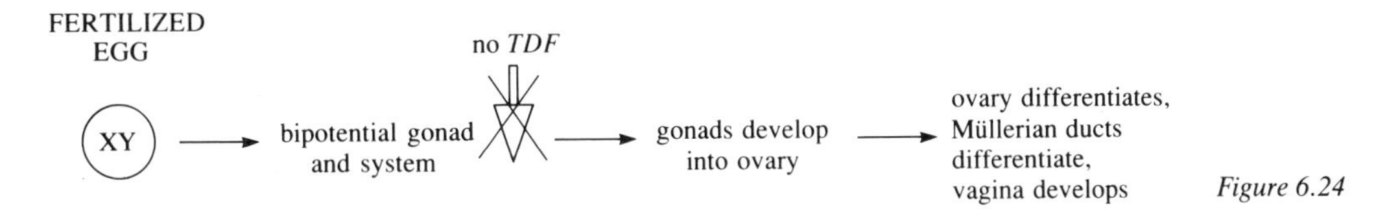

Figure 6.24

Question 6

(i) Hermaphrodite flowers.

(ii) Unisexual flowers with either anthers or ovules, borne on the same plant.

(iii) Unisexual flowers with either anthers or ovules, borne on separate plants.

Question 7 All three statements (a)–(c) could be correct explanations of the observation.

Question 8 Statement (a) is a possible explanation.

(b) All the cells would be XY or XO in which case they would all have the 1X:2A male-determining ratio.

(c) All the structures in such a fly would be male, although it might be sterile as consequence of loss of some of the genes on the Y chromosome.

Question 9 Your flow diagram should look something like Figure 6.18.

Question 10 Statement (c) provides conclusive evidence for TSD. (a) and (b) support the notion but do not provide evidence on their own. (d) suggests factors in addition to temperature may play a role in sex determination. (e) suggests that there is a period during development when temperature is important for sex determination.

Question 11 Freemartins may arise because some substance (possibly a product of TDF) travels from the male fetus to the female fetus causing masculinization of the ovaries.

CHAPTER 7

Question 1 The order is:

(e) Cells of neural ectoderm increase in height.

(f) and (a) The constriction of the apices of the neural ectoderm cells and the elevation of neural ectoderm are directly dependent events.

(b) Neural folds fuse at dorsal midline.

(d) The anterior neuropore closes.

(c) Finally, the posterior neuropore closes.

Question 2

(a) Colchicine will affect (i) elongation of neural ectoderm cells.

(b) Cytochalasin will affect (ii) elevation of neural folds.

(c) Papaverine will affect (ii) elevation of neural folds.

Question 3 The *curly-tail* phenotype appears to result from an imbalance in the rate of cell division between dorsal (neural tube) and ventral structures (endoderm). Mitomycin, since it inhibits cell division, may help to restore the balance in relative rates of cell division.

Question 4 Maternal diet appears to affect the risk of producing an affected infant and it may do so in a number of ways. Vitamin A is known to affect the rate of cell division in the mouse *curly-tail* mutant. So it is possible that there is a preventative effect of vitamins with respect to neural tube defects in humans.

Question 5

(a) False. The epidermis, a stratified squamous epithelium, contains some dead differentiated cells without nuclei.

(b) True.

(c) False. The epidermis contains no blood vessels or nerves.

(d) False. Epidermal cell growth inhibitors do not influence cell differentiation.

(e) True.

(f) True.

(g) False. Oncogene products affect primarily cell division, not differentiation.

(h) False. Cancer can result from alterations to receptor sites as well as to control substances.

Question 6

(a) The expected qualitative pattern of distribution of the substances is shown in Figure 7.14.

(i) TGFα will be highest in the dividing cells in the basal layer and will decrease progressively in the other layers.

(ii) TGFβ will rise in the differentiating cell layers, where cell divisions do not occur, but it will probably fall off again in the outer layers where cells are differentiated, particularly in the non-nucleated cells of the cornified layer.

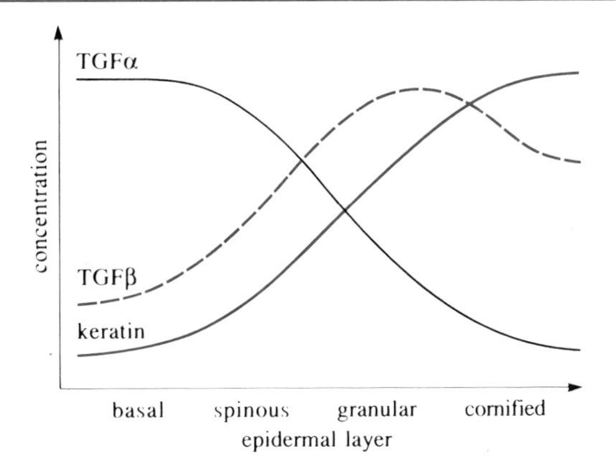

Figure 7.14

(iii) Keratin will increase progressively towards the outer layers since it is a stable differentiation product. However, the absence of nuclei in the cornified cells will result in a levelling off of concentration because of reduced production rates.

(b) Qualitative expectations are shown in Figure 7.15

(i) TFGα will be at a high level throughout the tumour, all the mutant cells over-producing it. Concentration will probably fall off slightly towards the dermis, which has a blood supply, so substances will diffuse out of the tumour and be removed by the blood. This is a small detail you may not have thought of.

(ii) TGFβ, a product of differentiating keratinocytes, will be present in low concentration uniformly throughout the tumour, representing basal production level, again falling off slightly towards the dermis, as with TGFα.

(iii) Keratin will show a similar distribution to TGFβ, but probably at an even smaller concentration.

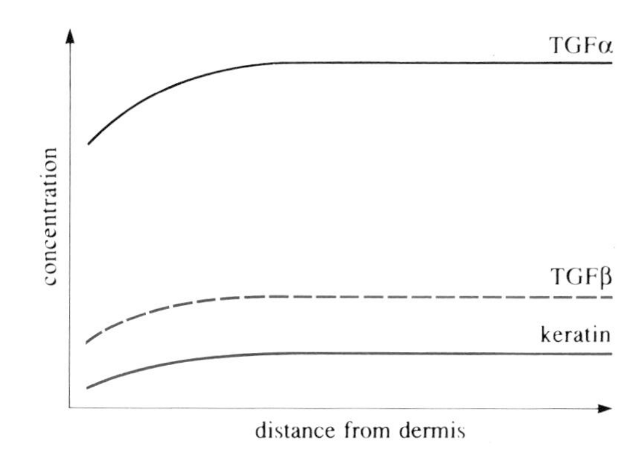

Figure 7.15

Question 7 The correct order is (e), (d), (c), (b), (a), (f).

FURTHER READING

Alberts, B. *et al.* (1989) *The Molecular Biology of the Cell*, 2nd edn, Garland Publishing Co. ISBN 0–8240–3695–6.

French, V., Ingham, P., Cooke, J. and Smith, J. (1988) Mechanisms of segmentation, *Development*, 104 (supplement), The Company of Biologists. ISBN 0–948601–15–9.

Gilbert, S. F. (1991) *Developmental Biology*, 3rd edn, Sinauer Associates Inc., Sunderland MA, USA. ISBN 0–87893–245–3.

Malacinski, G. (1990) *Cytoplasmic Organization Systems*, McGraw-Hill. ISBN 0–07–039749–X.

Walbot, V. and Holder, N. (1987) *Developmental Biology*, Random House. ISBN 0–394–33736–0.

ACKNOWLEDGEMENTS

The series *Biology: Form and Function* (for Open University Course S203) is based on and updates the material in Course S202. The present Course Team gratefully acknowledges the work of those involved in the previous Course who are not also listed as authors in this book, in particular: Ian Calvert, Lindsey Haddon, Sean Murphy and Jeff Thomas.

Grateful acknowledgement is made to the following sources for permission to reproduce material in this book:

FIGURES

Figures 1.4, 1.15: Berrill, N.J. (1971) *Developmental Biology*, figures 6.9 and 11.7, McGraw Hill, permission to reproduce from McGraw Hill, Book Company Europe Ltd; *Figures 1.16, 1.17:* Schaller, C.H. 'Bioassay of head activator and head inhibitor in *Hydra* from three different sections of the adult *Hydra* mutant maxi and mini', copyright © Professor C.H. Scaller; *Figures 1.18, 1.19, 1.20, 1.21:* copyright © Professor W. Muller, Zoologischen Institute DER, Universitat Heidelberg; *Figure 1.32:* Walpot, V. and Holder, N. (1987) *Developmental Biology*, Random House Inc.; *Figure 1.33:* Slack, J.M.W. (1984) 'Formation of basil body plan', *Egg to Embryo*, figure 1.1, Cambridge University Press.

Figure 2.3: Greenwalt, G.S and Moor, R.M. (1989) 'EM of nest of primordial follicles', *Journal of Reproduction and Fertility*, pp.561–71, figure 2; *Figure 2.5:* Johnson and Everitt (1980) *Essential Reproduction*, figure 4.4, Blackwell Scientific Publications; *Figures 2.6, 2.7:* Graham and Waring (1984) *Developmental Control in Animals and Plants*, figures 2.2.1 and 2.2.2, Blackwell Scientific Publications; *Figures 2.8, 2.11, 2.12, 2.13:* Gilbert S.F. (1988) *Developmental Biology*, Sinauer Associates Inc., figure 2.9 (photograph by D.E. Chandler, unable to trace).

Figure 3.2: Garrod, D.R. (1973) *Cellular Development*, figure 30, Chapman and Hall; *Figure 3.3:* Buckley, A. and Konigsberg, I.R. (1974) *Developmental Biology*, **37**, pp. 193–212, figure 1, Academic Press Ltd; *Figure 3.34:* Lawrence, P.A. (1973) 'A clonal analysis of segment development in *Oncopeltus* (Hemiptera)', *Journal of Embryology and Experimental Morphology*, **30**(3), Company of Biologists Limited; *Figure 3.4:* Elsdale, T. (1969) 'Pattern formation and homeostasis', *CIBA Foundation Symposium of Homeostatic Regulators*, figure 1; *Figures 3.6, 3.7, 3.8, 3.13 3.14, 3.18, 3.20:* Gilbert, S.F. (1988) *Developmental Biology*, Sinauer Associates Inc.; *Figure 3.10:* Gustafson, T. and Wolpert, L. (1967) 'Cellular movement and contact in sea-urchin morphogenesis', *Biological Review*, **42**(3), Cambridge University Press; *Figure 3.11:* Gustafson, T. and Wolpert, L. (1963) 'The cellular basis of morphogenetics', *International Review of Cytology*, **15**, figures 13a and b, Academic Press Inc. (New York) Ltd; *Figure 3.13:* Holtfreter, J. 'Cells into organs', *Journal of Experimental Zoology*, **94**, pp.261–318, **95**, pp.171–212, Prentice Hall and Wistar Institute of Anatomy and Biology; *Figure 3.15:* Steinberg, M.S. (1964) 'The problems of adhesive selectivity in cellular interactions', *Cellular Membranes in Development*, ed. M. Locke; *Figure 3.17:* (photograph by D. Stocum, unable to trace); *Figure 3.18:* Gilbert, S.F. (1988)

Developmental Biology, Sinauer Associates Inc.; *Figure 3.25:* Graham, G.F. and Naring, P.F. (1976) *The Developmental Biology of Plants and Animals*, figure 3.4.14, copyright © (1976) Blackwell Scientific Publications; *Figure 3.27:* Gurdon, J.B. (1987) 'Embryonic induction molecular prospects', *Development*, **99**, p.289, figure 3, copyright © (1987) Company of Biologists Ltd; *Figures 3.32, 3.33:* Lawrence, P.A. (1970) *Polarity and Patterns in the Post-Embryonic Development of Insects*, vol. 7, Academic Press Inc. (London) Ltd; *Figures 3.36, 3.37:* Lawrence, Crick and Munro (1972) *Journal of Cell Science*, **11**, p.815, figure 3, Company of Biologists Ltd.

Figures 4.4, 4.5: Raff, R.A. and Kaufman, T.C. (1983) *Embryos, Genes and Evolution*, Macmillan Publishing Co. Inc.; *Figure 4.10:* Jackle, M., Tautz, D., Schum, R., Seifert, E. and Lehmann, R. (1986) Cross-regulatory interactions among the gap genes of *Drosophila*, reprinted by permission from *Nature*, **324**(December), pp.18–25, copyright © (1986) Macmillan Magazines Ltd; *Figure 4.19:* Hadorn, E. (1968) *Transdetermination in Cells*, copyright © (1968) Scientific American Inc.; *Figures 4.24, 4.25, 4.26, 4.27, 4.29, 4.30:* French, V., Bryant, P.J. and Bryant, V.S. (1976) 'Pattern regulation in epimorphic fields', *Science*, **193**, American Association for Advancements of Science; *Figures 4.31, 4.32, 4.34, 4.39, 4.40:* Walbot, V. and Holder, N. (1987) *Developmental Biology*, Random House Inc., reproduced with permission of McGraw-Hill Inc, New York.

Figure 6.9: Gilbert, S.F. (1988) *Developmental Biology*, Sinauer Associates Inc.; *Figure 6.21:* (1982) reprinted by permission from *Nature*, **296**(April), p.29, copyright © (1982) Macmillan Magazines Ltd.

INDEX

Note Entries in **bold** are key terms, page numbers in *italics* refer to figures and tables, and inclusive page numbers refer to main sections.